Soft Material Robotic Systems

Annika Raatz · Tobias Kraus · Gianluca Rizzello ·
Jan Peters · Cora Maria Sourkounis
Editors

Soft Material Robotic Systems

Recent Advances and Future Developments

Springer

Editors
Annika Raatz
Institute of Assembly Technology and Robotics
Leibniz University Hannover
Garbsen, Niedersachsen, Germany

Gianluca Rizzello
Soft Robotic Systems and Control Lab
Saarland University
Saarbrücken, Germany

Cora Maria Sourkounis
Institute of Assembly Technology and Robotics
Leibniz University Hannover
Garbsen, Niedersachsen, Germany

Tobias Kraus
Colloid and Interface Chemistry
INM—Leibniz-Institute for New Materials and Saarland University
Saarbrücken, Germany

Jan Peters
Institute of Assembly Technology and Robotics
Leibniz University Hannover
Garbsen, Niedersachsen, Germany

ISBN 978-3-032-22452-1 ISBN 978-3-032-22453-8 (eBook)
https://doi.org/10.1007/978-3-032-22453-8

This work was supported by Gottfried Wilhelm Leibniz Universität Hannover.

This Springer imprint is published by the registered company Springer Nature Switzerland AG
The registered company address is: Gewerbestrasse 11, 6330 Cham, Switzerland

Foreword

The field of soft robotics is at a critical moment. Seven years removed from the first International Conference on Soft Robotics (RoboSoft) in 2018, I've noticed our community doing a bit of soul-searching. What are the "killer" applications? When will we see soft robots out in the real world, and what will they be doing? I get asked versions of these questions all the time. I usually respond with optimistic visions of soft robots in medical, wearable, manipulation, and locomotion applications. Given my own obsession with soft morphing robots, I often go on (and on) about the potential energetic savings to be gained by optimizing a robot's morphology and control policy to match its environment during multi-environment locomotion.

But I get the sense that these responses don't always land. A colleague recently pushed back. "But what will your morphing locomotion robots actually *do*?" they asked. "Is the morphing really necessary?"

I suppose that answering that question is the scientific journey I'm on right now. And answering the broader questions that will allow us, as a research community, to transition our technologies into engineered solutions that support human flourishing is the journey we are all on. We are at a critical juncture. Soft robotics doesn't carry the novelty buzz it once did. So, who are we now, as a soft robotics community? What are we meant to do? And how are we making the world a better place?

Fortunately, we have an edge. Soft robotics is inherently interdisciplinary. In a scientific landscape defined by disciplinary boundaries and silos, the field of soft robotics is breaking the mold. Research at disciplinary intersections, where tools from one field are applied in novel ways to another, is the heart of our field. This is our greatest strength. It drives discovery and makes collaboration and community central to our work.

The chapters in this book capture that spirit. This collection spans research contributions, reviews, and perspectives on fundamental materials science, innovative actuation strategies, modeling frameworks, design methodologies, and application-driven system design. A recurring theme is the co-design of robot morphology, material properties, and control policy. Taken together, these chapters exemplify a field that is maturing, deeply interdisciplinary, and collaborative at its core.

Enjoy the collection—and our ongoing work to better understand the world (and beyond), develop creative solutions to critical problems, foster collaborative approaches, and support one another along the way.

July 2025

Rebecca Kramer-Bottiglio
John J. Lee Associate Professor
of Mechanical Engineering
Yale University
New Haven, CT, USA

Introduction

Soft robotics has evolved from an emerging concept into a rapidly expanding interdisciplinary field that promises transformative impact across science, engineering, and medicine. This book brings together a diverse collection of contemporary research that reflects the breadth and maturity the field has reached—and outlines where it may be heading next.

The motivation behind this collective volume lies in the recognition that soft material systems are not only reshaping robotic design but also pushing the boundaries of what robots can physically achieve in unstructured, complex, or human-centered environments. The softness, compliance, and adaptability of materials serve not merely as mechanical features but as integral components in sensing, actuation, intelligence, and interaction.

A Collective Effort Rooted in a Major Research Programme

This book has its roots in the German Priority Programme SPP2100, a six-year initiative funded by the German Research Foundation (DFG) with a total budget of more than 13 million euros. Over two funding periods, 16 research projects and more than 50 researchers contributed to developing fundamental and applied knowledge on soft robotics. True to the field's nature, the programme's first funding period fostered strong interdisciplinary collaboration between the 12 participating projects—across robotics, mechanical engineering, mechatronics, materials science, chemistry, biology, and medicine. The second funding period with four additional projects built on these foundational insights, with the new call for proposals focusing on applications and contributions that leveraged the knowledge accumulated.

But the reach of the book extends beyond national boundaries. A number of internationally recognized researchers—many of whom have been close collaborators or friends of the SPP2100 programme—have contributed chapters. Their involvement reflects not only the global character of the soft robotics community but also the collaborative, open, and well-connected spirit that has characterized the

SPP2100 initiative from the beginning. It was supported by a community characterized by shared vision, ongoing dialogue, focused inquiry, teamwork, interdisciplinary openness, and enthusiasm for experimentation.

Structure and Scope

The book is divided into five parts, each capturing a central aspect of soft material robotic systems:

1. *Materials for Soft Robots*
   This part explores the latest developments in smart and functional materials, from humidity-responsive polymers and azobenzene-based photomechanical compounds, to triboelectric and piezoelectric elastomers, soft magnetic materials, and flexible printed electrodes. These foundational materials define the limits and possibilities of what soft robots can sense, do, and become.
2. *Design of Soft Robots*
   A critical challenge in soft robotics lies in how to design and fabricate systems whose behavior is often nonlinear, underactuated, and highly coupled with their environment. The chapters here span from automated pneumatic logic circuits and programmable curved rods, to tendon-driven manipulators, tensegrity structures, and strain-limiting materials—demonstrating how creative engineering approaches enable control through morphology, material, and structure.
3. *Modelling and Control of Soft Robots*
   Unlike traditional robots, soft robots defy easy modeling and require new paradigms for control. This section addresses exactly that, offering insights into dielectric elastomer modeling, stiffness modulation, continuum mechanics, and software toolboxes tailored for soft robotic platforms. These contributions reflect the deep technical complexity required to make soft systems both predictable and controllable.
4. *Applications of Soft Robots*
   The real-world potential of soft robotics comes to life in these chapters—from deep-sea sampling devices developed in collaboration with a volcanologist, to bio-inspired grippers and miniaturized medical robots designed to navigate the human body. These applications highlight how soft systems can operate safely, efficiently, and robustly in domains where traditional rigid robots fall short.
5. *The Future of Soft Robotics*
   The final chapter, based on a panel held at the 2025 Symposium on the Future of Soft Robotics in Hannover, offers a forward-looking view of the field. The discussion addresses challenges such as scalability, durability, and managing inflated expectations, while also emphasizing the creative potential and growing relevance of soft robotics in healthcare, environmental monitoring, and beyond. Importantly, it highlights that the future of the field lies not in replacing traditional robotics, but in integrating soft technologies into broader systems. Above

all, the panel underscores that interdisciplinary collaboration among engineers, chemists, biologists, and medical researchers will be essential for meaningful progress.

A Field Defined by Interdisciplinarity

What unites the contributions in this volume is not a single methodology or point of view, but a shared interdisciplinary ethos. Soft robotics demands expertise that spans scales—from molecular interactions in smart polymers to full-body control of multi-segmented robots and from the subtleties of human–robot interaction to the harsh conditions of the deep ocean. The editors would like to express their sincere gratitude to all the brilliant researchers with whom they had the privilege to collaborate over the past 6 years, to the authors who generously contributed their work to this volume, to the dedicated students whose efforts powered many of the individual projects, and to the enduring friendships that grew alongside the science. We hope this collection serves both as a snapshot of recent progress and as a catalyst for future collaboration. As soft robotics continues to evolve, it will require not only technical innovation but also sustained investment in interdisciplinary research, education, and open exchange. In that spirit, this book is dedicated to the researchers—established and emerging—who are helping shape the future of robotics, one soft system at a time.

Annika Raatz
Institute of Assembly Technology
and Robotics
Garbsen, Germany
raatz@match.uni-hannover.de

Contents

Applications of Soft Robots

The Future of Soft Robotics

Materials for Soft Robots

Advanced Rheology of Humidity-Responsive Functional Polymers for Soft Robotics

Dominik Fauser, Stephan Pflumm, Sabine Ludwigs, and Holger Steeb

Abstract Functional polymers are advanced soft materials capable of altering their shape and mechanical behavior in response to external stimuli. Functional materials science seeks to mimic adaptive behavior of biological systems for applications in soft material robotics, such as smart skin or sensing actuators. A natural example of this adaptability is found in the leaves of Ramonda myconi, that undergo reversible shape changes in response to moisture fluctuations. To tailor-make, understand and predict the complex behavior of synthetically-made functional polymers and devices thereof, advanced rheological characterization techniques are essential. These methods determine viscoelastic properties under controlled environmental conditions, capturing time-dependent responses through oscillatory measurements. By analyzing the complex modulus in the frequency domain, the shear and Young's modulus can be systematically evaluated under varying temperature and humidity. A major focus of this chapter is the rheological characterization and prediction of moisture-sensitive materials, including shape memory polymers, hydrogels and moisture responsive bilayer structures. The influence of humidity on material programmability and mechanical response is explored, demonstrating the predictive capabilities of rheological analysis. This study provides a comprehensive framework for the characterization of functional polymers, paving the way for their optimized integration into soft robotics, smart skins, and bioinspired sensor-actuator systems.

D. Fauser · H. Steeb (✉)
University of Stuttgart, Institute of Applied Mechanics (MIB), Stuttgart, Germany
e-mail: holger.steeb@mechbau.uni-stuttgart.de

D. Fauser
e-mail: dominik.fauser@mib.uni-stuttgart.de

S. Pflumm · S. Ludwigs
University of Stuttgart, IPOC-Functional Polymers, Institute of Polymer Chemistry (IPOC), Stuttgart, Germany
e-mail: stephan.pflumm@ipoc.uni-stuttgart.de

S. Ludwigs
e-mail: sabine.ludwigs@ipoc.uni-stuttgart.de

D. Fauser · H. Steeb
University of Stuttgart, Stuttgart Center for Simulation Science (SC SimTech), Stuttgart, Germany

A. Raatz et al. (eds.), *Soft Material Robotic Systems*,
https://doi.org/10.1007/978-3-032-22453-8_1

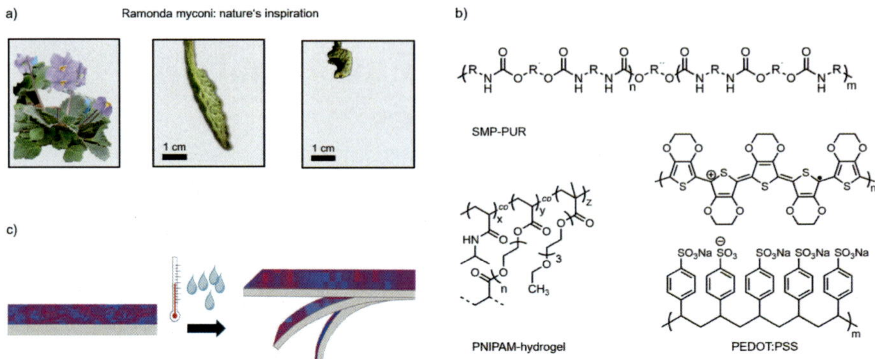

Fig. 1 **a** Photograph of the plant Ramonda myconi and response of a cut-off leaf from the wet to dry state (scale bar 1 cm). Adapted with permission under the terms of the CC-BY 4.0 license. Copyright 2018, Kampowski, Demandt, Poppinga, and Speck, published by Frontiers [6]. **b** Structural formula of PU-SMP (DiAPLEX, SMP Technologies Inc.) (upper center), PNIPAM-based hydrogel (lower left) and PEDOT:PSS (lower right). **c** Schematic representation of the actuation mechanism in a bilayer actuator, illustrating the temperature- and moisture-responsive behavior of the composite structure

1 Introduction

Functional polymers are materials capable of altering their shape [1] and mechanical behavior in response to external stimuli such as heat [2], moisture [2, 3], light [4], magnetic fields, or electric fields. Inspired by biological systems, the field of functional materials science seeks to replicate natural adaptive mechanisms to develop responsive materials for applications such as soft robotics [5].

An illustrative example of nature's dynamic adaptability is found in Ramonda myconi, a resurrection plant that exhibits reversible shape changes in response to variations in moisture content [6] (see Fig. 1a). The leaves of Ramonda myconi undergo complex folding motions upon dehydration, with both intact and excised leaves shrinking significantly, and fully recover their initial form upon rehydration, demonstrating an efficient water-driven actuation mechanism. The leaves can be regarded as multifunctional and hierarchically assembled layered structures.

Figure 1b shows three exemplary functional polymer materials, poly(urethane) shape memory polymer (PU-SMP), poly(N-isopropylacrylamide) (PNIPAM)-based hydrogels and poly(ethylenedioxythiophene)-poly(styrenesulfonate) (PEDOT:PSS), all of which exhibit moisture responsiveness, effectively mimicking the natural examples. This moisture sensitivity can replicate the behavior of Ramonda myconi, as demonstrated in Fig. 1c, where a bilayer architecture is designed to undergo curvature changes in response to variations in humidity and temperature.

To understand the intricate material behavior exhibited by functional polymers for soft robotics, it is imperative to employ advanced rheology characterization techniques (see Fig. 2). These methods determine viscoelastic material parameters under

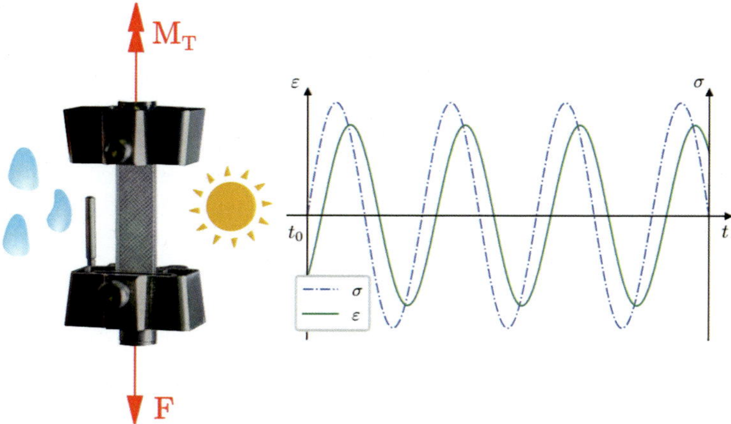

Fig. 2 Schematic representation of isothermal frequency measurements using harmonic torsion and/or harmonic compression/tension excitation, performed under constant temperature and relative humidity conditions

systematically varied external influences in controlled experiments. The viscoelastic behavior is rigorously analyzed through harmonic excitation in oscillatory shear- and tensile-mode measurements, as demonstrated by Dingler et al. [1], Fauser and Steeb [2] and Pflumm et al. [3]. The time-dependent response of the material, obtained from these oscillatory measurements, is captured by the complex modulus and complex compliance in the frequency domain [7].

For each type of deformation, the material response is characterized by associated complex moduli–such as the complex-valued bulk modulus K^*, shear modulus G^*, and Young's modulus E^*–with corresponding complex compliances obtained from their inverses. These material parameters are inherently dependent on external influences such as moisture and temperature. The complex representation of moduli, $M^* = \{K^*, G^*, E^*\}$, physically distinguishes between stored (retrievable) and dissipated energy via its real and imaginary components ($M^* = M' + i\,M''$). This formulation provides a comprehensive framework for predicting and engineering functional materials with tunable mechanical properties, essential for soft robotics applications.

A key approach to describing these material responses is through time-dependent rheological models, where model parameters vary with moisture content, as detailed in Fauser et al. [8]. Furthermore, studies have demonstrated that the programmed shapes of SM polymers (SMPs) are highly dependent on environmental moisture levels [2]. To advance the application of these materials in soft robotics, it is crucial to understand how moisture influences their programmability and mechanical performance.

In addition, advanced rheology provides an effective tool for assessing the moisture dependence of hydrogels in situ, as demonstrated by Pflumm et al. [3]. This approach offers valuable insights into the potential applications of hydrogels in soft robotics and smart skins. Similarly, Dingler et al. [1] showcased the predictive capabilities of rheological characterization in a moisture-dependent actuator, consisting of an active, moisture-responsive layer and a passive layer. Their study successfully combined experimental measurements with analytical composite beam theory to analyze and predict complex shape transformations.

In this chapter, advanced rheology is employed to investigate the influence of moisture on various functional polymer systems, including a PU-SMP, the in-situ characterization of PNIPAM-based hydrogel patches, and a moisture-dependent bilayer structure consisting of a PEDOT:PSS layer and a poly(dimethylsiloxane) (PDMS) layer. The chapter is structured as followed. Section 2 introduces the fundamentals of rheological characterization under varying environmental influences, along with the evaluation methods. The effects of moisture on the PU-SMP system are then described in Sect. 3. In Sect. 4, the in-situ measurement of the hydrogels is discussed. The subsequent section, Sect. 5, provides a detailed exposition of the two-layer actuator structure. Finally, Sect. 6 summarizes the key findings of the chapter.

2 Fundamentals of Rheological Characterization

Harmonic characterization of viscoelastic materials are performed by an applied sinusoidal stress, where the corresponding measured strain of the linear excitation follows with a phase shifted sinusoidal response as described in [9]. From harmonic shear stress controlled experiments, the ratio of shear stress τ to shear strain γ gives the complex shear modulus,

$$G^* = \frac{\tau(t)}{\gamma(t)} = |G^*|(\cos\delta_S + i\sin\delta_S), \tag{1}$$

where the real part describes the storage modulus $G' = |G^*|\cos\delta_S$, the imaginary part the loss modulus $G'' = |G^*|\sin\delta_S$. The absolute value is given by $|G^*| = \sqrt{G'^2 + G''^2}$. The viscous losses can be determined with the loss factor $\tan\delta_S = G''/G'$. The complex Young's modulus,

$$E^* = \frac{\sigma(t)}{\varepsilon(t)} = |E^*|(\cos\delta_A + i\sin\delta_A), \tag{2}$$

is derived with the axial stress σ and axial strain ε. In (2) the real-part is obtained by $E' = |E^*|\cos\delta_A$ and the imaginary part by $E'' = |E^*|\sin\delta_A$. The absolute value and loss factor of the complex Young's modulus is given by $|E^*| = \sqrt{E'^2 + E''^2}$ and $\tan\delta_A = E''/E'$.

2.1 Dynamical Mechanical Thermal Humidity Analysis (DMTHA)

In the case of a thermorheologically simple polymer, the chain mobility and the molecular rearrangement process exhibit comparable behavior with respect to temperature and time [10, 11]. Consequently, Time-Temperature Superposition (TTS) can be applied to the polymer under consideration, i.e., the frequency measurements performed at different temperatures can be shifted horizontally in frequency space with a shift factor to generate master curves at a reference temperature. By varying the RH at constant levels and measuring the isothermal frequency measurements (see Fig. 2), the time-, temperature- and moisture-dependent material behavior can be determined as described by [2], by generating master curves at reference temperature. This method is referred to as Dynamic Mechanical Thermal Humidity Analysis (DMTHA).

It follows that the determined harmonic generic storage modulus

$$M'(f, \Theta, \varphi) = M'(\alpha_T(\Theta) f, \Theta_0, \varphi) \tag{3}$$

and loss modulus

$$M''(f, \Theta, \varphi) = M''(\alpha_T(\Theta) f, \Theta_0, \varphi) \tag{4}$$

of the generic complex modulus $M^* = \{K^*, G^*, E^*\}$ at a temperature Θ and a RH level φ can be represented by a horizontal shift of the frequency $\alpha_T(\Theta) f$ of the determined harmonic modulus at a reference temperature Θ_0.

2.2 Master Curve Generation

From isothermal frequency measurements (see Fig. 3a), a master curve is generated to describe the time-dependent material behavior at the selected reference temperature Θ_0 in the reduced angular frequency range. To reduce the bias of manual master curve generation, we developed an algorithmic approach to calculate the frequency shift between two isothermal measurements.

We consider two isothermal measurements at temperatures Θ_k and Θ_{k+1}, where $\Theta_k = \Theta_0 + k\,\Delta\Theta$ with $k = 0, 1, ..., p$ and Θ_0 represents the reference temperature used for generating the master curve. For clarity, the Algorithm 1 is described only for the case where $\Theta > \Theta_0$. However, the procedure works similarly for measurements below the reference temperature, $\Theta < \Theta_0$, by simply reversing the direction of the incremental temperature change (i.e., subtracting instead of adding $\Delta\Theta$). Based on the previous shift parameter $\alpha_T(\Theta_k)$, Algorithm 1 computes the next shift parameter $\alpha_T(\Theta_{k+1})$. To enable a finer resolution of the overlapping range of the storage moduli, we first interpolate the measurement data at the two temperatures Θ_k and Θ_{k+1} with linear splines, resulting in two functions $\tilde{M}_k(\omega)$ and $\tilde{M}_{k+1}(\omega)$, respectively. We

Algorithm 1 Calculating Shift Parameter $\alpha_T(\Theta_{k+1})$

Require: Θ_k, Θ_{k+1}: Measurement temperatures
Require: $\alpha_T(\Theta_k)$: Previous shift parameter
Require: $\{\omega_n\}_{n=1,\ldots,N}$: Measurement frequencies
Require: $\{M'_{k,n}\}_{n=1,\ldots,N}$: Experimental determined storage modulus at frequency ω_n and temperature Θ_k
Require: $\{M'_{k+1,n}\}_{n=1,\ldots,N}$: Experimental determined storage modulus at frequency ω_n and temperature Θ_{k+1}

$\tilde{M}_k(\omega) \leftarrow \text{interpolate}\left[\{(\omega_n, M'_{k,n})\}_{n=1,\ldots,N}\right]$
$\tilde{M}_{k+1}(\omega) \leftarrow \text{interpolate}\left[\{(\omega_n, M'_{k+1,n})\}_{n=1,\ldots,N}\right]$
shift $\leftarrow 0$, counter $\leftarrow 0$
for $m = 0, \ldots, M$ **do**
  $\hat{M}_m \leftarrow ((M-m)\, M'_{k,1} + m\, M'_{k+1,N})/M$
  $\overline{\omega} \leftarrow \text{argmin}_\omega |\hat{M}_m - \tilde{M}_k(\omega)|$
  $\overline{\overline{\omega}} \leftarrow \text{argmin}_\omega |\hat{M}_m - \tilde{M}_{k+1}(\omega)|$
  shift $\leftarrow$ shift $+ \overline{\omega}/\overline{\overline{\omega}}$
  counter $\leftarrow$ counter $+ 1$
end for
$\alpha_T(\Theta_{k+1}) \leftarrow \alpha_T(\Theta_k) \cdot \text{shift}/\text{counter}$

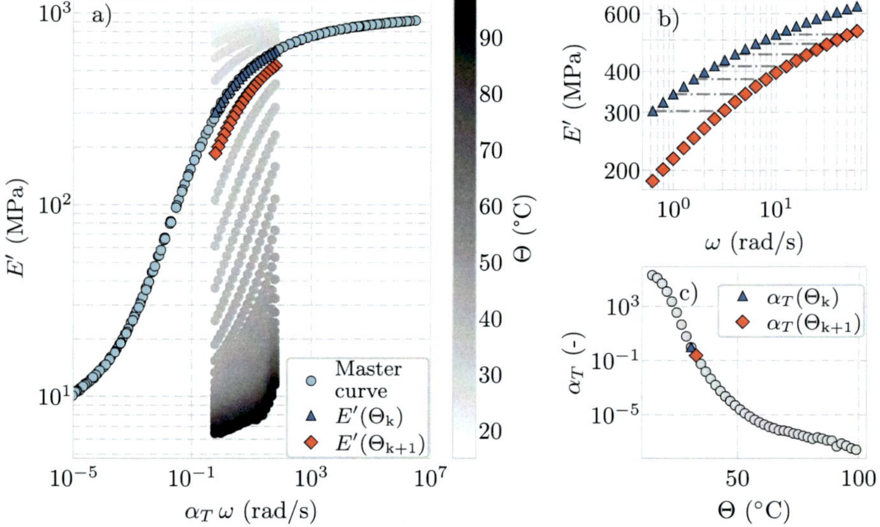

Fig. 3 **a** Isothermal harmonic uniaxial tensile measurements of PU-SMP at different temperatures in the dry state ($\varphi = 0\%$ RH) are represented by gray circles. The reference frequency-measurement at the reference temperature (Θ_0) is highlighted with blue triangles, while the corresponding master curve at Θ_0 is shown with light blue circles. The frequency measurement at $\Theta_{k+1} = \Theta_0 + \Delta\Theta$ is represented by red diamonds. **b** Comparison of storage modulus values at the two observed temperatures, highlighting the overlapping region where self-similar values appear. **c** The frequency shift between the self-similar storage moduli at both temperatures is used to determine the shifting parameter, as depicted in the plot

uniformly sample the overlapping range of the storage moduli and determine for each sample $\hat{M}_m$ the corresponding frequencies $\overline{\omega}$ and $\overline{\overline{\omega}}$, such that $\hat{M}_m = \tilde{M}_k(\overline{\omega}) = \tilde{M}_{k+1}(\overline{\overline{\omega}})$, see Fig. 3b. The optimal shift at this sampling point is then given by $\overline{\omega}/\overline{\overline{\omega}}$. The shift $\alpha_T(\Theta_{k+1})$ at temperature Θ_{k+1}, is calculated as the mean of the optimal shift at all sampling points times $\alpha_T(\Theta_k)$, i.e. the shift at temperature Θ_k, see Fig. 3c. Repeating this procedure for all isothermal measurements in incremental steps toward the reference temperature results in the full master curve.

3 Humidity Triggered Shape Memory Polymers

In this section we apply the DMTHA to the thermo-responsive PU-SMP. A key property of PU is their intrinsic SM behavior [2], which is governed by the mobility of their molecular chains. Above Θ_G, the increased chain mobility allows PU to undergo large deformations. When this deformation is fixed and the material is cooled below Θ_G, the reduction in chain mobility effectively "freezes" the deformation, programming the material into a new shape. Upon reheating above Θ_G, the programmed state is released, and the material returns to its initial configuration. This SM behavior is strongly influenced by the viscoelastic properties of PU both above and below Θ_G, which can be characterized and predicted through rheological measurements [2, 8]. Effective material parameters, such as complex moduli and creep compliances, provide crucial insights into materials behavior of PU under varying environmental conditions.

The interplay between temperature and moisture significantly affects PU's mechanical response. Through DMTHA, it has been shown that increasing RH levels lead to systematic shifts in time-dependent mechanical properties. One key observation is the moisture-induced shift in the glass transition temperature, which decreases from approximately 55 °C in dry conditions to 22.12 °C at 70% RH. This reduction in Θ_G is accompanied by an acceleration of relaxation times, while the viscous losses remain constant. Master curves are generated according to the procedure outlined in Sect. 2.2 and in Algorithm 1. Comparing master curves obtained at different RH levels, a distinct time-frequency shift is observed, emphasizing the strong dependency of PU's mechanical response on environmental moisture.

This humidity-dependent transition behavior can be effectively modeled using rheological methods as described by Fauser et al. [8], enabling precise predictions of PU's performance under varying environmental conditions. Since moisture lowers Θ_G, the thermal fixation of PU structures becomes humidity-dependent, requiring a thermal-humid mechanical cycle for accurate programming (see Fig. 4). As a result, PU structures that are programmed into a temporary shape can be deactivated by increasing RH at ambient temperatures (Fig. 4c), eliminating the need for thermal activation. This property makes PU highly attractive for applications in soft robotics and adaptive materials, where moisture-driven actuation can be leveraged for energy-

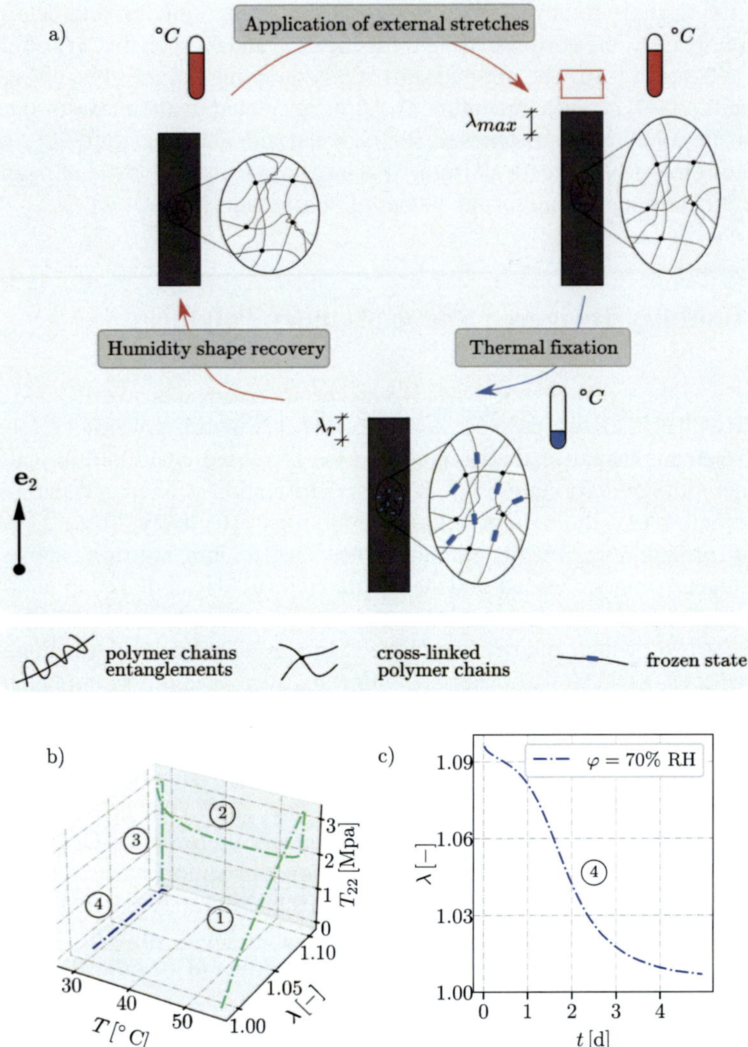

Fig. 4 **a** Graphical abstract of the Thermo-Humidity-Mechanical Cycle (THMC) from [2], licensed under CC-BY 4.0. **b** The experimental result of the THMC is illustrated in the plot. In the first step, at a temperature above the glass transition temperature ($\Theta > \Theta_G$) and in dry conditions ($\varphi = 0\%$RH), the PU-SMP sample is stretched to a stretch value of $\lambda = 1.10$. In the second step, the stretch is maintained while the temperature is reduced to $\Theta = 27\ °C$ ($\Theta < \Theta_G$), and the axial Cauchy stress T_{22} is continuously measured. During the third step, the applied force is released ($F = 0$ N), allowing the sample to move freely. Finally, in the fourth step, the humidity is increased to 70% RH, leading to water absorption from the surrounding air. The absorption process depends on the diffusion time, influencing the material's behavior. The corresponding stretch recovery over time is depicted in **c**. The recovery process is governed by the diffusion time, as the absorbed water lowers the glass transition temperature ($\Theta_G(\varphi)$). This shift enables the material to gradually return to its original shape. Figures adapted from Fauser and Steeb [2]

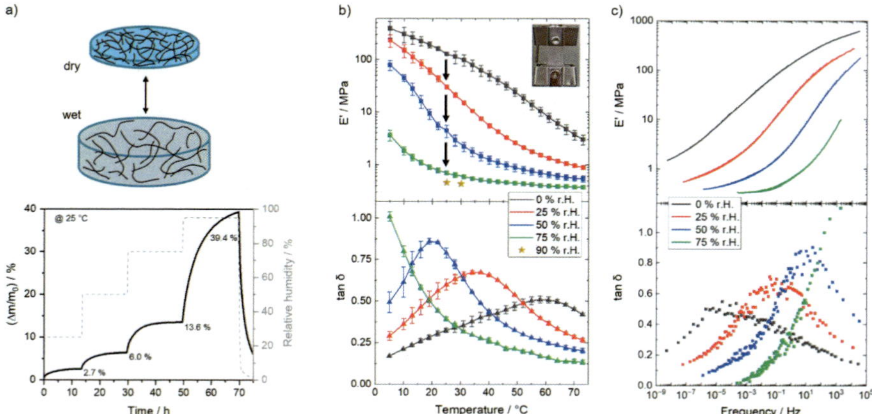

Fig. 5 **a** (Upper) Schematic representation of the reversible swelling behavior of PNIPAM-based hydrogel patches. (Lower) Relative swelling degree as a function of time at 25 °C and varying RH levels. **b** (Upper) Storage modulus E' and (Lower) Loss factor tan δ of the complex Young's modulus as a function of temperature at different RH levels. **c** Master curves generated as a function of reduced frequency, derived from isothermal frequency measurements at varying RH levels. Adapted from Pflumm et al. [3], licensed under CC-BY 4.0

efficient shape transformations. These adaptive PU behaviors are reminiscent of the shape transformations observed in Ramonda myconi leaves, where moisture fluctuations drive reversible morphological changes.

4 Humidity-Responsive Hydrogel Patches

Hydrogels, such as PNIPAM-based materials, as described in Pflumm et al. [3], exhibit high stretchability and tunable water uptake and release, making them highly adaptable to evolving environmental conditions. These materials can absorb significant amounts of water, reaching up to 39.4% equilibrium water content at 95% RH (cf. Fig. 5a) and 708% in water bath at 5 °C. Their mechanical properties vary considerably with moisture content (see Fig. 5b,c). In their dry state, they become brittle and rigid, whereas under ambient conditions, they maintain an optimal balance between flexibility and stability. As the RH increases, their stiffness decreases, as shown by the storage modulus E' 1 Hz and 25 °C, which decreases from 388.9 MPa in the dry state ($\varphi = 0\%$ RH) to 0.5 MPa at 90% RH. Similarly, the glass transition temperature decreases from 59 °C in dry conditions to 21 °C at 50% RH, demonstrating a strong dependence on moisture content.

By applying DMTHA, master curves can be constructed at a chosen reference temperature and RH level (see Fig. 5c). Higher water content accelerates relaxation processes, leading to greater energy dissipation within the glass transition zone. This behavior differs from PU-SMP, as described in Sect. 3, where no significant variation in the loss factor was observed at different RH levels.

In contrast, temperature affects the equilibrium swelling degree, facilitating water release at elevated temperatures. This phenomenon enables adaptive drug release kinetics, as demonstrated by Pflumm et al. [3]. The complex material behavior of PNIPAM-based hydrogels under varying external influences makes them particularly suitable for applications in personalized medicine, smart actuators, and soft robotics. Their flexibility and exceptional water absorption capacity closely resemble the adaptive behavior of Ramonda myconi, making them promising candidates for bioinspired designs where environmental responsiveness plays a crucial role.

5 Humidity-Responsive Bilayer Actuators

Simple bilayer structures composed of a hydrophilic PEDOT:PSS as the active layer and a hydrophobic PDMS as the passive layer exhibit reproducible bending behavior in response to environmental RH (see Fig. 6a,b), as described by [1, 12–14]. PEDOT:PSS absorbs water through diffusion, leading to volumetric expansion (see Fig. 6c) [15]. This expansion generates internal stresses within the bilayer structure due to the differential swelling between the active and passive layers, resulting in bending. Unlike the one-way shape change observed in PU-SMPs, as described in Sect. 3, the humidity-driven deformation of the PEDOT:PSS-PDMS bilayer is fully reversible. The structure continuously adapts to changes in environmental RH, making it a promising candidate for applications requiring dynamic and autonomous responsiveness, such as soft actuators and biomimetic devices.

Advanced rheological characterization was conducted to analyze the mechanical behavior of the individual layers under varying RH conditions. Harmonic tensile measurements revealed that the complex Young's modulus of PEDOT:PSS decreases with increasing RH from 130 MPa at $\varphi = 10\%$ to 54 MPa at 90% RH (see Fig. 6c), indicating its sensitivity to moisture absorption. The volumetric strain increases from $\varepsilon_{\text{vol}} = 0.8\%$ at 10% RH to 6.1% at 90% RH. In contrast, the PDMS passive layer exhibited no significant dependence on RH, confirming its stability in humid environments. The material parameters obtained from DMTHA were subsequently utilized to predict the curvature behavior of the bilayer structure at varying RH levels. This prediction was based on Timoshenko's composite beam theory, expressed as

$$\kappa = \frac{\varepsilon_{\text{vol}}}{d} \frac{6(1+m)^2}{3(1+m)^2 + (1+mn)\left(m^2 + \frac{1}{mn}\right)} \tag{5}$$

where d represents the total thickness of the bilayer, $m = d_{\text{PEDOT:PSS}}/d_{\text{PDMS}}$ denotes the thickness ratio between the active (PEDOT:PSS) and passive (PDMS) layers, and $n = E_{\text{PEDOT:PSS}}/E_{\text{PDMS}}$ corresponds to the ratio of their (static) Young's moduli. By

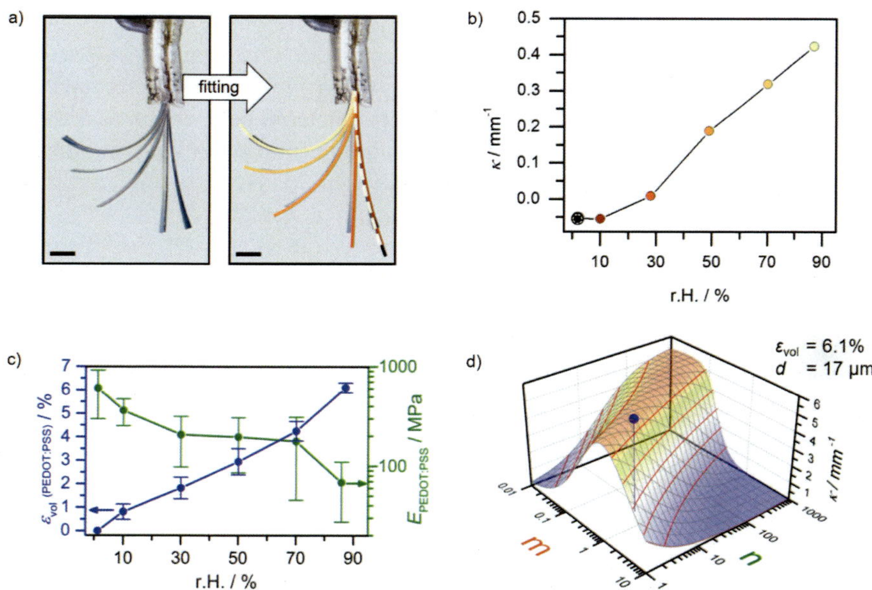

Fig. 6 **a** Actuator motion of a PEDOT:PSS/PDMS bilayer (thickness: $d = 54\,\mu\text{m}$) in response to changes in RH (scale bar: 1 mm). Circle segments were fitted to the image to extract the curvature $\kappa(\varphi)$ as a function of RH, as shown in **b**. The color coding in **a** corresponds to that in **b**. **c** The volumetric strain ε_{vol} is shown in blue and the storage modulus of the Young's modulus E' of PEDOT:PSS as a function of RH is shown in green. **d** Curvature behavior of a bilayer beam for varying thickness ratios m and stiffness ratios n, derived using Timoshenko's equation (5). Adapted from Dingler et al. [1], licensed under CC-BY 4.0

applying these material parameters, the bending response of the bilayer could be accurately predicted, demonstrating how structural curvature can be tailored by adjusting the thickness ratio of the active and passive layers (see Fig. 6d). This approach provides a foundation for designing adaptive actuators whose deformation can be precisely controlled in response to environmental humidity variations.

6 Summary and Outlook

In this chapter, we explore the application of advanced rheological characterization to understand and predict the material behavior of functional materials for soft robotics. Using three examples–thermoplastic polyurethane shape memory polymers, intelligent hydrogels, and a bilayer structure composed of a hydrophilic active and hydrophobic passive layer–we demonstrate how functional materials can replicate complex natural behaviors, such as the moisture-responsive deformation of Ramonda myconi.

Advanced rheological characterization involves harmonic excitation of uniaxial tension/compression or shear measurements, conducted under varying external influences like temperature and humidity. This approach allows for the determination of the time-, temperature- and moisture-dependent material behavior, providing valuable insight into the material's response to environmental conditions. We show that Dynamic Mechanical Thermal Humidity Analysis (DMTHA) is an effective method for understanding and predicting complex material behaviors.

By employing DMTHA for material characterization and utilizing rheological models that can be extended into three-dimensional constitutive equations for finite deformation simulations, complex material behaviors in structures can be effectively designed and predicted. This methodology advances the field of soft robotics and opens new possibilities in personalized medicine, facilitating the development of materials and actuators that dynamically adapt to environmental stimuli such as temperature and humidity fluctuations.

Acknowledgements This research was funded by the Deutsche Forschungsgemeinschaft (DFG, German Research Foundation) under grant no. 498339709.

References

1. Dingler, C., Müller, M., Wieland, M., Fauser, D., Steeb, H., Ludwigs, S.: From understanding mechanical behavior to curvature prediction of humidity-triggered bi- layer actuators. Adv. Mater. **33**, 2007982 (2021)
2. Fauser, D., Steeb, H.: Influence of humidity on the rheology of thermoresponsive shape memory polymers. J. Mater. Sci. **57**, 9508–9524 (2022)
3. Pflumm, S., Wiedemann, Y., Fauser, D., Safaraliyev, J., Lunter, D., Steeb, H., Ludwigs, S.: Autonomous adaption of intelligent humidity-programmed hydrogel patches for tunable stiffness and drug release. Adv. Mater. Technol. **8**, 2300937 (2023)
4. Mainik, P., Hsu, L.Y., Zimmer, C.W., Fauser, D., Steeb, H., Blasco, E.: DLP 4D printing of multi-responsive bilayered structures. Adv. Mater. Technol. **8**, 2300727 (2023)
5. Farhan, M., Klimm, F., Thielen, M., Rešetič, A., Bastola, A., Behl, M., Speck, T., Lendlein, A.: Artificial tendrils mimicking plant movements by mismatching modulus and strain in core and shell. Adv. Mater. **35**, 2211902 (2023)
6. Kampowski, T., Demandt, S., Poppinga, S., Speck, T.: Kinematical, structural and mechanical adaptions to desiccation in poikilohydric ramonda myconi (Gesnericeae). Front. Plant Sci. **9**, 1701 (2018)
7. Lakes, R.S.: Viscoelastic Materials. Cambridge University Press (2009)
8. Fauser, D., Beddrich, J., Wohlmuth, B., Steeb, H.: A fractional moisture-dependent viscoelasticity model for thermoplastic polymers. J. Rheol. **70**(4):1–16 (2026)
9. Fauser, D., Rodríguez Agudo, J.A., Madadi, H., Haeberle, J., Renner, J., Steeb, H.: Complex Poisson's ratio for viscoelastic materials: direct and indirect measurement methods and their correlation. Proc. R Soc. A **481**, 20240543 (2025)
10. Schwarzl, F., Staverman, A.J.: Time-temperature dependence of linear viscoelastic behavior. J. Appl. Phys. **23**, 838–843 (1952)
11. Williams, M.L., Landel, R.F., Ferry, J.D.: The temperature dependence of relaxation mechanisms in amorphous polymers and other glass-forming liquids. J. Am. Chem. Soc. **77**, 3701–3707 (1955)

12. Pelrine, R., Kornbluh, R., Pei, Q., Joseph, J.: High-speed electrically actuated elastomers with strain greater than 100%. Science **287**, 836–839 (2000)
13. Ilievski, F., Mazzeo, A.D., Shepherd, R.F., Chen, X., Whitesides, G.M.: Soft robotics for chemists. Angew. Chem. Int. Ed. **50**, 1890–1895 (2011)
14. Rus, D., Tolley, M.T.: Design, fabrication and control of soft robots. Nature **521**, 467–475 (2015)
15. Wieland, M., Dingler, C., Merkle, R., Maier, J., Ludwigs, S.: Humidity-controlled water uptake and conductivities in ion and electron mixed conducting polythiophene films. ACS Appl. Mater. Interfaces. **12**, 6742–6751 (2020)

Azobenzene-Based Photomechanical Materials for Soft Robotics: A Focus on Visible Light-Activated Liquid Crystal Networks

Thore Klüwer, Sven Schultzke, and Anne Staubitz

Abstract This chapter explores the development and application of azobenzene-based liquid crystal networks (LCNs) for soft robotics, with a particular emphasis on visible light-activated materials. The unique properties of tetra-*ortho*-fluorinated azobenzenes, including their photoisomerization behavior as well as their extremely long thermal half-life times in the switched state are discussed in detail. Tetra-*ortho*-fluorinated azobenzenes can also be incorporated in liquid crystalline networks. This chapter highlights the significance of photomechanical and photothermal effects in these materials and their relevance to soft robotic applications. The focus is on a novel, visible light-activated liquid crystalline monomer, which overcomes the limitations of traditional UV-light-activated systems. The implications of this advancement for soft robotics are explored, along with the challenges and future directions in this field.

Keywords Azobenzene · Tetra-*ortho*-fluorination · Liquid crystalline networks · Photomechanical effect

1 Introduction

Soft robotics [1] has emerged as a transformative field, leveraging soft, flexible, and adaptive materials to create robots capable of interacting with delicate environments. A key challenge in soft robotics is the development of intrinsically functional materials that can be actuated without external mechanical or hydraulic systems. Photomechanical materials, particularly those incorporating azobenzene, [2] have gained significant attention due to their ability to undergo macroscopic deformations induced by light.

T. Klüwer · S. Schultzke · A. Staubitz (✉)
Otto-Diels-Institute for Organic Chemistry, University of Kiel, Kiel, Germany
e-mail: staubitz@oc.uni-kiel.de

A. Raatz et al. (eds.), *Soft Material Robotic Systems*,
https://doi.org/10.1007/978-3-032-22453-8_2

Azobenzene, a molecule that undergoes reversible *E* to *Z* isomerization upon light irradiation, has been widely studied for its potential in photomechanical applications [2]. The unsubstituted parent molecule has a relatively short thermal half-life time for the *Z*-isomer (around 1.4 days at 35 °C) [3] and can only be switched with UV-light. Tetra-*ortho*-fluorinated azobenzenes however, can be switched with visible light [4]. This is a significant advantage if the application is underwater or in contact with living beings [5].

Certain azobenzenes can be incorporated into liquid crystalline polymer networks (LCNs), which then exhibit a bending deformation under light irradiation (photomechanical or occasionally photothermal effect), [6] making them ideal candidates for soft robotic actuators [7]. However, the use of UV light or even blue light, which is harmful to biological tissues [8] and has limited penetration depth, [9] has restricted their broader application. Recent advancements in visible light-activated azobenzene derivatives have addressed these limitations, enabling safer and more efficient actuation [10].

This chapter focuses on the development of a visible light-activated liquid crystalline monomer [10]. The monomer combines liquid crystalline properties and an azobenzene with visible light responsiveness, enabling the creation of photomechanical LCNs that can be actuated under safe and efficient conditions. The implications of this innovation for soft robotics are explored, along with the underlying mechanisms and challenges.

2 The Chemical Components and Physical Processes of Light Bendable Polymer Films

2.1 *Azobenzene and its Photoisomerization*

2.1.1 Azobenzene as a Photomechanical Switch

Azobenzene is a photoswitchable molecule that undergoes a reversible *E* to *Z* isomerization upon irradiation with light (Fig. 1). Both the *E*-and the *Z*-isomer show two electronic transitions: a strong symmetry-allowed $\pi\pi^*$ transition ($\lambda_{max} \approx 320$ nm [*E* to *Z*]; $\lambda_{max} \approx 270$ nm [*Z* to *E*]) as well as a weaker symmetry-forbidden $n\pi^*$ transition for the *E*-isomer ($\lambda_{max} \approx 450$ nm [*E* to *Z*]) and a symmetry-allowed $n\pi^*$ transition for the *Z*-isomer ($\lambda_{max} \approx 450$ nm [*Z* to *E*]) [11]. Due to the poorly separated $n\pi^*$ transitions, the *E* to *Z* isomerization relies on UV-light, which excites the well-separated $\pi\pi^*$ transition. However the downside is that UV-light is harmful to living beings [8] and has a very low penetration depth in materials and liquids [9].

The isomerization process involves a significant change in the molecular geometry, with the *E*-isomer being planar and thermally stable, while the *Z*-isomer is nonplanar and metastable. The distance between the *para*-position carbon atoms changes

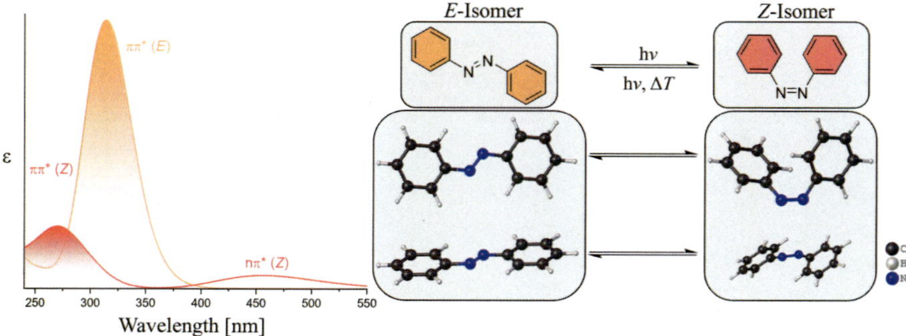

Fig. 1 Light-induced switching of azobenzene: calculated spectra. The *E* to *Z* isomerization results in a reduction in molecular length by approximately one-third and a transition from a planar to a twisted geometry [13]. Adapted from the M.Sc. thesis of Hinrich Reller. Copyright Hinrich Reller [15]

from 9 Å in the *E*-isomer [12] to 5.5 Å in the *Z*-isomer, [13] leading to a substantial alteration in the molecular shape from an elongated, rod-like molecule to a more bent, three-dimensional structure [14]. This is significant, because, as discussed later, the geometrical change on the nanoscale can be translated into anisotropic volume changes in a material and thus a macroscopic volume change resulting in a bending movement [6]. One challenge with azobenzene is that the metastable *Z*-isomer slowly relaxes back to the *E*-isomer via a thermal pathway (approx. 1.4 days at 35 °C) [3]. This background reaction cannot be prevented, limiting applications since material in the *Z*-state is constantly converted to the *E*-state.

2.1.2 Tetra-*Ortho*-Fluorinated Azobenzenes: Properties and Syntheses

Tetra-*ortho*-substituted azobenzenes, particularly those incorporating fluorine atoms, exhibit remarkable photochemical and thermal properties [4]. The substitution of all four *ortho*-positions with fluorine strongly enhances the thermal stability of the *Z*-isomer, resulting in significantly extended thermal half-lives compared to other azobenzene derivatives [4]. Notably, *ortho*-fluorine substituents stabilize the *Z*-isomer to such an extent that thermal half-life time values of up to 2 years at room temperature have been reported [4]. The mechanism of thermal relaxation is still discussed and many factors appear to play a role, a discussion of which is beyond the scope of this chapter [16]. Tetra-*ortho*-fluorination in combination with the incorporation of the azobenzene moiety in a macrocycle has led to the longest reported half-life times for azobenzenes of 120 years in DMSO at 25 °C (extrapolated) [17] (Fig. 2).

Another effect is the clearer and wider separation between the nπ* transition of the *E*-and the *Z*-isomer, enabling the use of this transition for molecular switching [4]. Therefore, switching in the visible range becomes possible.

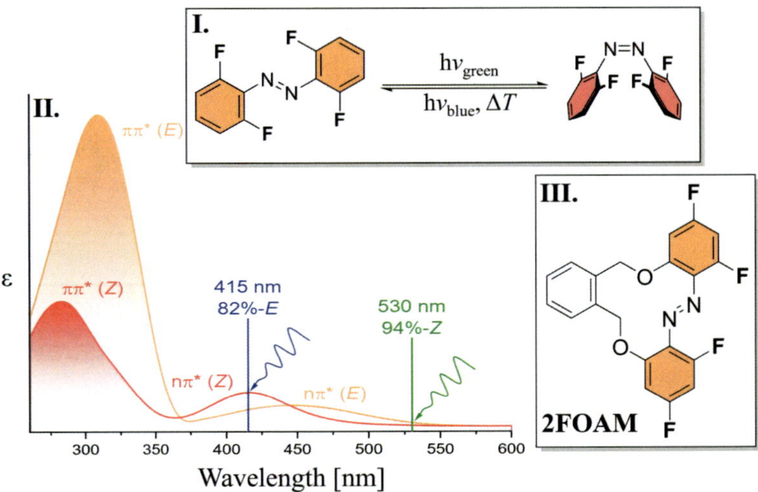

Fig. 2 **I**. Isomerization of tetra-*ortho*-fluorinated azobenzene induced by light **II**. Illustration of the measured UV-Vis spectra of the respective *E*-and *Z*-isomer **III**. The fluorinated oxygen-bridged azobenzene macrocycle (**2FOAM**) exhibits a thermal half-life time of 120 years (extrapolated to 25 °C in DMSO). Adapted from the M.Sc. thesis of Hinrich Reller. Copyright Hinrich Reller [15]

Synthetically, the most common method is an oxidative coupling of 2,6-difluoroanilines, in which the other desired substituents are already in place or which contain functionalizable groups. Further strategies for the synthesis of azo motifs include the diazotization of aromatic amines and subsequent coupling with electron-rich aromatic nucleophiles. In another method, known as the Mills reaction, [18] aromatic nitroso derivatives react with anilines to form the corresponding azo compounds [19].

2.2 *Liquid Crystals and their Properties*

2.2.1 Liquid Crystalline Molecules

Liquid crystals (LCs) are materials that exhibit properties intermediate between liquids and crystals, combining the fluidity of liquids with the ordered structure of crystals. They are classified into several types, including nematic, smectic, and cholesteric phases, each with distinct ordering characteristics (Fig. 3). Nematic liquid crystals, in particular, are characterized by their long-range orientational order; they typically contain a rigid, rod-like molecule to impart the crystal-like properties as well as a flexible linker. The phase transitions of pure LCs are thermotropic; however, there are a number of alignment techniques [20] to arrange the LCs with respect to a desired geometry. Liquid crystals can be used in a vast range of applications, [21]

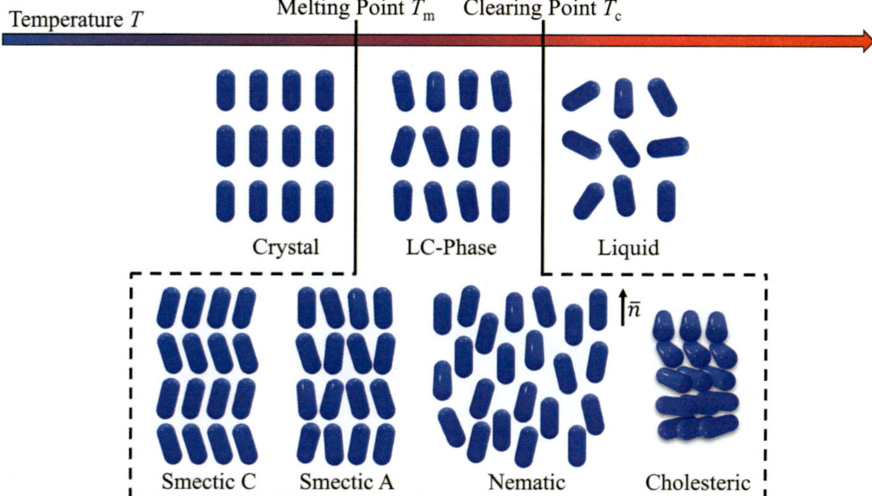

Fig. 3 Schematic representation of the liquid crystalline phase categorized into several types depending on the respective alignment of mesogens. Adapted from the Ph.D. thesis of Sven Schultzke. Copyright 2024 Sven Schultzke [23]

but in this chapter, we focus on photoresponsive liquid crystals with an azobenzene core.

2.2.2 Liquid Crystals Fixed in Liquid Crystalline Networks (LCNs)

When the liquid crystal molecules are synthesized with a polymerizable terminal group, the liquid crystalline phase present at the temperature of polymerization can be fixed in place and the chosen alignment geometry is by and large retained, although the post-polymerization LC phase can have a slightly different ordering (i. e. different LC phase type). The properties of these "solidified liquid crystals depend on the degree of cross-linking"; low degrees of cross-linking lead to liquid crystalline elastomers, [22] whereas high degrees result in liquid crystalline networks.

2.3 Photomechanical and Photothermal Effects in Azobenzene Based LCNs

Azobenzenes are well-suited for functional LCs, due to their rigid, rod-like structure, which bends upon irradiation. When azobenzene is incorporated into liquid crystalline polymer networks, the molecular-level shape change is amplified, resulting in macroscopic bending or deformation of the material [6]. Upon irradiation, the *E* to *Z*

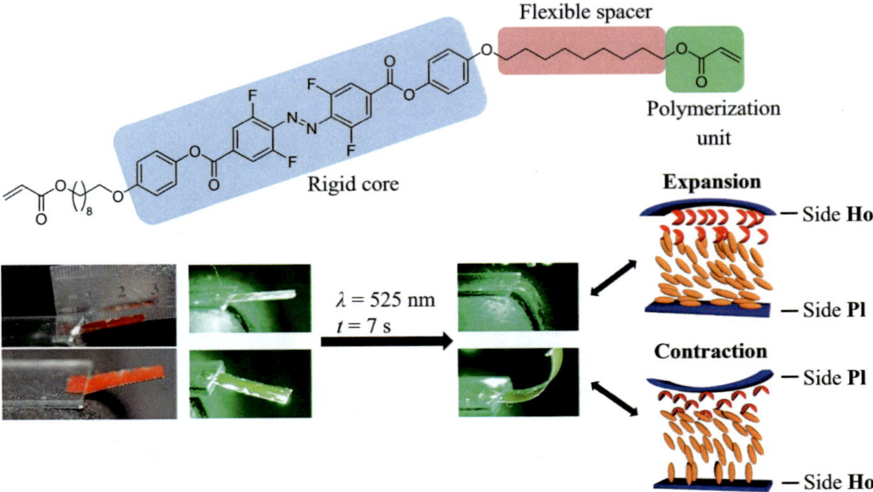

Fig. 4 Illustration of the liquid crystalline monomer that can be polymerized into a photomechanical film. At the bottom, the mechanistic explanation for the direction of bending of the film is shown: In a homeotropic alignment (Ho), isomerization increases lateral expansion, because the azobenzenes are perpendicular to the plane of the film. *E* to *Z* isomerization increases the thickness of the azobenzenes to approximately 3 Å [13]. On the contrary, in a planar alignment (Pl) the lateral expansion decreases, because the azobenzenes are parallel to the plane of the film and decrease their length due to *E* to *Z* isomerization. Adapted with permission from [10], licensed under CC-BY 4.0

isomerization of azobenzene induces a change in the molecular orientation, leading to a contraction or expansion of the material [24]. This effect is highly dependent on the alignment of the azobenzene molecules: If the azobenzenes are perpendicular to the surface normal, irradiation on this side leads to a surface contraction; however, if they are aligned parallel to the surface normal an expansion is observed (Fig. 4) [10, 25].

In addition to the photomechanical effect, photothermal effects also play a role in the deformation of LCNs. The absorption of light by azobenzene does not always lead to a switching event, but the depletion of the excited states (S_2 or S_1) can also occur via vibrational relaxation pathways [29]. In addition, the thermal *Z* to *E* relaxation forms heat. These exothermic processes can cause thermal expansion of the material, which leads to bending, if only those layers heat up, which are irradiated. While photothermal effects can complicate the interpretation of photomechanical behavior, they can also be harnessed to enhance the actuation of soft robotic devices. In practice, the two effects are relatively difficult to separate. One method is switching under water, as water serves as an effective heat sink [26].

3 Visible Light-Activated Liquid Crystalline Monomers

Traditional azobenzene-based LCNs are activated by UV light, which has several limitations, including potential harm to biological tissues and limited penetration depth [5, 8, 9].To address these challenges, visible light-activated azobenzene derivatives have gained attention, such as tetra-*ortho*-fluorinated azobenzenes. However, despite the flat crystal structure of *E*-tetra-*ortho*-substituted azobenzenes, [27] it did not exhibit any liquid crystallinity (Fig. 5I) [28]. There are methods to overcome this issue by blending in these light switchable azobenzenes with proven liquid crystal mixtures, leading to bending effects upon irradiation [28]. The azobenzene loading in these mixtures is relatively low, which is both an advantage, as the tetra-*ortho*-substituted azobenzene tends to be the non-commercial and therefore cost intensive part. But a disadvantage is the dilution of the dye and lower effectiveness in bending.

To address these issues, a novel liquid crystalline monomer was developed, which combines the photoswitching properties of azobenzene with visible light responsiveness (Fig. 5II) [10]. The design principle was an extension of the rigid core; crystal structures of model compounds proved however, that this molecule does not possess an entirely flat core but rather, there are interactions between the fluorine atoms and neighboring mesogens. The monomer exhibits a highly fluid nematic phase, but upon polymerization it is fixed into a layered smectic C phase [10]. The resulting films could be actuated with visible light and exhibited reversible bending behavior, with bending angles of up to 40° within 30 s. Not only are these very high in comparison with other known systems (see however porous films: [29]); they are also observed in the thickest films that were prepared up to date (60 μm). This is of practical importance, as too thin films are easily mechanically broken.

Furthermore, bending experiments were carried out in water (Fig. 6), with water serving as the cooling medium due to the higher heat convection coefficient compared to air. The successful implementation suggests that the cause of the bending is predominantly photomechanical rather than photothermal. The thermal half-life time of the polymer was estimated to be roughly 17 h, whereas the monomer exhibited values of 49 d at 25 °C in toluene. The much shorter half-life time of the polymer

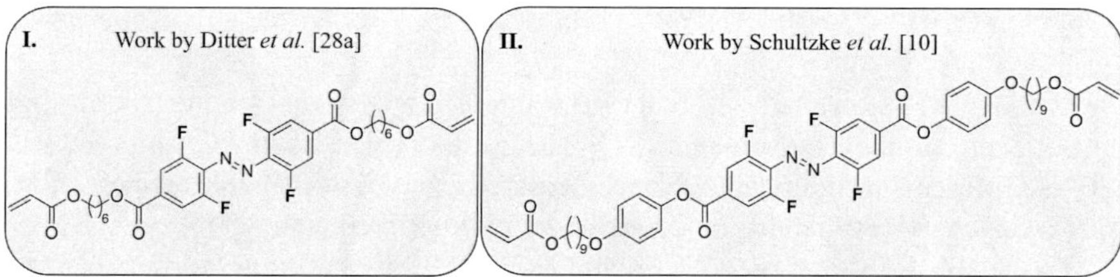

Fig. 5 **I**. Design of a tetra-*ortho*-fluorinated azobenzene, which did not exhibit liquid crystalline properties, used as a crosslinker in combination with LC monomers [28] **II**. Further development of the design based on previous reports, combining the tetra-*ortho*-fluorinated azobenzene motif with a phenyl and acrylate unit. The expansion of the rigid core results in a reintroduction of liquid crystallinity. From [10], licensed under CC-BY 4.0

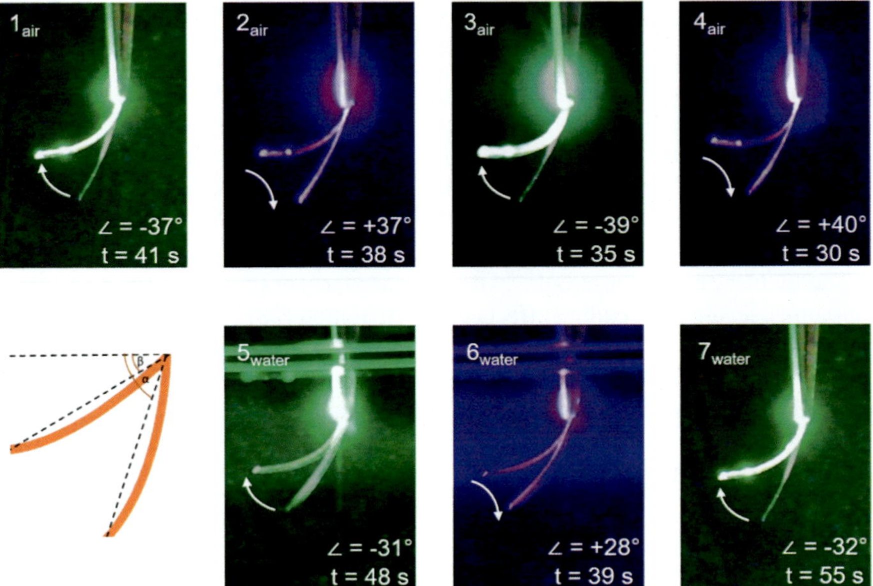

Fig. 6 The study of Schultzke et al. [10] examines how the bending angles of a polymer film consisting exclusively of the tetra-*ortho*-fluorinated azobenzene monomer change in air (top row) and water (bottom row) depending on the illumination wavelength (525 nm or 420 nm). The angles shown represent the difference between the initial and final positions ($\angle =$ starting angle—ending angle; see schematic for reference). The sign indicates the direction of change upon illumination, while the time value specifies how long the film was exposed to light to reach this state. Reprinted with permission from [10], licensed under CC-BY 4.0

might be attributed to mechanical strain [10]. However, in the literature, much shorter half-life times of approx. 4 h for light switchable azobenzene containing LCNs were observed [30].

4 Relevance to Soft Robotics

Soft robotics relies on materials that can be actuated by external stimuli, such as light, to perform complex movements [31]. The development of visible light-activated LCNs represents a significant advancement in this field, as it enables the creation of soft robotic devices that can be controlled with high precision.

The visible light-activated LCNs and LCEs exhibit several properties that make them particularly suitable for soft robotics: First, the bending is reversible. This enables repetitive actuation. It is important to note that thermal relaxation can differ significantly between solution and polymer environments, making it difficult to predict these differences prior to incorporating an azobenzene switch into a LCN or LCE. In the case of the discussed example, [10] the half-life time of bending in the film was 17 h, whereas the monomer in solution exhibited a thermal half-life time of 49 days. Also, while thermal relaxation in solution follows a simple first order kinetic, this is no longer the case for the polymer and therefore, this issue remains to be explored further.

A very relevant advantage of the visible light switchable film is the actuation under water: aquatic environments are relatively common in soft robotics [32] because of the application in biologically relevant systems.

5 Challenges and Future Directions

Despite the significant advancements in visible light-activated LCNs, several challenges remain. These include the need for improved thermal stability of the Z-isomer, which can limit the duration of actuation. For this, not only are switches with extensive half-life times required; what also needs to be understood is the anisotropy of the change in modulus upon switching and how mechanical strain induced is released mechanistically.

Additionally, the upscaling of the material synthesis and the integration of these materials into soft robotic devices require further research. With the photomechanical film discussed in this chapter, one challenge is always the processing, which is delicate and at present can only deliver relatively small pieces of polymer film. A very promising improvement in this regard is 3D-printing (or 4D-printing as a functional material is employed) [33]. For example, with this technique a photomechanical liquid crystal elastomer (LCE) actuator was printed that was operative in biologically relevant phosphate-buffered saline (PBS) [33]. A reversible contraction of 7% of the initial length was achieved under a 1 g load, with half the maximum contraction reached in just 3 s at 37 °C (human body temperature). Unlike azobenzene containing hydrogels, which are also discussed as bistable [34] or actuating materials [35] in biological contexts, these actuators were salt-insensitive and retained functionality in PBS, where efficient heat dissipation minimized photothermal effects, enabling a predominantly photochemical response. A 5–6% photochemical contraction persisted after UV light was removed and reversibly reset with blue light.

Furthermore, at the moment, most materials are not very sustainable in the sense that they can only be used once in the shape they have been prepared. One significant advancement is the use of mere physical cross-links (hydrogen bonds), [33] but more such techniques are very desirable.

Future directions in this field may include the development of new photoswitchable molecules with improved switching properties and the exploration of hybrid materials that combine the benefits of LCNs with other stimuli-responsive materials. Materials with the additional function of self-healing or shape-memory effects would be of interest. Eventually, also the engineering side will need to be addressed: How can light be delivered in a controlled manner under dynamic bending and in a device? How can large forces be generated, or, even better, adaptable forces? Eventually, the use of photomechanical materials will be an interdisciplinary endeavor, which is a highly promising undertaking.

6 Conclusion

The development of visible light-activated liquid crystalline polymer networks (LCNs) represents a significant breakthrough in the field of soft robotics. These materials combine the photoswitching properties of azobenzene with the anisotropic mechanical properties of liquid crystals, enabling precise and efficient actuation under visible light. Although tetra-*ortho*-fluorinated azobenzenes were thought to be not liquid crystalline, the elongation of the rigid core led to a novel liquid crystalline monomer and, further, polymer with visible light responsiveness. Not only are the bending angles very high; bending can also be achieved under water with only an insignificant loss in bending magnitude. This material has much potential for the design of soft robotic devices, because it enables contactless bending with a high half-life time. As research in this field continues to advance, the potential applications of these materials in areas such as artificial muscles, grippers, and locomotion systems are expected to expand, paving the way for a new generation of soft robotic technologies.

Acknowledgments The research concerning the visible light switchable photomechanical film was funded by the German Research Foundation (DFG) within the priority program SPP 2100 "Soft Material Robotic Systems", Subproject STA1195/5-1, "Insect feet inspired concepts soft touch grippers with dynamically adjustable grip strength".

Competing Interests The authors declare no competing interests.

References

1. (a) Appiah, C., Arndt, C., Siemsen, K., Heitmann, A., Staubitz, A., Selhuber-Unkel, C.: Living materials herald a new era in soft robotics. Adv. Mater. **31**, n/a (2019). https://doi.org/10.1002/adma.201807747; (b) Wallin, T.J., Pikul, J., Shepherd, R.F.: 3D printing of soft robotic systems. Nature Rev. Mater. **3**, 84–100 (2018). doi: 10.1038/s41578-018-0002-2; (c) Whitesides, G.M.: Soft robotics. Angew. Chem. Int. Ed. **57**, 4258–4273 (2018). doi: 10.1002/anie.201800907; (d) Sachyani Keneth, E., Kamyshny, A., Totaro, M., Beccai, L., Magdassi, S.: 3D printing materials for soft robotics. Adv. Mater. **33**, 2003387 (2021). doi: 10.1002/adma.202003387
2. (a) Mahimwalla, Z., Yager, K.G., Mamiya, J.-i., Shishido, A., Priimagi, A., Barrett, C.J.: Azobenzene photomechanics: prospects and potential applications. Polym. Bull. **69**, 967–1006 (2012). https://doi.org/10.1007/s00289-012-0792-0; (b) Pang, X., Lv, J.-a., Zhu, C., Qin, L., Yu, Y.: Photodeformable Azobenzene-containing liquid crystal polymers and soft actuators. Adv. Mater. **31**, 1904224 (2019). doi: 10.1002/adma.201904224
3. Talaty, E.R., Fargo, J.C.: Thermal cis–trans-isomerization of substituted azobenzenes: a correction of the literature. Chem. Commun., 65–66 (1967). https://doi.org/10.1039/C19670000065

4. (a) Bleger, D., Schwarz, J., Brouwer, A.M., Hecht, S.: O-Fluoroazobenzenes as readily synthesized Photoswitches offering nearly quantitative two-way isomerization with visible light. J. Am. Chem. Soc. **134**, 20597–20600 (2012). https://doi.org/10.1021/ja310323y; (b) Knie, C., Utecht, M., Zhao, F., Kulla, H., Kovalenko, S., Brouwer, A.M., Saalfrank, P., Hecht, S., Bleger, D.: Ortho-Fluoroazobenzenes: visible light switches with very long-lived Z isomers. Chem. Eur. J. **20**, 16492–16501 (2014). doi: 10.1002/chem.201404649
5. Lerch, M.M., Hansen, M.J., van Dam, G.M., Szymanski, W., Feringa, B.L.: Emerging targets in Photopharmacology. Angew. Chem. Int. Ed. **55**, 10978–10999 (2016). https://doi.org/10.1002/anie.201601931
6. (a) Barrett, C.J., Mamiya, J.-i., Yager, K.G., Ikeda, T.: Photo-mechanical effects in azobenzene-containing soft materials. Soft Matter **3**, 1249–1261 (2007). https://doi.org/10.1039/B705619B; (b) Yu, Y., Nakano, M., Ikeda, T.: Directed bending of a polymer film by light. Nature **425**, 145–145 (2003). doi: 10.1038/425145a
7. (a) Yamada, M., Kondo, M., Mamiya, J.-i., Yu, Y., Kinoshita, M., Barrett, C.J., Ikeda, T.L: Photomobile polymer materials: towards light-driven plastic motors. Angew. Chem. Int. Ed. **47**, 4986–4988 (2008). https://doi.org/10.1002/anie.200800760; (b) Kizilkan, E., Strueben, J., Staubitz, A., Gorb, S.N.: Bioinspired photocontrollable microstructured transport device. Science Robot. **2** (2017). doi: 10.1126/scirobotics.aak9454
8. Opländer, C., Hidding, S., Werners, F.B., Born, M., Pallua, N., Suschek, C.V.: Effects of blue light irradiation on human dermal fibroblasts. J. Photochem. Photobiol. B Biol. **103**, 118–125 (2011). https://doi.org/10.1016/j.jphotobiol.2011.02.018
9. Nababan, B., Ulfah, D., Panjaitan, J.P.: Light propagation, coefficient attenuation, and the depth of one optical depth in different water types. IOP Conf. Ser.: Earth Environ. Sci. **944**, 012047 (2021). https://doi.org/10.1088/1755-1315/944/1/012047
10. Schultzke, S., Scheuring, N., Puylaert, P., Lehmann, M., Staubitz, A.: A photomechanical film in which liquid Crystal Design shifts the absorption into the visible light range. Adv. Sci. **10**, 2302692 (2023). https://doi.org/10.1002/advs.202302692
11. (a) Bandara, H.M.D., Burdette, S.C.: Photoisomerization in different classes of azobenzene. Chem. Soc. Rev. **41**, 1809–1825 (2012). https://doi.org/10.1039/C1CS15179G; (b) Vetráková, L., Ladányi, V., Anshori, J.Al., Dvořák, P., Wirz, J., Heger, D.: The absorption spectrum of cis-azobenzene. Photochem. Photobiol. Sci. **16**, 1749–1756, (2017). doi: 10.1039/c7pp00314e
12. Bouwstra, J.A., Schouten, A., Kroon, J.: Structural studies of the system trans-azobenzene/trans-stilbene. I. A reinvestigation of the disorder in the crystal structure of trans-azobenzene, C12H10N2. Acta Cryst. C. **39**, 1121–1123 (1983). https://doi.org/10.1107/S0108270183007611
13. Mostad, A., Rømming, C., Hammarström, S., Lousberg, R., Weiss, U.: A refinement of the crystal structure of cis-Azobenzene. Acta Chem. Scand. **25**, 3561–3568 (1971)
14. Merino, E., Ribagorda, M.: Control over molecular motion using the cis–trans photoisomerization of the azo group. Beilstein J. Org. Chem. **8**, 1071–1090 (2012). https://doi.org/10.3762/bjoc.8.119
15. Reller, H.K.: Master Thesis, University of Kiel (2025)
16. Axelrod, S., Shakhnovich, E., Gómez-Bombarelli, R.: Thermal half-lives of Azobenzene derivatives: virtual screening based on intersystem crossing using a machine learning potential. ACS Central Science. **9**, 166–176 (2023). https://doi.org/10.1021/acscentsci.2c00897
17. Schultzke, S., Puylaert, P., Wang, H., Schultzke, I., Gerken, J., Staubitz, A.: Centennial isomers: A unique fluorinated Azobenzene Macrocyclus with dual stability over 120 years. Adv. Funct. Mater. **34**, 2313268 (2024). https://doi.org/10.1002/adfm.202313268
18. Mills, C.: XCIII.—Some new azo-compounds. J. Chem. Soc. Trans. **67**, 925–933 (1895). https://doi.org/10.1039/CT8956700925
19. Volaric, J., Buter, J., Schulte, A.M., Van den Berg, K.O., Santamaría-Aranda, E., Szymanski, W., Feringa, B.L.: Design and synthesis of visible-light-responsive Azobenzene building blocks for chemical biology. J. Org. Chem. **87**, 14319–14333 (2022). https://doi.org/10.1021/acs.joc.2c01777; (b) Jerca, F.A., Jerca, V.V., Hoogenboom, R.: Advances and opportunities in the

exciting world of azobenzenes. Nature Rev. Chem. **6**, 51–69 (2022). doi: 10.1038/s41570-021-00334-w; (c) Merino, E.: Synthesis of azobenzenes: the coloured pieces of molecular materials. Chem. Soc. Rev. **40**, 3835–3853 (2011). doi: 10.1039/C0CS00183J

20. Zhao, J., Zhang, L., Hu, J.: Varied alignment methods and versatile actuations for liquid crystal elastomers: A review. Adv. Intell. Sys. **4**, 2100065 (2022). https://doi.org/10.1002/aisy.202100065
21. Zhang, Z., Yang, X., Zhao, Y., Ye, F., Shang, L.: Liquid crystal materials for biomedical applications. Adv. Mater. **35**, 2300220 (2023). https://doi.org/10.1002/adma.202300220
22. Herbert, K.M., Fowler, H.E., McCracken, J.M., Schlafmann, K.R., Koch, J.A., White, T.J.: Synthesis and alignment of liquid crystalline elastomers. Nature Rev. Mater. **7**, 23–38 (2022). https://doi.org/10.1038/s41578-021-00359-z
23. Schultzke, S.: Dissertation, University of Bremen, Germany (Bremen) (2024)
24. Yu, H.: Recent advances in photoresponsive liquid-crystalline polymers containing azobenzene chromophores. J. Mater. Chem. C. **2**, 3047–3054 (2014). https://doi.org/10.1039/C3TC31991A
25. van Oosten, C.L., Harris, K.D., Bastiaansen, C.W.M., Broer, D.J.: Glassy photomechanical liquid-crystal network actuators for microscale devices. Eur. Phys. J. E. **23**, 329–336 (2007). https://doi.org/10.1140/epje/i2007-10196-1
26. Verpaalen, R.C.P., Pilz da Cunha, M., Engels, T.A.P., Debije, M.G., Schenning, A.P.H.J.: Liquid crystal networks on thermoplastics: reprogrammable photo-responsive actuators. Angew. Chem. Int. Ed. **59**, 4532–4536 (2020). https://doi.org/10.1002/anie.201915147
27. Hermann, D., Schwartz, H.A., Ruschewitz, U.: Crystal structures of Z and E ortho-Tetrafluoroazobenzene. Chemistry Select. **2**, 11846–11852 (2017). https://doi.org/10.1002/slct.201702185
28. (a) Ditter, D., Braun, L.B., Zentel, R.: Influences of Ortho-Fluoroazobenzenes on liquid crystalline phase stability and 2D (planar) actuation properties of liquid crystalline elastomers. Macromol. Chem. Phys. **221**, 1900265 (2020). https://doi.org/10.1002/macp.201900265; (b) Kumar, K., Knie, C., Bléger, D., Peletier, M.A., Friedrich, H., Hecht, S., Broer, D.J., Debije, M.G., Schenning, A.P.H.J.: A chaotic self-oscillating sunlight-driven polymer actuator. Nat. Commun. **7**, 11975 (2016). doi: 10.1038/ncomms11975
29. Kizilkan, E., Strueben, J., Jin, X., Schaber, C.F., Adelung, R., Staubitz, A., Gorb, S.N.: Influence of the porosity on the photoresponse of a liquid crystal elastomer. R. Soc. Open Sci. **3**, 150700/150701 (2016). https://doi.org/10.1098/rsos.150700
30. Liu, D., Broer, D.J.: New insights into photoactivated volume generation boost surface morphing in liquid crystal coatings. Nat. Commun. **6**, 8334 (2015). https://doi.org/10.1038/ncomms9334
31. (a) Shen, Z., Chen, F., Zhu, X., Yong, K.-T., Gu, G.: Stimuli-responsive functional materials for soft robotics. J. Mater. Chem. B **8** (2020). https://doi.org/10.1039/D0TB01585G; (b) Dou, W., Zhong, G., Cao, J., Shi, Z., Peng, B., Jiang, L.: Soft robotic manipulators: designs, actuation, stiffness tuning, and sensing. Adv. Mater. Tech. **6**, 2100018 (2021). doi: 10.1002/admt.202100018
32. Qu, J., Xu, Y., Li, Z., Yu, Z., Mao, B., Wang, Y., Wang, Z., Fan, Q., Qian, X., Zhang, M., Xu, M., Liang, B., Liu, H., Wang, X., Wang, X., Li, T.: Recent advances on underwater soft robots. Adv. Intell. Sys. **6**, 2300299 (2024). https://doi.org/10.1002/aisy.202300299
33. (a) Ceamanos, L., Mulder, D.J., Kahveci, Z., López-Valdeolivas, M., Schenning, A.P.H.J., Sánchez-Somolinos, C.: Photomechanical response under physiological conditions of azobenzene-containing 4D-printed liquid crystal elastomer actuators. J. Mater. Chem. B. **11**, 4083–4094 (2023). https://doi.org/10.1039/D2TB02757G; (b) Lugger, S.J.D., Ceamanos, L., Mulder, D.J., Sánchez-Somolinos, C., Schenning, A.P.H.J.: 4D printing of Supramolecular liquid crystal elastomer actuators fueled by light. Adv. Mater. Technol. **8**, 2201472 (2023). doi: 10.1002/admt.202201472
34. Colaco, R., Appiah, C., Staubitz, A.: Controlling the LCST-phase transition in Azobenzene-functionalized poly (*N*-Isopropylacrlyamide) hydrogels by light. Gels. **9**, 1–18 (2023). https://doi.org/10.3390/gels9020075

35. Mauro, M.: Gel-based soft actuators driven by light. J. Mater. Chem. B. **7**, 4234–4242 (2019). https://doi.org/10.1039/C8TB01893F

Crosslinked Elastomers for Triboelectricity and piezoelectricity-Driven Tactile Sensors and Arrays in Soft Robotic Systems

Injamamul Arief, Anik Kumar Ghosh, Andreas Fery, and Amit Das

Abstract Tactile sensing is essential for advancements in robotics, wearable electronics, and human–machine interfaces. Hybrid piezoelectric-triboelectric nanogenerators (HPTENGs) have emerged as self-powered tactile sensors capable of converting mechanical stimuli into usable electrical signals. Despite substantial advancements in materials and device engineering, the integration of commercial rubbers into HPTENGs offers both exciting opportunities and notable challenges. This review summarizes the recent findings within the framework of Deutsche Forschungsgemeinschaft (DFG) special priority program (SPP2100) highlighting fundamental principles, materials strategies, fabrication approaches, and unresolved research questions. By incorporating recent developments and critical insights, this report aims to facilitate the design of robust, energy-efficient, and cost-effective tactile sensors optimized for industrial and wearable soft robotic applications.

1 Introduction

Modern robotics, wearable systems, and healthcare devices increasingly require soft, reliable, and highly sensitive tactile sensors. Traditional rigid sensors typically lack the mechanical compliance necessary for seamless human–machine interactions, highlighting the necessity for elastomers, often referred to as "commercial rubbers," in advanced sensor technologies [1]. Elastomer-based tactile sensors can accommodate deformations, interface with rough surfaces, and recover from repeated

I. Arief (✉) · A. K. Ghosh · A. Fery · A. Das
Leibniz Institute of Polymer Research Dresden, Dresden, Germany
e-mail: arief@ipfdd.de

A. K. Ghosh
e-mail: anik-ghosh@ipfdd.de

A. Fery
e-mail: fery@ipfdd.de

A. Das
e-mail: das@ipfdd.de

A. Raatz et al. (eds.), *Soft Material Robotic Systems*,
https://doi.org/10.1007/978-3-032-22453-8_3

mechanical stress within a shorter timeframe. The superior mechanical compliance of soft robotic systems arises primarily from the exceptional mechano-adaptive and mechano-electric properties inherent to elastomers [2–6]. Furthermore, compliant robots equipped with engineered metasurfaces provide high sensitivity to various stimuli, including pressure and force. However, sensor prototypes derived from thermoplastics often encounter issues such as limited mechanical compliance, inadequate stretchability, poor long-term cyclic performance, and compatibility challenges [5]. Conversely, commercial rubbers demonstrate superior responsiveness, mechanical stability under sustained stress, and seamless integration capabilities with robotic segments [6].

To enable precise robotic control, tactile sensors are optimized specifically for physiological motion detection. Several sensing principles, including piezoresistive, piezoelectric, and triboelectric nanogenerators, have been explored for developing pressure-sensitive and tactile integrated sensors. In this project, we specifically employed hybrid piezoelectric-triboelectric-piezoresistive modules, alongside individual sensing components, to develop integrated tactile systems for soft robotics. Recent advancements suggest that combining triboelectric nanogenerators (TENGs), piezoelectric nanogenerators (PENGs), and piezoresistive strain sensors provides an effective approach for developing tactile-sensitive robotic limbs [2–4, 7, 8]. TENGs generate charge via surface contact and separation, whereas PENGs take advantage of electric polarization under mechanical strain [9]. The integration of these mechanisms produces hybrid piezoelectric-triboelectric nanogenerators (HPTENGs), which exhibit significantly enhanced power outputs, sensitivity, and broader frequency responses compared to single-mode sensor devices [10–12]. In this context, commercial rubbers offer significant advantages, including cost-effectiveness, ease of processing, and excellent mechanical performance, making them highly suitable for sensor matrices [13].

Despite these advantages, ensuring consistent and sustainable performance, particularly over thousands of operational cycles in real-world applications, remains a challenge [14, 15]. Additionally, designing commercial elastomer-based metasurfaces for cutting-edge electronic skin fabrication requires further control of the processes involving nanotechnology, micropatterning and lithography-based processes. The following sections summarize significant research progress as part of this project and highlight knowledge gaps and upcoming and current challenges in soft robotic research.

2 Commercial Rubbers-Based Piezoresistive Strain Sensors for Soft Robotic Segments

2.1 Hydrocarbon Softener Enabled Ultrasoft Natural Rubber (NR) Composites for Soft Robots

Since there are certain limitations in producing ultra-soft NR by altering just the crosslinking densities, strategies were employed to use low molecular weight hydro-

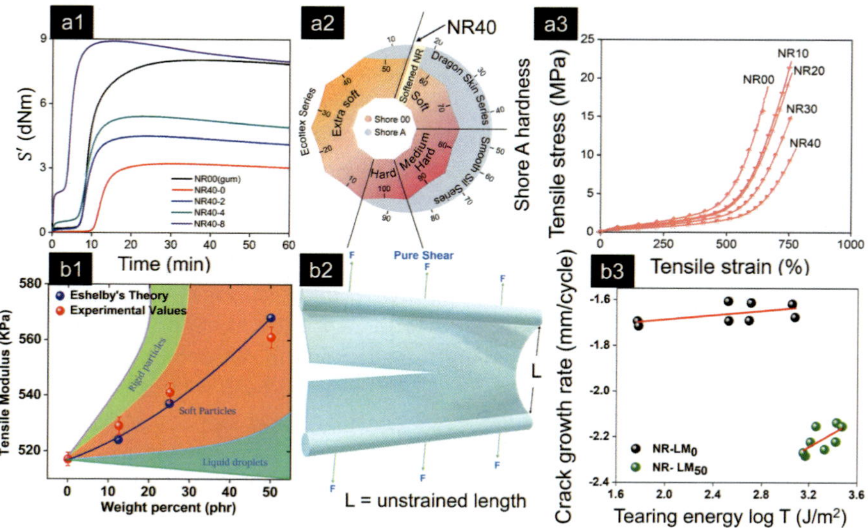

Fig. 1 (a1) Torque-time profiles illustrating the curing characteristics of softened NR composites, with S′ denoting torque in deciNewton meters (dNm). (a2) Softness of NR composites measured in Shore A hardness compared to silicone rubbers. (a3) Uniaxial stress-strain curves for unsubstituted and softened NR. Panels (a1)–(a3) adapted with permission from Banerjee et al., Appl. Mater. Today, 2021, 25, 101219 [3]. Copyright 2021 Elsevier. (b1) Tensile modulus at 100% elongation versus LM content (wt%), (the dashed line shows Eshelby's prediction, illustrating an increase in tensile moduli with soft particle inclusion). (b2) Schematic of the pure shear process for tear fatigue analysis. (b3) Comparison of fracture properties (crack growth rate vs. tearing energy) for unfilled NR and NR-LM50. Panels (b1)–(b3) adapted with permission from from Banerjee et al., ACS Appl. Mater. Interfaces, 2021, 13 (13), 15610–15620. Copyright 2021 American Chemical Society [5]

carbons that could act as inactive softening agents for crosslinked NR. For this, a hydrocarbon based softener was used that contained no unsaturated olefins and with high branching character that had a significant impact on the miscibility of elastomer and the additives [3]. The resulting composite showed remarkable softness following the addition of the softener (Fig. 1a1–a3). The alteration of these mechanical properties by the softener loading could also be associated to the cohesive forces and chain mobility of polymer networks [3, 16]. We demonstrated that commercial silicone-based soft robotic body parts can be designed using these state-of-the-art soft NR composites, marking a significant advancement in soft robotics.

The hydrocarbon-softened NR composites described here were specifically employed in the development of piezoresistive strain sensors embedded in soft robotic actuators, as detailed subsequently in Sec. 2.2. This utilization emphasizes the critical role of the hydrocarbon softener in achieving the necessary softness and mechanical adaptability required for effective integration in sensors and actuators.

2.2 Soft Piezoresistive Strain Sensors Embedded on Soft Robotic Actuators

Piezoresistive soft rubber composites are ubiquitous in strain sensing that manifests in a dramatic increment of electrical resistivity upon elongation [8, 17]. Embedding a piezoresistive strain sensor into a soft robotic arm has been a challenging task in terms of surface compatibility, shape, and dynamics of the soft robotic components. We demonstrated a super-elastic, ultrasoft NR composite containing multiwalled carbon nanotubes (MWCNTs) in presence of a softener. The resulting conducting elastomer offered low electrical percolation, ultra-softness, elastic modulus in the kilopascal (kPa) range, ultra-stretchability and high tensile strength. We demonstrated the functionality of the sensor with respect to custom developed soft robotic actuator movement. The initial and on-strain values of resistance corroborated with the parameters obtained from free-standing sensing analyses. The proof-of-concept sensor-integrated soft robot studies pave the path to future development of proprioceptive sensing robots and soft robotic segments [3].

2.3 Development of Super-Tough and Ultra-Stretchable Liquid Metal-Embedded Natural Rubber for Soft Robotic Body Parts

Commercial elastomers offer significant advantages but still present challenges, including complex functionalization processes and relatively poor dielectric responses [3, 5]. Additionally, typical industrial elastomers rarely achieve softness comparable to biological tissues or skin. To introduce enhanced thermal and electrical conductivities into soft elastomeric systems, recent studies have explored blending functional fluids into elastomers. A prominent example is the incorporation of liquid metals (LM) into liquid or gel rubbers, yielding composites with markedly improved properties [18]. Several studies demonstrated excellent compatibility with polydimethylsiloxane (PDMS). To overcome the limitations associated with PDMS, we developed liquid metal-embedded NR composites [19]. The optimized balance between flexibility and improved tear resistance makes these NR-LM composites highly attractive candidates for constructing soft robotic components, potentially replacing conventional PDMS-based composites (Fig.1b1–b3) [5]. Consequently, these LM-based NR composites are ideal for fabricating advanced soft robotic parts.

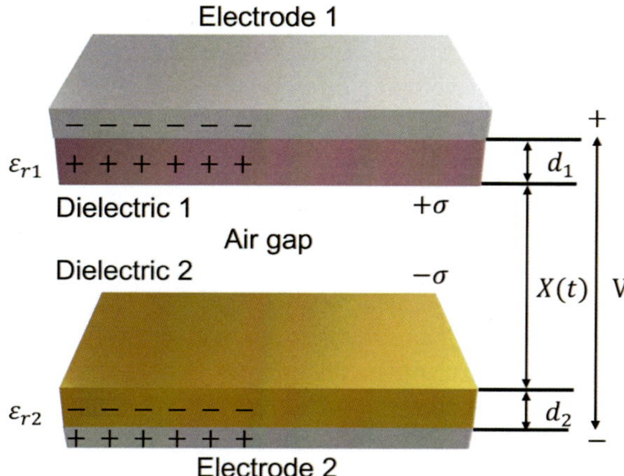

Fig. 2 Schematic illustration of the CE process in a TENG

3 Commercial Rubber-Based Triboelectric Energy Harvester for Tactile Application

Stretchable TENGs have been the subject of huge interest for sustainable power generation and power supply to flexible electronics [6]. The use of flexible elastomers for TENG fabrication is a relatively new field and is very desirable, as mechanically robust rubbers could offer a long service life. In this project, we have extensively investigated commercial rubber-based TENGs and their performance in terms of energy harvesting and sensing performance [15].

3.1 Robust, Stretchable TENGs Using Commercial Rubbers for Affordable Energy Harvesting

A typical TENG can convert mechanical motion/friction based static charge into electricity by electrostatic induction. Elastomers, particularly cross-linked rubbers, are an important class of materials which offer extreme robustness against friction and other various mechanical deformations under a wide range of temperature, deformation amplitude, and frequency (Fig. 2) [3–5]. We introduced high-performance stretchable TENGs comprising 100% amorphous commercial rubbers [6], highlighting their potential for energy harvesting applications. The active components of the generator had been prepared following the conventional rubber-processing method, i.e., solid-state mixing and vulcanization. We systematically investigated several commercial rubbers for their triboelectric output and categorized them based on triboelectric performance, providing a comprehensive evaluation of their efficiency. In addition to functionalized rubbers, several studies demonstrated the possibility of generating more charges, when the polymer films are textured or patterned to generate more friction between them. It was concluded that texturing exhibited a mostly positive effect on the triboelectric characteristics. This could be associated with stronger friction from the textured contact layers, resulting in a large number of charges developing on the surfaces [6, 20, 21].

3.2 Transfer-Printed TENGs Using Fluororubbers for Ultrasensitive Tactile Sensor and E-Skin

Self-powered tactile sensor-based electronic skin derived from TENG appeared to be one of the most worthwhile alternatives for sustainable energy harvesting. Unlike PDMS, fluoroelastomers (FKM) with particularly high F-content could address charge generation more effectively [20]. In addition to the micropatterned substrates, high-conductivity electrodes were essential to maximize the overall output of a TENG. Recent reports on laser-induced graphene (LIG) have opened up a plethora of possibilities in terms of cost-effective photothermal fabrication of graphene-based electrodes from a commercial carbon source [22]. We incorporated commercial FKM with high F-content ($\sim$ 70%) for the construction of single electrode TENG (STENG). For the first time, our work reported direct imprinting-cum-curing of both the active surface and LIG electrode on a tribonegative fluorinated rubber that exhibited impressive output power density of 715 mW/m^2 (Fig. 3a1–a3). Several demonstrations were conducted to validate the direct current (DC) STENG's usefulness in harvesting ambient mechanical energy [20]. The resulting TENG sensor was shown to have remarkable touch and motion sensitivity and therefore, could be embedded into soft robotic actuators for potential development in e-skin/bionic skin applications.

4 Triboelectric-Piezoelectric Hybrid Nanogenerators in Terms of Tactile Sensing

At their core, TENGs rely on contact electrification (CE), also known as triboelectrification, and electrostatic induction [1]. In contrast, PENGs convert mechanical stress into electrical charges through oriented dipoles in ferroelectric phases [8]. Hybridizing these principles yields HPTENGs with enhanced charge generation, often surpassing the voltage and current outputs of either approach alone [10, 11, 23]. Additionally, localized charge accumulation on a tribo-surface can influence the overall charge transfer process [24]. Enhanced charge retention is achievable by incorporating piezoceramic particles, such as barium titanate ($BaTiO_3$/BTO), into the TENG matrix [13]. This inclusion creates a synergistic interaction between the triboelectric and piezoelectric mechanisms, improving impedance matching and overall output performance (Fig.3b1–b3) [13]. At resonance, the PENG component offers high energy conversion efficiency that can be tuned within a TENG framework, resulting in reduced output phase differences and superior hybrid performance [13, 25]. Recent computational studies confirm this synergy between triboelectric and piezoelectric processes, highlighting opportunities for adaptive designs at micro- and nanoscales [26]. Commercial rubbers, akin to piezoresistive and triboelectric materials, offer distinct advantages for hybrid nanogenerator sensors. While traditional triboelectric or piezoelectric materials such as fluoropolymers and lead zirconate titanate (PZT) are costly and less flexible, commercial rubbers provide notable benefits including mechanical durability, chemical and thermal stability, industrial scalability, and ease of customization [3, 10, 13, 20, 24, 27, 28]. These factors highlight why commercial rubbers are increasingly favored in HPTENG applications, particularly in wearable and large-area sensing contexts.

4.1 Key Approaches To Enhance Tactile Sensitivity in Hybrid Systems

4.1.1 Micro-/Nanopatterning

Improving surface morphology through micropillars, microwells, or hierarchical porous layers significantly enhances triboelectric charge density and frictional contact [9, 20]. Enhanced surface charge density directly correlates to increased output charges [2–6]. Micropatterning remains a highly effective technique among various approaches [29], with structured textures extensively explored to boost charge generation [20]. The interplay between surface roughness and contact electrification must be carefully considered. Several systematic patterning techniques, including lithography, stamping, imprinting, and 3D printing, have been shown to significantly enhance triboelectric voltage outputs [2–4]. Advanced methods such as ultraviolet (UV) lithography and laser ablation further produce structured elastomeric surfaces, yielding considerable gains in open-circuit voltage [30, 31].

4.1.2 Fillers and Composite Approaches

Incorporating conductive and dielectric fillers such as single-walled carbon nanotubes (SWCNTs), multi-walled carbon nanotubes (MWCNTs), carbon black, or BTO into elastomer matrices significantly enhances triboelectric charge generation and modulates piezoelectric responses [13]. Such enhancements arise mainly from increased interfacial polarization, improved dielectric permittivity, and optimized pathways for charge transport [15, 25]. However, balancing filler loading is crucial to avoid structural embrittlement, phase segregation, or elasticity reduction, which compromise mechanical durability [3]. Effective optimization thus depends on careful control of filler dispersion, matrix compatibility, and processing conditions to maintain structural integrity alongside superior electroactive performance.

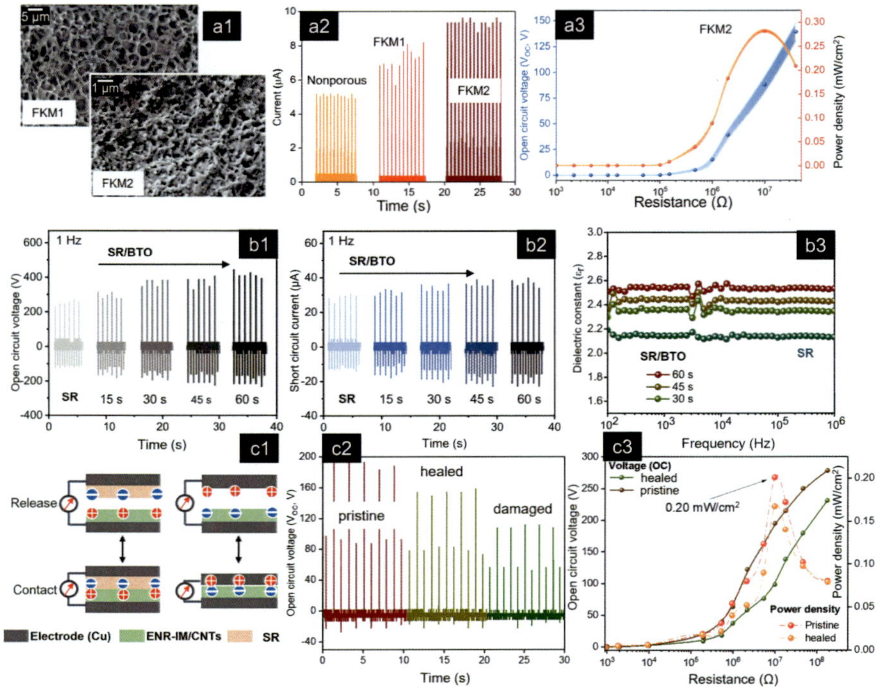

Fig. 3 Electrical characterization of rubber-based (FKM, hybrid BTO/silicone rubber) TENGs: (a1) SEM micrographs of transfer-printed microwells on the FKM surface. (a2) Current output shows a microstructured FKM-TENG with a higher short-circuit current (∼ 8 and ∼ 10 μA), compared to the pristine substrate (5 μA). (a3) Output voltage and power for FKM-TENG. (b1, b2) Tribo-voltage and current output increase with increasing dielectric BTO layer thickness in the hybrid BTO/ silicone rubber (SR) TENG. (b3) Frequency-dependent dielectric constants of BTO/SR with varying dielectric layer thickness. Panels (a1)–(b3) adapted from Arief et al., Mater. Horiz. 2022, 9, 1468, under the terms of the Creative Commons Attribution 3.0 Unported License (CC BY 3.0). (c1) Mechanism of CE in TENG with SR and ENR (epoxidized NR)-IM (imidazole)/4CNTs (4 wt%) under single and dual-electrode mode. (c2) Voltage before and after the healing process. (c3) Comparison of output power density and voltage as a function of variable resistance for PTENG (STENG) before and after healing. Panels (c1)–(c3) adapted from Mandal et al. [10], under the terms of the Creative Commons Attribution 4.0 International License (CC BY 4.0)

4.1.3 Incorporating Self-healing Mechanisms Into Hybrid Nanogenerator-Sensors

Several recent studies have focused on introducing dynamic crosslinking networks and reversible ionic bonding to facilitate self-healing elastomeric matrices [10, 24]. These self-healing mechanisms enable the restoration of mechanical performance and functionalities by repairing microtears or surface cracks, thereby prolonging the device lifespan under continuous operational stress. Ongoing research further explores thermally responsive self-healing strategies, where localized heating -facilitated by triboelectric energy harvesting induces polymer chain mobility and induce bond reformation [27]. The integration of these self-healing/repairing functionalities presents a unique opportunity for next-generation wearable and industrial sensors, where long-term reliability against mechanical fatigue are important. Self-healing rubber allows sensor modules to recover from mechanical tears or pronounced abrasions [10]. In particular, we have reported that reversible ionic bonds have demonstrated near-complete restoration of triboelectric outputs after damage (Fig. 3c1–c3) [27]. Such modules hold great promise for reliable sensing in scenarios like prosthetics, high-motion robotics, and industrial automation lines prone to abrasion.

5 Knowledge Gaps and Current Challenges in Soft Robotic-Integrated Sensors

5.1 Extended Durability, Environmental Compatibility and Device Integration

Although most studies assess the performance of HPTENGs over a few thousand cycles, advanced robotic and wearable applications may demand much longer operational lifetimes, often exceeding a million cycles and exposure to extreme temperatures [23] However, long-term degradation mechanisms, such as triboelectric charge dissipation, viscoelastic creep, and material fatigue, remain poorly understood. Further studies should incorporate accelerated aging tests and mechanical fatigue analyses to improve durability under dynamic stress and environmental conditions. Environmental factors such as moisture absorption, airborne contaminants, and chemical exposure significantly influence triboelectric charge retention over time. Humidity has shown to reduce triboelectric output by altering surface work functions and promoting charge dissipation [25]. While recent studies explore hydrophobic coatings and encapsulation strategies, their long-term efficacy remains unclear. In addition to these, the effects of temperature fluctuations on charge retention need further investigation to ensure reliability across diverse operating conditions. In addition, integrating HPTENGs into soft robotic skins, wearable electronics, and other flexible systems necessitates robust adhesion strategies that minimize mechanical compliance mismatches [24]. Key integration challenges include sensor delamination, substrate wrinkling, and signal drift under repeated motion. To address these, emerging approaches such as ionic bonding adhesives, flexible hybrid circuits, and surface-functionalized polymers are being explored to enhance long-term device stability.

Finally, investigation into the pressure sensitivity, durability, and response time of the triboelectricity driven tactile array matrix (TTAM) is crucial owing to its broad application in emulating tactile sensing. Preliminary data indicated that the pressure

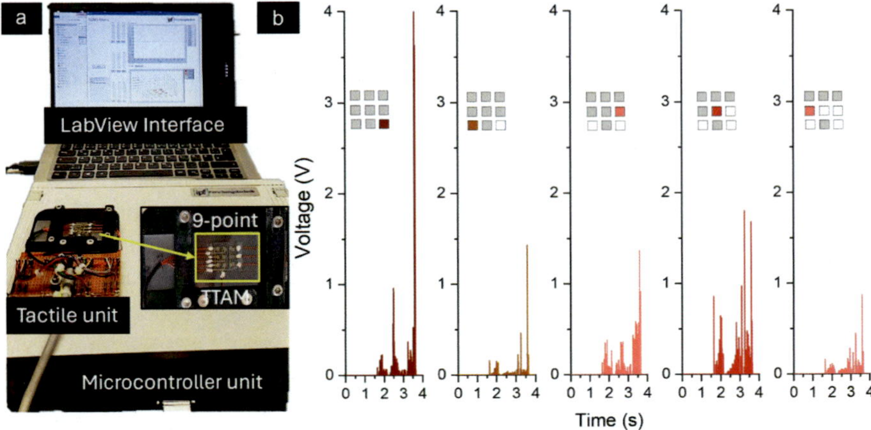

Fig. 4 Tactile sensor array based on commercial rubbers. **a** A 9-point TTAM based on the triboelectric principle has been shown to operate under mild tactile contact. The figure illustrates a state-of-the-art self-powered sensor system connected to an interface, demonstrating live contact force mapping. **b** Pressure mapping of the 9-point tactile sensor array; the output is expressed in terms of triboelectric voltage

sensitivity of the micropatterned array was higher than that of smooth sensor and is attributable to an increase in effective contact area. Following a successful evaluation of material parameters for optimal TENG performance and subsequent validation through numerical modeling, we developed the first proof-of-concept TTAM based on a 9-point tactile array composed of functional elastomers Fig. 4. However, further investigation into arrays is highly anticipated, as the large amount of acquired data from wide-area tactile arrays can be effectively fed into predictive algorithms and machine learning.

5.2 *Parameterized Finite Element Modeling (FEM)*

Despite numerous efforts aimed at optimizing textured surfaces for TENG applications, existing shape parameterization methods frequently overlook critical factors such as precise texture geometry and scale [29]. Additionally, computational simulations dedicated explicitly to texture optimization remain limited, though such simulations are essential for the theoretical validation of optimization processes and for reducing unnecessary experimental iterations [32]. To effectively address these limitations and enhance the performance of materials in sensing applications, it is crucial to integrate a diverse range of microstructures and material properties into a comprehensive parameterized simulation model. Our recent studies emphasize advancements in modeling contact electrification phenomena using parameterized

FEM [26, 32]. Nonetheless, current FEM and multiphysics simulations often oversimplify critical aspects of HTPENGs, particularly triboelectric charge transport, piezoelectric responses under non-uniform strain conditions, and viscoelastic relaxation. Consequently, existing models frequently fall short of accurately capturing charge retention and redistribution behaviors across prolonged cycles, leading to discrepancies between computational predictions and empirical observations. Future research should therefore adopt comprehensive, multi-scale modeling techniques integrating detailed descriptions of contact electrification, viscoelastic deformation, and real-time charge decay mechanisms to substantially improve predictive accuracy.

6 Conclusion

Given the rapidly expanding market potential of self-powered pressure and tactile sensors in robotics and electronic skin technologies, and Internet of Things (IoT) applications, significant progress in the modular TENG embedded within robotic actuators are highly anticipated. Moreover, the sustainable nature of these energy-harvesting modules, especially in robotic applications, provides strong commercial incentives that outweigh existing technical and deployment challenges. This review highlights the growing potential of HPTENGs, piezoresistive, and triboelectric sensors for tactile sensing, focusing on commercial rubbers that deliver sustained mechanical robustness, convenient scalability, and flexible processing. It emphasizes significant advances in these technologies, with particular attention to commercial elastomers renowned for their mechanical robustness, scalability, and processing versatility. Novel approaches, from hierarchical micropatterning to self-healing architectures, continue to push the boundaries of device performance and reliability. Yet, translating these laboratory findings into commercially viable products necessitates deeper insights into long-term reliability, environmental durability, and manufacturing standardization. Advanced computational modeling must evolve to more accurately predict complex material behavior, while machine learning frameworks should be fine-tuned to manage large volumes of sensory data streams under dynamic operational conditions. By integrating bio-inspired design principles, advanced filler technologies, and high-throughput fabrication methods, the next generation of elastomer-based HPTENGs will likely drive a transformative shift in self-powered sensors across robotics, consumer electronics, and beyond.

Acknowledgements I.A., A.K.G., A.F., and A.D. have been supported by DFG Project No. 404941515 under the SPP 2100 program, `Soft Material Robotic Systems`. We would like to acknowledge Prof. Annika Raatz and the SMaRT team from MATCH, Leibniz Universität Hannover, for their kind support under the SPP 2100 framework during our investigation of the motion sensitivity of an ultrasoft NR sensor mounted on a soft robotic component.

References

1. Luo, J., Wang, Z.L.: Recent progress of triboelectric nanogenerators: from fundamental theory to practical applications. EcoMat **2**(4), e12059 (2020)
2. Mandal, S., Hait, S., Simon, F., Ghosh, A., Scheler, U., Arief, I., Tada, T., Hoang, T.X., Wießner, S., Heinrich, G., Das, A.: Transformation of epoxidized natural rubber into ionomers by grafting of 1H-imidazolium ion and development of a dynamic reversible network. ACS Appl. Polym. Mater. **4**(9), 6612–6622 (2022)
3. Banerjee, S.S., Arief, I., Berthold, R., Wiese, M., Bartholdt, M., Ganguli, D., Mitra, S., Mandal, S., Wallaschek, J., Raatz, A., Heinrich, G., Das, A.: Super-elastic ultrasoft natural rubber-based piezoresistive sensors for active sensing interface embedded on soft robotic actuator. Appl. Mater. Today **25**, 101219 (2021)
4. Banerjee, S.S., Arief, I., Fery, A., Heinrich, G., Das, A.: Aspects of materials processing and engineering for the development of soft robotic devices. In: Soft Robotics, pp. 1–20. Bentham Science Publishers (2022)
5. Banerjee, S.S., Mandal, S., Arief, I., Layek, R.K., Ghosh, A.K., Yang, K., Kumar, J., Formanek, P., Fery, A., Heinrich, G., Das, A.: Designing supertough and ultrastretchable liquid metal-embedded natural rubber composites for soft-matter engineering. ACS Appl. Mater. Interfaces. **13**(13), 15610–15620 (2021)
6. Natarajan, T.S., Finger, S., Lacayo-Pineda, J., Bhagavatheswaran, E.S., Banerjee, S.S., Heinrich, G., Das, A.: Robust triboelectric generators by all-in-one commercial rubbers. ACS Appl. Electron. Mater. **2**(12), 4054–4064 (2020)
7. Wang, Z.L., Wang, A.C.: On the origin of contact-electrification. Mater. Today **30**, 34–51 (2019)
8. Wang, Y., Cao, X., Wang, N.: Recent progress in piezoelectric-triboelectric effects coupled nanogenerators. Nanomaterials **13**(3), 385 (2023)
9. Zhang, J., He, Y., Boyer, C., Kalantar-Zadeh, K., Peng, S., Chu, D., Wang, C.H.: Recent developments of hybrid piezo-triboelectric nanogenerators for flexible sensors and energy harvesters. Nanoscale Adv. **3**(19), 5465–5486 (2021)
10. Mandal, S., Arief, I., Chae, S., Tahir, M., Hoang, T.X., Heinrich, G., Wießner, S., Das, A.: Self-repairable hybrid piezoresistive-triboelectric sensor cum nanogenerator utilizing dual-dynamic reversible network in mechanically robust modified natural rubber. Adv. Sens. Res. **3**(10), 2400036 (2024)
11. Ye, Z., Liu, T., Du, G., Shao, Y., Wei, Z., Zhang, S., Chi, M., Wang, J., Wang, S., Nie, S.: Bioinspired superhydrophobic triboelectric materials for energy harvesting. Adv. Funct. Mater. **35**(2), 2412545 (2025)
12. Li, W., Lu, L., Kottapalli, A.G.P., Pei, Y.: Bioinspired sweat-resistant wearable triboelectric nanogenerator for movement monitoring during exercise. Nano Energy **95**, 107018 (2022)
13. Meena, K.K., Arief, I., Ghosh, A.K., Liebscher, H., Hait, S., Nagel, J., Heinrich, G., Fery, A., Das, A.: 3D-printed stretchable hybrid piezoelectric-triboelectric nanogenerator for smart tire: onboard real-time tread wear monitoring system. Nano Energy **115**, 108707 (2023)
14. Kundu, A., Arief, I., Mandal, S., Meena, K.K., Krause, B., Staudinger, U., Mondal, T., Wießner, S., Das, A.: Elastomeric sensor-triboelectric nanogenerator coupled system for multimodal strain sensing and organic vapor detection. ACS Appl. Mater. Interfaces. **16**(39), 53083–53097 (2024)
15. Nuthalapati, S., Chakraborthy, A., Arief, I., Meena, K.K., Kaja, K.R., Kumar, R.R., Kumar, K.U., Das, A., Altinsoy, M.E., Nag, A.: Wearable high-performance MWCNTs/PDMS nanocomposite-based triboelectric nanogenerators for haptic applications. IEEE J. Flex. Electron. **3**(9), 393–400 (2024)
16. Shtarkman, B.P., Razinskaya, I.N.: Plasticization mechanism and structure of polymers. Acta Polym. **34**(8), 514–520 (1983)
17. Jonscher, A.K., Loh, C.K.: Poole-Frenkel conduction in high alternating electric fields. J. Phys. C: Solid State Phys. **4**(11), 1341 (1971)

18. Pan, C., Markvicka, E.J., Malakooti, M.H., Yan, J., Hu, L., Matyjaszewski, K., Majidi, C.: A liquid-metal-elastomer nanocomposite for stretchable dielectric materials. Adv. Mater. **31**(23), 1900663 (2019)
19. Bartlett, M.D., Fassler, A., Kazem, N., Markvicka, E.J., Mandal, P., Majidi, C.: Stretchable, high-k dielectric elastomers through liquid-metal inclusions. Adv. Mater. **28**(19), 3726–3731 (2016)
20. Arief, I., Zimmermann, P., Hait, S., Park, H., Ghosh, A.K., Janke, A., Chattopadhyay, S., Nagel, J., Heinrich, G., Wießner, S., Das, A.: Elastomeric microwell-based triboelectric nanogenerators by in situ simultaneous transfer-printing. Mater. Horiz. **9**(5), 1468–1478 (2022)
21. Meena, K.K., Arief, I., Ghosh, A.K., Knapp, A., Nitschke, M., Fery, A., Das, A.: Transfer-printed wrinkled PVDF-based tactile sensor-nanogenerator bundle for hybrid piezoelectric-triboelectric potential generation. Small **n/a**, 2502767 (2025)
22. Stanford, M.G., Li, J.T., Chyan, Y., Wang, Z., Wang, W., Tour, J.M.: Laser-induced graphene triboelectric nanogenerators. ACS Nano **13**(6), 7166–7174 (2019)
23. Chen, G., Wang, J., Xu, G., Fu, J., Gani, A.B., Dai, J., Guan, D., Tu, Y., Li, C., Zi, Y.: The potential application of the triboelectric nanogenerator in the new type futuristic power grid intelligent sensing. EcoMat **5**(11), e12410 (2023)
24. Li, C., Guo, H., Wu, Z., Wang, P., Zhang, D., Sun, Y.: Self-healable triboelectric nanogenerators: marriage between self-healing polymer chemistry and triboelectric devices. Adv. Funct. Mater. **33**(2), 2208372 (2023)
25. Fan, B., Liu, G., Fu, X., Wang, Z., Zhang, Z., Zhang, C.: Composite film with hollow hierarchical silica/perfluoropolyether filler and surface etching for performance-enhanced triboelectric nanogenerators. Chem. Eng. J. **446**, 137263 (2022)
26. Verners, O., Šutka, A., Arief, I., Das, A., Mālnieks, K., Lungevičs, J.: The effect of surface texture components on the contact electrification of triboelectric materials: a theoretical study. Mater. Sci. Eng., B **317**, 118140 (2025)
27. Liao, W., Liu, X., Li, Y., Xu, X., Jiang, J., Lu, S., Bao, D., Wen, Z., Sun, X.: Transparent, stretchable, temperature-stable and self-healing ionogel-based triboelectric nanogenerator for biomechanical energy collection. Nano Res. **15**(3), 2060–2068 (2022)
28. Wajahat, M., Kouzani, A.Z., Khoo, S.Y., Mahmud, M.A.P.: A review on extrusion-based 3D-printed nanogenerators for energy harvesting. J. Mater. Sci. **57**(1), 140–169 (2022)
29. Fang, Z., Chan, K.H., Lu, X., Tan, C.F., Ho, G.W.: Surface texturing and dielectric property tuning toward boosting of triboelectric nanogenerator performance. J. Mater. Chem. A **6**(1), 52–57 (2018)
30. Kim, D., Tcho, I.W., Jin, I.K., Park, S.J., Jeon, S.B., Kim, W.G., Cho, H.S., Lee, H.S., Jeoung, S.C., Choi, Y.K.: Direct-laser-patterned friction layer for the output enhancement of a triboelectric nanogenerator. Nano Energy **35**, 379–386 (2017)
31. Feng, H., Li, H., Xu, J., Yin, Y., Cao, J., Yu, R., Wang, B., Li, R., Zhu, G.: Triboelectric nanogenerator based on direct image lithography and surface fluorination for biomechanical energy harvesting and self-powered sterilization. Nano Energy **98**, 107279 (2022)
32. Verners, O., Das, A.: Comparison of contact electrification mechanisms of selected polymers and surface-functionalized molecules. J. Phys. Chem. B **127**(46), 10035–10042 (2023)

BY NC ND

Development and Characterization of Soft Magnetic Materials with Anisotropy in the Mechanical Properties and Multi-stimulated Compliance

Nina Sindersberger, Darshan K. Gowda, Stefan Odenbach, and Klaus Zimmermann

Abstract A hybrid material, resulting in a thermosensitive elastomer (TSE), which enables new applications with controllable material properties for the use in soft robotic systems was developed. Recent advances in soft robotics demonstrate robust and versatile performance in gripping and manipulation. By manufacturing such compliant structures using thermosensitive hybrid materials made of polydimethylsiloxane (PDMS) and embedded polycaprolactone (PCL) particles, the system design could obtain higher versatility and additional functionality. In certain investigations, soft magnetic carbonyliron (CI) particles are also included in order to prove the compatibility of the two filler particle types within the PDMS matrix and to utilize a possible symbiotic effect of the particle mixtures. By means of mechanical tests, the TSE enables shape changes through the simultaneous application of external heat and stress/force. With a low melting point in the temperature range of 58–60 °C, PCL as thermoplastic filler offers good application potential. One of the most important and application-oriented phenomena of PCL is the shape memory effect (SME). By adding soft magnetic CI particles, an accelerated heat distribution within the samples was recognized, which results in a faster occurrence of the corresponding thermally induced effect and an additional magnetically induced effect. Microcomputed tomography (μCT) and scanning electron microscopy (SEM) examinations indicate a homogeneous distribution of PCL and CIP within the TSE. Moreover, the influence of an external alternating magnetic field can result in magnetic heating. The

N. Sindersberger (✉) · K. Zimmermann
Institute of mechanics of compliant systems, Technical University of Ilmenau, Ilmenau, Germany
e-mail: nina.sindersberger@oth-regensburg.de

K. Zimmermann
e-mail: klaus.zimmermann@tu-ilmenau.de

D. K. Gowda · S. Odenbach
Institute of mechanical science and engineering, TUD Dresden University of Technology, Dresden, Germany
e-mail: stefan.odenbach@tu-dresden.de

A. Raatz et al. (eds.), *Soft Material Robotic Systems*,
https://doi.org/10.1007/978-3-032-22453-8_4

multiple controllable properties of the developed material open up fields of application in the area of gripping and locomotion technology. The focus is thereby on the adaptive mechanical flexibility of the technical functional elements.

Keywords Thermosensitive hybrid materials · Shape Memory Effect · Magnetic particles · Shape adaption

1 Introduction

The increasing tendency to integrate extensive functions in a single smart material instead of separated components with distinct material properties, processing of energy and information, leads to their steadily growing importance. The development results in a wide range of smart materials. In general, these materials are able to adaptively interact with the environment or can be specifically controlled by physical, chemical or mechanical internal or external stimuli [8]. Furthermore, a distinction is made between materials whose (micro −) structure has been modified and consequently their properties as well as composite materials whose material properties are characterized by the combination of the individual components. The latter are the focus of this work.

Photosensitive hybrid materials exhibit physical or chemical changes, e.g. changes in molecular structure, in volume or in elastic properties, as response to an optical signal [12, 33]. Such photosensitive materials are often utilized as sensors [5] or even as drug delivery systems [2] or actuators [11]. The advantages of the stimulation with light are the controllable spectral range of the light source and thus the adjustable properties, the precise spatial and temporal control as well as an on-demand interruption and resumption of the activation. Photosensitive materials can respond quickly and sensitively to changes in the incident light with a high degree of deformation [30]. Whereas composites such as magnetosensitive elastomer (MSE) or magnetoactive polymer (MAP) belong to a category of smart materials whose mechanical behavior can be controlled by applying an external magnetic field [15, 22]. However, the areas of application are very similar, since MSE are also employed as sensors [6, 10], actuators [1, 9] or for biomedical engineering applications [34], MSE ordinarily consist of non-magnetic elastic polymer matrices, in which soft or hard magnetic filler particles are embedded. The common size of these filler particles is generally limited to the nm to μm range [16]. This kind of smart material shows multi-functional field-dependent properties under the influence of an external magnetic field. Examples are the elastic and plastic properties of MSE, which may be isotropic or anisotropic and are magnetically controllable [24]. These externally controllable material properties are induced by the interaction of both components (magnetic filler and elastomer), thereby the compound is exceeding the usability and possibilities of each solitary component. Beyond that these properties can be set during MSE manufacturing by prior application of magnetic fields, mechanical stresses and preliminary treatment of the filler particles [3, 4]. Smart materials consisting of an elastic matrix and electrically conductive fillers, which causes the actual insulator to become electrically conductive, resulting in the so-called electroactive polymers (EAP) [16]. Consequently, EAP can significantly change their size and/or shape by applying voltage. In general, a material can be referred to as active if it is able to convert electrical/magnetical/chemical energy directly into mechanical energy through the response of the material [13]. Due to these controllable material properties, there is a great interest especially in the fields of (soft) robotics and medical engineering, etc. [26, 34]. The various fabrication methods are also increasingly being researched. Not only thermoplastics but also MAP and EAP can now be formed into a certain shape using 3D printing technology [17, 28]. This enables hybrid materials to be produced more quick and even scalable in different designs and to be tuned individually according to specific requirements.

1.1 Objectives

After numerous studies in the extensive field of smart materials mentioned above, the focus of this work is on a practical aspect of thermal induced effects within a polymer compound and is not restricted on pure polymers. The developed TSE, consisting of a combination of an elastic PDMS matrix with thermoplastic PCL particles and enables a change in the material properties of the resulting hybrid material or element through a temperature influence. Temperature-dependent property changes have advantages compared to the previously mentioned stimulation, e.g. an unhindered light influence cannot be guaranteed in every application. Similarly, the generation of strong magnetic or electric fields often requires a large setup compared to thermal activation. In addition, the aspect must be considered that magnetic or electric fields may not be implemented in every application, especially regarding biomedical applications inside or outside the human body. Furthermore, these magnetic or electric fields are often not or not easily focusable or precise spatial and temporal controllable. Therefore, it is possible to unintended influence surrounding non-involved structures or parts of the application. Thus, these methods of activation are also often problematic for applications in soft robotics, which are of interest for this work. A further advantage of the TSE is that the resulting composites have to be heated up to a maximum temperature of 60 °C, caused by the low melting temperature of the PCL filler particles used. Additionally, these hybrid materials exhibit a Shape Memory Effect SME, which is of increasing interest in the field of soft robotics. Materials with a SME are able to change their shape in a predefined and controlled manner when exposed to a suitable stimulus, such as light or temperature [23, 25].

1.2 Technical Approach

The main focus is the systematic investigation of the resulting hybrid material, which includes mechanical analysis such as the measurement of stress-strain curves through tensile and compression tests of the TSE. Thereby, it is one of the key parameters in controlling their change of stiffness. Additionally, the SME is examined and thermal as well as structural analysis of TSE with CI particles were carried out. A particle combination of thermoplastic and magnetic particles is being discussed in order to expand application possibilities through a suitable material configuration and to be able to exploit the advantages of the resulting systems. Therefore, it was necessary to investigate the compatibility of the material mixtures within the matrix in order to improve the understanding of the theory. For hybrid materials, polymer blends and composites with different fillers, the distribution of particles is of significance to determine their macroscopic properties. Respectively, it is important to analyze the distribution during and after the temperature treatment due to the thermosensitivity of the developed material. The resulting hybrid material enables new applications with a controllable material characteristic for the use in soft robotic systems, especially on the development of adaptive mechanical compliance of technical functional elements, such as actuators and sensors. In order to get an idea of the application potential, applications such as modified hard gripping structures or the extension of an existing gripping system (robot hand for in-hand manipulation, CO-Design of Prof. Dr. Oliver Brock from TU Berlin [20]) have already been investigated and successfully implemented.

2 Experimental Methods and Characterization

The developed material requires a stable, repeatable and optimized manufacturing process as well as knowledge of synthesizable material combinations. Verification of the manufacturing process and the homogeneous distribution of the particles as well as analysis of the internal structure and surface of all sample sets were carried out using micro-computed tomography (μCT) and scanning electron microscopy (SEM), as illustrated in Fig. 1.

The polymer blend of PDMS in which PCL particles are embedded shows a good sample quality when this material combination contains up to max. 20 wt% PCL particles, while the composites containing additionally CIP have reached a maximum mixing ratio of about 15 wt% PCL particles and 50 wt% CIP. In order to measure the effects of temperature on physical (e.g. mechanical properties) and chemical properties, it is necessary to achieve a uniform heat distribution within the samples, especially at temperatures around the melting point of PCL. Therefore, a well-insulated and temperature-controlled heating chamber, shown in Fig. 2, has been developed, built and verified experimentally through a FEM thermostatic analysis. The extension of the Zwick/Roell Z005 tensile and compression machine [7, 18]

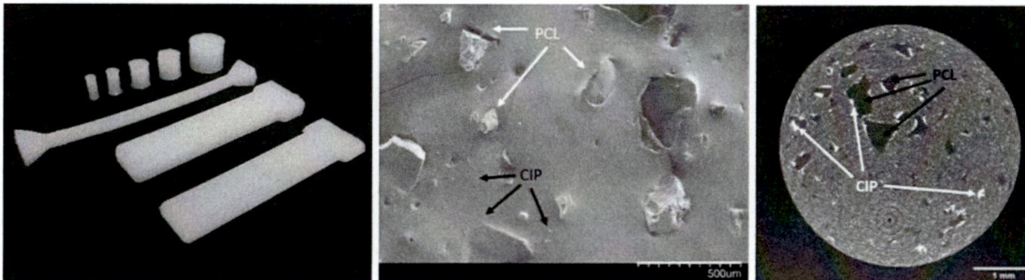

Fig. 1 The image on the left side shows variously shaped thermosensitive hybrid material samples for different investigations including mechanical testing, scanning electron microscopy (SEM), and micro-computed tomography (μCT). The middle panel presents an SEM image of the material's surface, revealing polycaprolactone particles along with embedded soft-magnetic carbonyl iron particles (CIP). The image on the right displays the results of a micro-computed tomography scan of these thermosensitive hybrid materials, highlighting the distribution of both particle types within the structure

incorporating this heating chamber enables the sample temperature to be controlled during the tests.

Before the results of the tensile and compression tests can be correctly interpreted, it is important to precondition the samples due to the Mullins Effect and Payne Effect. The stress-strain curve of the polymer blend depends on its historical maximum load, consequently, if the same load test is performed several times in sequence, a softening of the material can be recognized. The preconditioning procedure consists of repetitive steps of heating, loading and cooling and is intended to ensure that standardized conditions are established for the samples allowing the results of different samples to be compared with each other adequately. An example for the preconditioning process within the tension test at room temperature is given in Fig. 3 for PDMS with 20 wt% PCL.

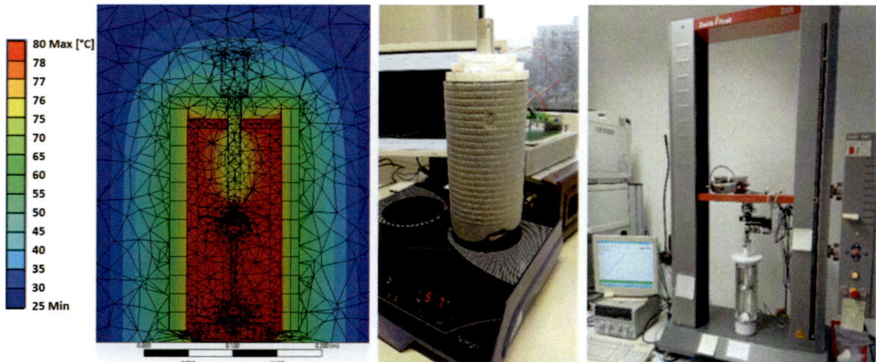

Fig. 2 Thermostatic simulations (FEM) of the heat distribution within the heating chamber for tension and compression tests (left), construction and experimental verification of the set up (middle), measurement set up for tension and compression tests (right)

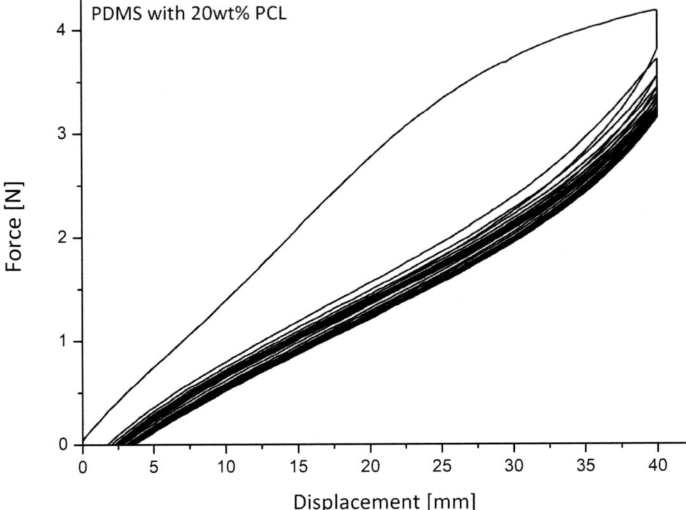

Fig. 3 Force-displacement-characteristic and preconditioning process of the thermosensitive hybrid material (PDMS with 20 wt% PCL) within the tension test at room temperature

During the preconditioning procedure it can be observed that the force at the maximum strain decreases between each loading cycle. The hysteresis can be explained by the typical material behavior of the elastomer. This characteristic is displayed more clearly when the maximum load is hold up for a few seconds before the sample is unloaded. The difference between the force curves and the maximum force becomes smaller with the subsequent load cycles, thus the measurement result becomes more stable. For this reason and for better comparability with the compression samples, the third loading cycle is utilized for the evaluation of both tests. This applies to all subsequent static or dynamic mechanical and thermal investigations.

As shown in Fig. 4, a change in temperature leads to a change in the mechanical properties, such as the stiffness ("modulus of elasticity" tension/compression) and dynamics ("eigenfrequency", damping) of the material. The melting of the PCL particles in a small temperature range around the PCL melting point (switching characteristic) within the elastomer matrix causes a decrease in the stiffness of the TSE of up to 60% depending on the particle concentration. This revealed a difference between tensile and compressive stress, with the force required to stretch the samples being lower than the force required to compress them [35].

In addition, soft magnetic particles (e.g. CIP) influence the heat transfer rate and the specific heat capacity of the material mixtures. The reaction time of the material to temperature changes can be reduced by up to 30% by adding CIP, depending on the geometry and dimensions of the samples [14, 19].

Furthermore, a shape memory effect (SME) was observed. The melting of PCL particles enables the reshaping of the material (e.g. compression, bending, "texturing") for temporary local or global morphological changes. After cooling, the shape

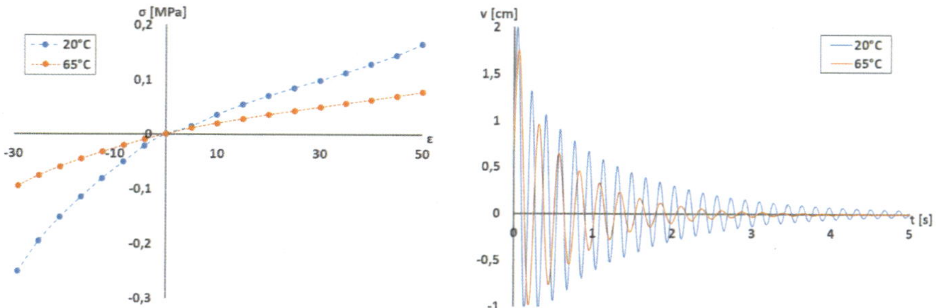

Fig. 4 Temperature-dependent mechanical response of the thermosensitive hybrid material containing 20wt% PCL. The left panel shows the results of tensile and compressive tests at different temperatures of selected samples. The right panel presents the dynamic analysis, including eigenfrequency and damping behavior of TSE

is maintained in the modified state by external deformation until reheating above the PCL melting point. This process is completely reversible and repeatable. However, the material retains its elastic behavior even in the deformed states, as illustrated in Fig. 5. The SME results from the internal stresses between the molecular chains of the elastomer and the PCL particles, while the external shape results from the balance of all internal forces [29]. Consequently, the material mixtures can be described as both thermosensitive elastomers and shape memory polymers.

Neither a shift in the melting temperature nor similarly striking differences between the various material mixtures and the pure PCL particles were found. As a result, it can be assumed that the thermoplastic particles only have physical bonds to the matrix.

These results were confirmed by thermogravimetric examination methods such as differential scanning calorimetry (DSC), shown in Fig. 6 and thermogravimetric analysis (TGA), illustrated in Fig. 7.

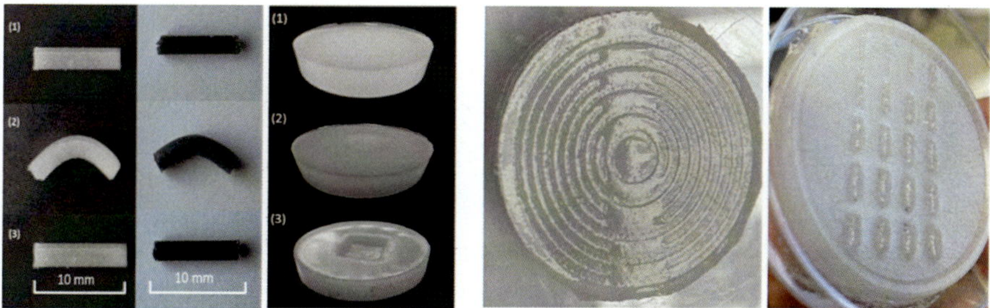

Fig. 5 Demonstration of the shape memory behavior of thermosensitive elastomers. The images on the left show samples with and without embedded carbonyl iron particles (CIP), illustrating reversible shape changes under thermal influence and a clear recovery to the original geometry after repeated the temperature stimulation. The middle and right images demonstrate shape adaptation (shape imprinting). The imprinted shape can remain stable as long as the material is not exposed to elevated temperatures again

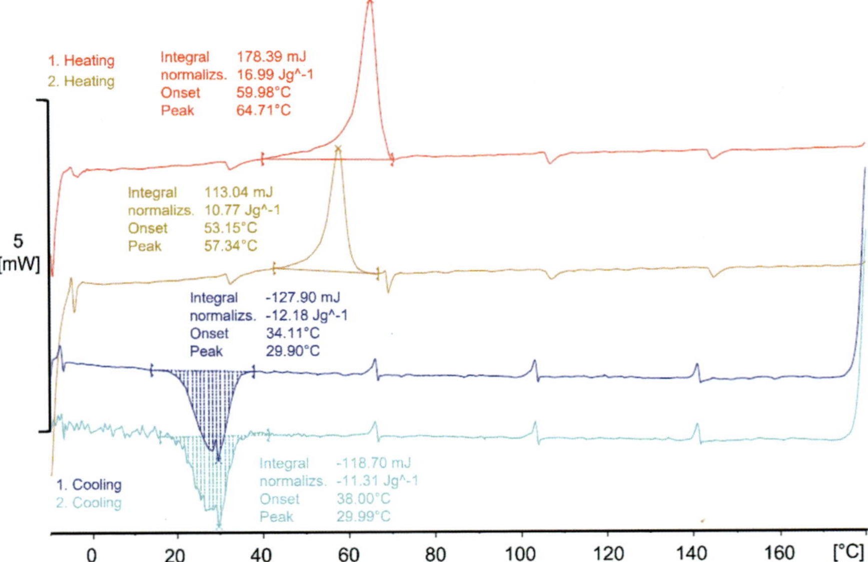

Fig. 6 Differential Scanning Calorimetry (DSC) analysis of the thermosensitive hybrid material for selected samples (PDMS with 20 wt% PCL) . The temperature-dependent phase behavior and corresponding thermal transitions of the material are illustrated

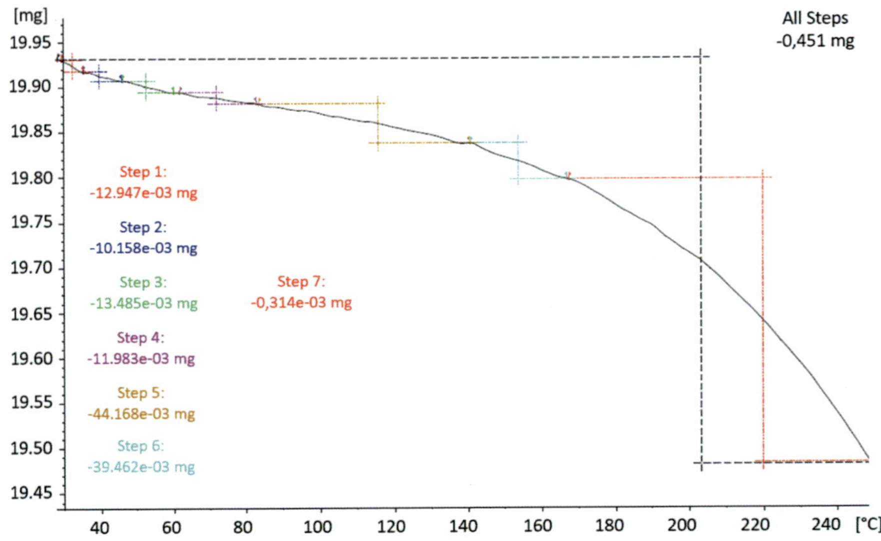

Fig. 7 Thermogravimetric analysis (TGA) of the thermosensitive hybrid material for selected samples (PDMS with 20wt% PCL). The graph shows the sample weight as a function of temperature

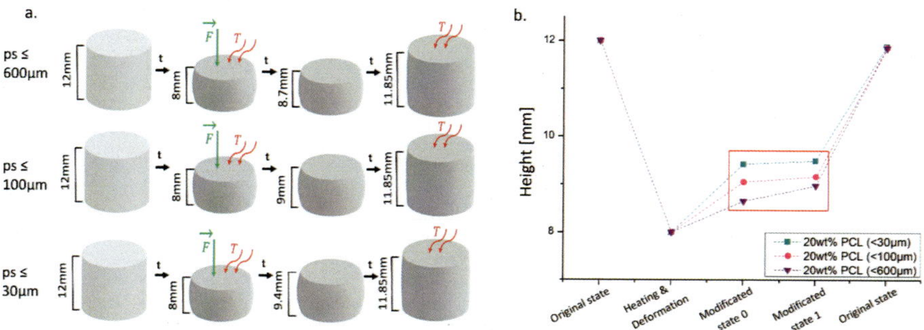

Fig. 8 Experimental procedure and results of shape memory effect (SME) tests on the thermosensitive hybrid materials containing polycaprolactone (PCL) particles of different sizes. The left image (a) illustrates the deformation and recovery sequence of cylindrical samples (initial height: 12 mm) subjected to mechanical loading and thermal stimulation, comparing particle sizes: ps $\leq$ 600 μm, ps $\leq$ 100 μm and ps $\leq$ 30 μm. The right image (b) presents the corresponding quantitative results, showing the sample height at different states (original, during heating and deformation, and after modification) for TSE with 20 wt% PCL of varying particle sizes, highlighting the influence of particle size on shape memory performance

Further investigations have revealed that the shape memory effect is more or less pronounced depending on the average particle size, compared Fig. 8. Thus, the extent of the SME of the material mixture with PCL particles < 30 µm is weaker than the SME with PCL particles < 600 µm. One possible approach explaining why the particle size is relevant in this case could be the fact that a sufficiently large area of the soft segment must be present at various points in the network in order to force the matrix to fix its shape. It appears irrelevant whether the rigid segment is formed exclusively by the polymer matrix itself or by additional materials, such as the soft magnetic CI particles.

Stretching the samples may cause air inclusions, which are subsequently the reason why the PCL particle no longer fills the space between the network alone. The SME would therefore be weakened. Whereas this effect does not occur when the samples are subjected to compression, as the network is not expanded but compacted. The possibility of creating space for possible air inclusions is in this case eliminated.

In addition, TSE with PDMS of different shore hardness (A0 and A13) were utilized. After the cooling process and removal of the external load, the indentation into the sample was measured at different points. The mean values of the results are illustrated in Fig. 9., which indicates that the TSE of lower shore hardness requires a lower intrinsic height to enable an imprint that corresponds to the actual height of the imprinted structures. A clear correlation between the quality and accuracy of the SME with the respective shape adaption possibilities and the sample geometry as well as the shore hardness of the matrix can be observed.

A suitable combination of silicone-based composite materials with thermoplastic particles and silicone-based composite materials with soft or hard magnetic particles results in a thermally/magnetically controllable element. Figure 10. shows a beam made of two different layers, the bending of which results from the effect of a static

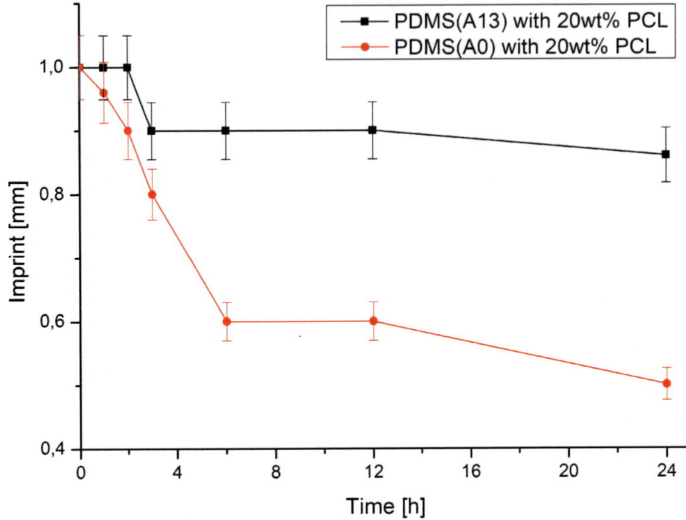

Fig. 9 Height of the imprinted shape as function of the thermosensitive elastomer containing 20wt% PCL with different shore hardness of the PDMS matrix

magnetic field and a temperature above 60 °C. The deformation of the beam is reversible and can be repeated as often as required in all directions.

Not only the targeted control of the thermosensitive hybrid material is possible through the effect of an external magnetic field, but also the heating of the samples from the inside of the sample. Magnetic heating is mainly based on the principle that the filler particle system absorbs energy when it is exposed to an alternating external magnetic field [21].

3 Application Possibilities

Due to its high flexibility and compliance, the TSE consisting of PDMS with 20 wt% PCL is suitable as a soft material, generally utilized for gripping fragile and sensitive objects of irregular shapes and sizes, as illustrated in Fig. 11. In addition to the numerous advantages, many challenges exist as soon as soft materials are employed for gripping objects. The challenges include not only the necessary rigidity but also sufficient gripping force, which is required to lift and hold objects and move them to their intended position. Especially pneumatic soft actuators expand radially when pressurized, which reduces the efficiency of the actuator performance [27, 31, 32].

A two-finger gripper was developed as part of a collaboration between TU Ilmenau and TU Berlin, drawing inspiration from a co-design project. The gripping surfaces that come into direct contact with the object consists of pure, blue-colored silicone.

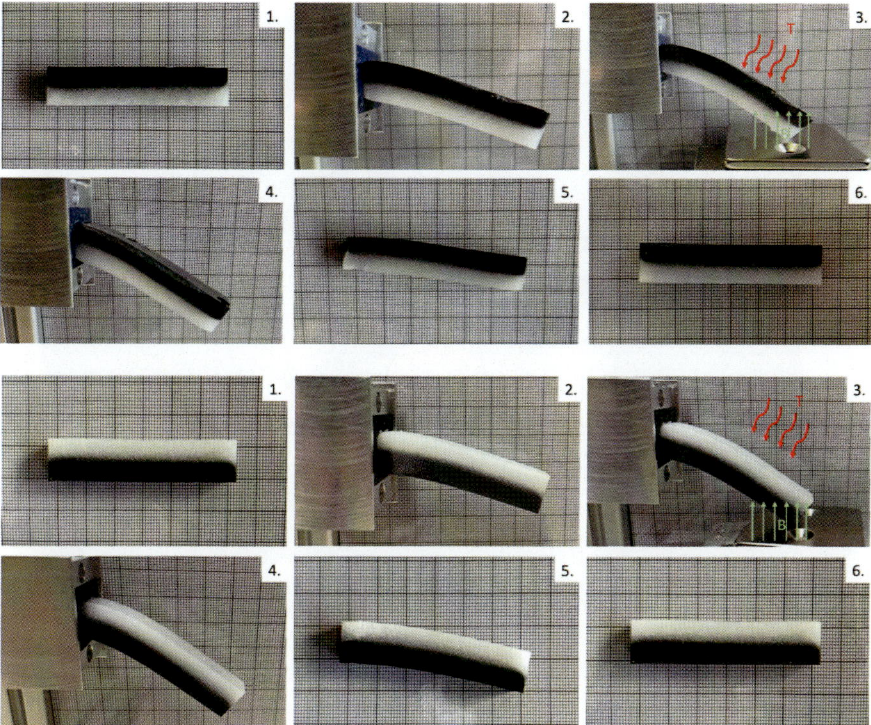

Fig. 10 Bending of a TSE beam; 1. Original state (o.s.), 2. o.s. of the fixed beam, 3. Applying an external magnetic field and a temperature of 70 °C, 4. Removing the external magnetic field and cooling down to 25 °C, 5. Modified state, 6. o.s. after heating the beam up and cooling down again

Since this part has negligible impact on the overall deformation behavior, PCL particles are omitted, resulting in reduced stiffness. At the mounting end of the gripper, a rigid base is integrated, featuring a threaded section for attachment and a tube connection for the pressure regulation system. To ensure structural stability, thin layers of temperature-resistant polycarbonate are applied to both the inner and outer sides of the base. The gripper is installed on a motorized linear actuator, enabling adjustable movement along the z-axis, and is part of a multi-axis test platform that ensures precise alignment with test objects or measurement devices. Pneumatic actuation is controlled using a manually operated mini pump and valves, while thermal energy is provided through a heating wire embedded in the TSE.

To evaluate the potential forces that can be exerted by the gripper design, the gripper and a force sensor are brought into direct contact, with no initial load applied to the sensor in the resting state. The pneumatic activation does not cause a gripping motion but instead generates a force directly. The force sensor is aligned perpendicularly to the expected force vector, allowing it to measure forces across all spatial directions. The resulting total force is recorded. The corresponding experimental setup is illustrated in Fig. 12. Test results compare the original PneuFlex gripper

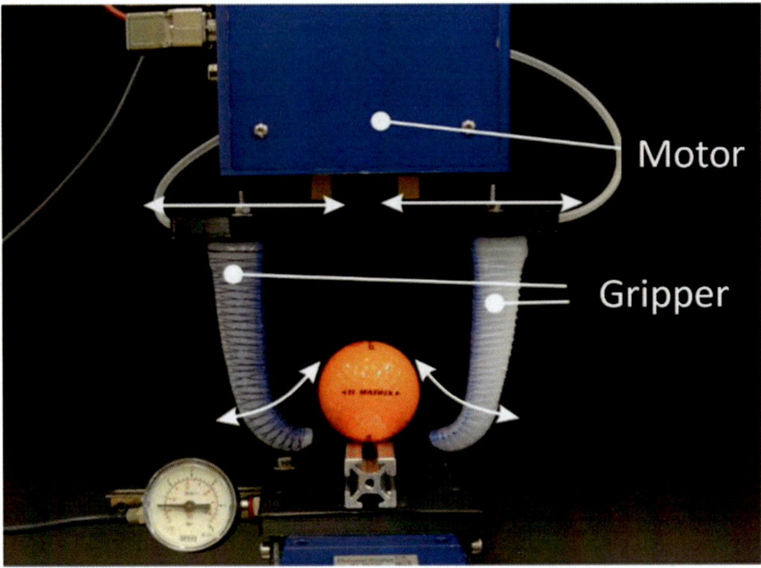

Fig. 11 The shown setup of the designed two-finger gripper consists of a motor-driven actuation system connected to flexible gripper elements, demonstrating controlled gripping of an object

with the TSE-based gripper in both its unmodified and modified forms. Overall, the data indicates that all gripper versions demonstrate similar performance. Both the original and the TSE variants show a slightly degressive pressure-force relationship, attributed to the internal pressure developed by the polymer mixture. This effect becomes more noticeable at elevated pressure levels.

Due to its considerably higher stiffness, the TSE gripper shows a less favorable pressure-to-force ratio compared to the original PneuFlex gripper. As a result, achieving the same gripping force with the unmodified TSE gripper requires a higher input pressure. The modified TSE gripper displays a slightly progressive force-pressure characteristic. Modifications to its shape influence not only the gripping motion but also the resulting gripping force. While the underlying mechanisms are not experimentally investigated in detail, it is assumed that the increased external surface area and reduced wall thickness of the modified gripper allow for more efficient conversion of air pressure into force, with lower energy losses.

To measure the forces at a defined point during the gripper's motion, the sensor is aligned with the movement trajectory of the gripper tip in a subsequent experiment, as depicted in Fig. 12. According to the force measurements shown in the figure, the curved tip of the modified TSE gripper contacts the sensor at a pressure of 0.1 bar. Initially, the pressure-force relationship displays a progressive trend, followed by a degressive curve. In contrast, the original silicone gripper makes contact at 0.4 bar and exhibits a primarily degressive response. The overall behavior of both gripper types is consistent with prior experiments, with the key distinction being the pronounced degressive pattern above 0.8 bar in both cases. It is assumed that

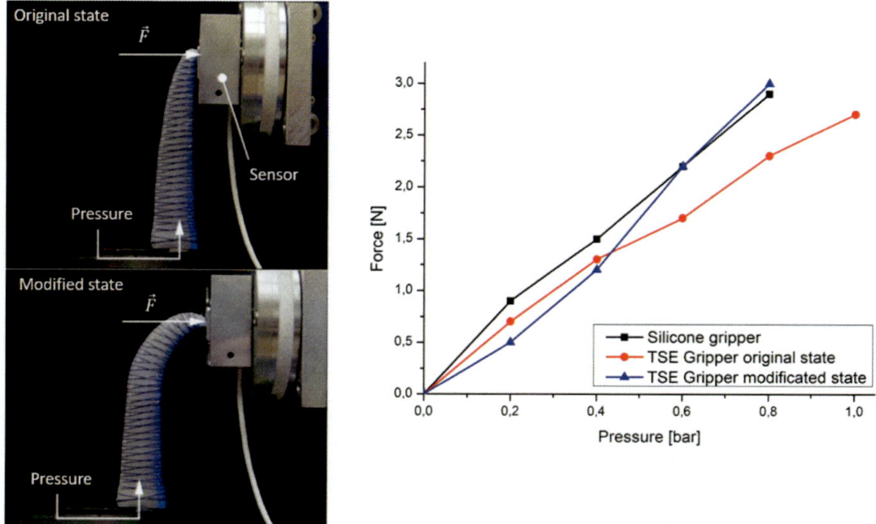

Fig. 12 Force development of the original silicone gripper (TU Berlin) compared to the original as well as modified state of the TSE gripper (TU Ilmenau)

the pneumatic pressure contributes to both motion and force generation. As a result, the force output is lower compared to earlier tests. This is due to the fact that larger elastic deformations consume more energy, causing the force component to diminish proportionally with increased deflection. Additionally, the contact area between the sensor and the gripper tip shifts during deformation. As the gripper bends, the share of frictional force increases while the direct compressive force decreases, altering both the direction and magnitude of the resulting force. In summary, the TSE-based gripper designs used in the experiments are capable of exerting forces of up to 3 N. Moreover, the degree of deformation significantly influences the effective gripping force.

Due to their properties, the TSE as shape memory polymers represent a potential opportunity to realize or support applications in soft robotics, such as the co-design project in a cooperation between TU Ilmenau and TU Berlin.

Other possibilities are soft robotic components with the ability to mimic different morphologies, illustrated in Fig. 13, which can be produced as well as an effect that resulting from the manipulation of the surface of the developed TSE resulting in the change in the friction coefficient of the material. The manipulation in this case includes the change of the surface at elevated temperature, which leads to a reduction in friction, as well as the permanent change of the surface due to the imprinted new morphology. In order to achieve a corresponding surface, it is necessary to imprint different shapes in the macro and micro range into the new materials. This can be advantageous when gripping and transporting various objects. To utilize these advantages, various samples with different geometries and dimensions as well as the appropriate choice of force and temperature influence on the material were investigated [29].

In addition to the modification of the soft robot hand at TU Berlin, another demonstrator was developed and experimentally verified as part of the project. The sole of the humanoid robot NAO, see Fig. 14, was modified using the new materials, which resulted in an improvement in the stability of the robot by adapting it to any surface. Heating the material leads to a temporary local change in the sole during operation. This results in a change in the coefficient of friction as well as an adaptation to different external structures or surfaces. As a result, the humanoid robot is not only able to move on any surface without undesirable interruptions, but also to maintain stability even under difficult conditions [29].

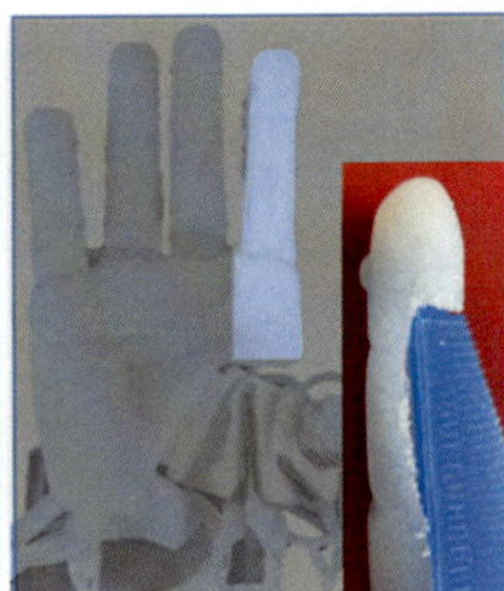

Fig. 13 Supplementing a conventional gripper by using thermosensitive hybrid materials (left) as well as shape adaption and morphology changes (contact area) to generate new morphological features for gripping of objects (right)

Fig. 14 Humanoid robot NAO with a modified shoe sole made of the thermosensitive hybrid material

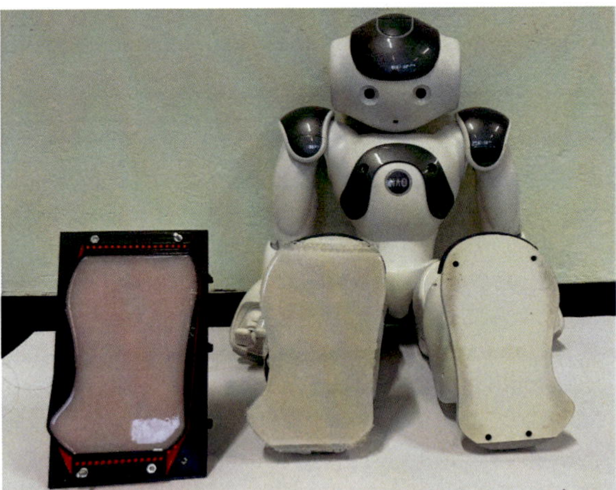

4 Conclusion

By means of various experimental analyses, the material properties for an immiscible but homogen thermosensitive elastomer (TSE) consisting of polydimethylsiloxane (PDMS) and polycaprolactone (PCL) was determined. This polymer blend can also be described as a shape memory polymer (SMP) according to the experimental results. Utilizing thermogravimetric analysis (TGA) and differential scanning calorimetry (DSC), the TSE proved to be thermally stable for the desired field of intended application in soft robotics. The mechanical experiments revealed a clear Payne and Mullins Effect of the samples, making preconditioning unavoidable for further investigations and applications. The higher the proportion of PCL and CI particles, the higher the shore hardness of the TSE becomes. The shore hardness value is almost identical before and after the temperature influence, since the PCL particles generally have a higher influence on the elasticity of the TSE. This also results in a limit to the mixing ratio of 20 wt% PCL particles in the matrix due to a lower miscibility and an unstable cross-linked end product. Both the weight percentage of the particles and the shore hardness of the matrix have an influence on the deformation and bending behavior of the TSE. Furthermore, the higher the weight percentage of PCL particles, the more significant is the influence of temperature on the thermal and mechanical effects such as the stress change, the different stiffness and deformation behavior of the TSE, whereby the important temperature range is between 60 and 65 °C. In some experiments, CI particles were included as an additional filler. An amount of 20 wt% PCL and an additional 50 wt% CI particles represents the maximum possible percentage by weight of fillers. The implementation has different consequences, on the one hand the addition of CIP accelerates the heat conduction within the material, but on the other hand when an external magnetic field is applied, the stiffness increases, which is a contrary effect to the softening of TSE under the influence of temperature. Alternatively, the magnetic attraction can be used as an external force to deform a sample during the melting process of the PCL particles. The deformation increases if the samples are subjected to a tensile force compared to a compressive force. Not only the static mechanical properties but also the dynamic mechanical properties of TSE can be influenced by temperature, such as the damping increase due to a higher temperature with a decrease in the resonance frequency as consequence. The SME of the TSE is the most significant under the condition that the PCL particles are larger than 100 μm. The SME can be utilized not only for deformations of the entire sample geometry, but also for shape adaption or morphology changes of the surface. The TSE of lower shore hardness require a lower intrinsic height to enable an imprint that corresponds to the actual height of the imprinted structures. However, with respect to the time the imprint remains in the various samples without reheating the sample to its original state, it is evident that the height of the imprint of the TSE of lower shore hardness decreases. Therefore, a clear correlation between the quality and accuracy of the SME with the respective shape adaption possibilities and the sample geometry as well as the shore hardness of the matrix can be observed. The TSE could be realized as an addition to hard structured grippers and as soft components such as a finger pulp of the RBO3 hand or as an entire gripping system. Various aspects such as deformation behavior, shape adaptations, morphology changes and stability improvements were investigated within these application possibilities [29]. As a result, the TSE can be implemented in many different soft robotics applications. The increase of temperature between 60 and 65 °C represents an field of application that does not affect the properties of other components made of silicone, metal or ceramic. This makes the material particularly attractive for implementation in existing systems.

5 Outlook

The TSE provide a good foundation for enabling new application possibilities in the field of soft robotics. Manipulation can involve the modification of the surface at elevated temperature, as already mentioned, which could result in a decrease of friction, as well as the permanent change of the surface due to the imprinted new morphology. This could be advantageous for gripping and during transporting various objects.

In previous investigations, the focus has been on ensuring the most homogeneous possible distribution of the PCL particles in the matrix. The corresponding properties are highly standardized and can be adapted to almost all requirements. However, for selected applications, it could be desirable to change the mechanical properties in specific directions.

According to the experiences and results with the developed TSE, the design of soft robotic structures based on particle-matrix interactions and controllable structural properties by thermal and magnetic fields could offer interesting investigation results. These fields, in their controlled temporal sequence, frequency, magnitude and time-dependent behavior, provides the necessary energy inputs for tasks such as grasping and locomotion and are used for information transfer.

In addition to manipulation, future applications are also seen in the field of locomotion. Active elements such as a controllable worm skin (apedal locomotion), elements such as an adaptive sole with a controllable friction coefficient (pedal locomotion) for humanoid robots, as illustrated in Fig. 11 are possible fields of investigation. The purpose is to improve the walking behavior of the humanoid robot by modifying the friction according to the different requirements of any surface. A possible stabilization of the walking behavior on different surfaces is then detected by the force sensors already integrated into the system of the robot. Minimizing the force therefore involves stabilizing the robot's body [29].

In summary, the TSE offers many ideas for implementation as actuators and/or sensors in the field of soft robotics and thus represents an important expansion of possibilities.

Acknowledgements Funded by the Deutsche Forschungsgemeinschaft (DFG, German Research Foundation) under grant no. 404586657.

Literatures

1. Ahamed, R., Choi, S.-B., Ferdaus, M.M.: A state of art on magneto-rheological materials and their potential applications. J. Intell. Mater. Syst. Struct. **29**(10), 2051–2095 (2018)
2. Alvarez-Lorenzo, C., Bromberg, L., Concheiro, A.: Light-sensitive intelligent drug delivery systems. Photochem. Photobiol. **85**(4), 848–860 (2009)
3. Becker, T.I., Böhm, V., Chavez Vega, J., Odenbach, S., Raikher, Y.L., Zimmermann, K.: Magnetic-field-controlled mechanical behavior of magneto-sensitive elastomers in applications for actuator and sensor systems. Arch. Appl. Mech. **89**(1), 133–152 (2019)
4. Bellan, C., Bossis, G.: Field dependence of viscoelastic properties of magnetorheological elastomers. Int. J. Mod. Phys. B. **16**(17n18), 2447–2453 (2012)
5. Denisyuk, Y.N.: Three–dimensional imaging by means of a reference–free selectogram recorded in a thick–layered light–sensitive material. Opt. Eng. **35**(2), 564 (1996)
6. García-Arribas, A., Gutiérrez, J., Kurlyandskaya, G.V., Barandiarán, J.M., Svalov, A., Fernández, E., Lasheras, A., de Cos, D., Bravo-Imaz, I.: Sensor applications of soft magnetic materials based on magneto-impedance, magneto-elastic resonance and magneto-electricity. Sensors. **14**(5), 7602–7624 (2014)
7. Gast, S., Prem, N., Schale, F., Zeidis, I., Zimmermann, K.: A contribution to the mechanics of a multi-layered compliant system with applications for soft robotics. Probl. of Mech., Int. Sci. J. IFToMM. **82**(2), 7–16 (2021)
8. Greco, F., Mattoli, V.: Introduction to active smart materials for biomedical applications. In: Ciofani, G., Menciassi, A. (eds.) Piezoelectric Nanomaterials for Biomedical Applications, Nanomedicine and Nanotoxicology, pp. 1–27. Springer, Berlin Heidelberg, Berlin, Heidelberg (2012)
9. Hou, X., Liu, Y., Wan, G., Xu, Z., Wen, C., Yu, H., Zhang, J.X.J., Li, J., Chen, Z.: Magneto-sensitive bistable soft actuators: experiments, simulations, and applications. Appl. Phys. Lett. **113**(22), 221902 (2018)
10. Hu, T., Xuan, S., Ding, L., Gong, X.: Stretchable and magneto-sensitive strain sensor based on silver nanowire-polyurethane sponge enhanced magnetorheological elastomer. Mater. Des. **156**, 528–537 (2018)
11. Huang, Y., Yu, Q., Su, C., Jiang, J., Chen, N., Shao, H.: Light-responsive soft actuators: mechanism, materials, fabrication, and applications. Actuators. **10**(11), 298 (2021)
12. Ji, W., Wu, Q., Han, X., Zhang, W., Wei, W., Chen, L., Li, L., Huang, W.: Photosensitive hydrogels: from structure, mechanisms, design to bioapplications. Sci. China Life Sci. **63**(12), 1813–1828 (2020)
13. Kim, K.J., Tadokoro, S.: Electroactive Polymers for Robotic Applications: Artificial Muscles and Sensors. Springer-Verl, London (2007)
14. Monkman, G., Striegl, B., Sindersberger, D., Prem, N.: Electrical properties of Magnetoactive boron-Organo-silicon oxide polymers. Macromol. Chem. Phys. **221**, 1900342 (2020). https://doi.org/10.1002/macp.201900342
15. Odenbach, S.: Magnetic Hybrid-Materials: Multi-Scale Modelling, Synthesis, and Applications. De Gruyter, Berlin (2022)
16. Prem, N., Vega, J.C., Böhm, V., Sindersberger, D., Monkman, G.J., Zimmermann, K.: Properties of polydimethylsiloxane and magnetoactive polymers with electroconductive particles. Macromol. Chem. Phys. **219**(18), 1800222 (2018)
17. Prem, N., Sindersberger, D., Monkman, G.J.: Mini-extruder for 3d magnetoactive polymer printing. Adv. Mater. Sci. Eng. **2019**, 1–8 (2019)

18. Prem, N., Schale, F., Zimmermann, K., Gowda, D.K., Odenbach, S.: Synthesis and characterization of the properties of thermosensitive elastomers with thermo-plastic and magnetic particles for application in soft robotics. J. Appl. Polym. Sci., 14 (2021). https://doi.org/10.1002/app.51296
19. Prem, N., Schale, F., Sindersberger, D., Zimmermann, K.: Thermosensitive elastomers for shape adaption of soft robotic systems. In: ACTUATOR 2022; International Conference and Exhibition on New Actuator Systems and Applications, pp. 1–4, Mannheim, Germany (2022)
20. Puhlmann, S., Harris, J., Brock, O.: Rbo hand 3: a platform for soft dexterous manipulation. IEEE Trans. Robot. **38**(6), 3434–3449 (2022)
21. Rongchen, L.: Synthese und experimentell-messtechnische von Elastomer-Verbundmaterialien mit magnetischen und thermoplastischen Partikeln Masterarbeit. TU Ilmenau, Ilmenau (2022)
22. Ivaneyko, D.,Toshchevikov, V., Saphiannikova, M., Heinrich, G.:Mechanical properties of magneto-sensitive elastomers: unification of the continuum-mechanics and microscopic theoretical approaches. Soft Matter. **10**, 2213–2225 (2014)
23. Scalet, G.: Two-way and multiple-way shape memory polymers for soft robotics: an overview. Actuators. **9**(1), 10 (2020)
24. Schümann, M., Morich, J., Günther, S., Odenbach, S.: The evaluation of anisotropic particle structures of magnetorheological elastomers by means of pair correlation function. J. Magn. Magn. Mater. **502**, 166537 (2020)
25. Schönfeld, D., Chalissery, D., Wenz, F., Specht, M., Eberl, C., Pretsch, T.: Actuating shape memory polymer for thermoresponsive soft robotic grip-per and programmable materials. Molecules. **26**(3) (2021)
26. Servant, A., Methven, L., Williams, R.P., Kostarelos, K.: Electroresponsive polymer-carbon nanotube hydrogel hybrids for pulsatile drug delivery in vivo. Adv. Healthc. Mater. **2**(6), 806–811 (2013)
27. Shintake, J., Cacucciolo, V., Floreano, D., Shea, H.: Soft robotic grippers. Adv. Mater., e1707035 (2018)
28. Sindersberger, D., Diermeier, A., Prem, N., Monkman, G.J.: Printing of hybrid magneto active polymers with 6 degrees of freedom. Mater. Today Commun. **15**, 269–274 (2018)
29. Sindersberger, N.: Synthesis, Characterization and Application of Soft Thermoresponsive Hybrid Materials and their Field-Based Controllable Properties. Universitätsbibliothek, TU Ilmenau (2025)
30. Stoychev, G., Kirillova, A., Ionov, L.: Light–responsive shape–changing polymers. Adv. Opt. Mater. **7**(16), 1900067 (2019)
31. Walker, J., Zidek, T., Harbel, C., Yoon, S., Strickland, F.S., Kumar, S., Shin, M.: Soft robotics: a review of recent developments of pneumatic soft actuators. Actuators. **9**(1), 3 (2020)
32. Wu, Z., Li, X., Guo, Z.: A novel pneumatic soft gripper with a jointed endoskeleton structure. Chin. J. Mech. Eng. **32**(1) (2019)
33. Xia, L.-W., Xie, R., Ju, X.-J., Wang, W., Chen, Q., Chu, L.-Y.: Nano-structured smart hydrogels with rapid response and high elasticity. Nat. Commun. **4**, 2226 (2013)
34. Yarali, E., Baniasadi, M., Zolfagharian, A., Chavoshi, M., Arefi, F., Hos-sain, M., Bastola, A., Ansari, M., Foyouzat, A., Dabbagh, A., Ebrahimi, M., Mirzaali, M.J., Bodaghi, M.: Magneto–/ electro–responsive polymers toward manufacturing, characterization, and biomedical/ soft robotic applications. Appl. Mater. Today. **26**, 101306 (2022)
35. Zimmermann, K., Zeidis, I., Gast, S., Prem, N., Gowda, D.K., Odenbach, S.: An approach to the modeling and simulation of multi-layered and multi-stimulable material for application in soft robots. In: 16th International Conference Dynamical Systems—Theory and Applications, pp. 265–266. Book of Abstract, Lodz (2021)

Flexible Printed Electrodes for Soft Robotic Applications

Lingyu Liu, Mario De Lorenzo, Thomas Kister, Makara Lay, Uwe Marschner, Andreas Richter, Tobias Kraus, and E.-F. Markus Vorrath

Abstract Dielectric elastomer (DE) components—such as actuators (DEAs), switches (DESs), transistors (DETs), and oscillators (DEOs) enable soft robots with integrated material intelligence by combining actuation and control at the material level. These components can be manufactured via printing methods that produce multifunctional elastomer structures capable of actuation, sensing, and signal processing. DEAs typically use pre-stretched elastic membranes like VHB 4905 or silicone, combined with stretchable carbon-based electrodes applied through printing. Existing printable electrodes often degrade, limiting device longevity. We explored alternative conductive composites with stable conductivity under strain and low stiffness. For DESs, piezoresistive electrodes are required: materials whose conductivity changes under load. To enhance switching performance, we developed functional inks with high and stable gauge factors. By combining DEA and DES materials, we fabricated DETs on a single substrate. Interconnecting multiple DEA-DES units forms DEOs capable of generating oscillations. These soft, stretchable systems with switchable piezoresistivity enable autonomous behaviors like walking, grasping, and sensing—without external control units. Our work shows how soft electronic materials support distributed control and multifunctionality in biomimetic soft robotics.

L. Liu (✉) · T. Kister · M. Lay · T. Kraus
INM-Leibniz Institute for New Materials, Saarbrücken, Germany
e-mail: lingyu.liu@leibniz-inm.de

T. Kister
e-mail: thomas.kister@leibniz-inm.de

M. Lay
e-mail: makara.lay@leibniz-inm.de

T. Kraus
e-mail: tobias.kraus@leibniz-inm.de

T. Kraus
Saarland University, Colloid and Interface Chemistry, Saarbrücken, Germany

---

Lingyu Liu and Mario De Lorenzo these authors contributed equally to this chapter.

A. Raatz et al. (eds.), *Soft Material Robotic Systems*,
https://doi.org/10.1007/978-3-032-22453-8_5

1 Introduction

Pelrine et al. introduced dielectric elastomer actuators (DEAs) in the1990sat the Stanford Research Institute (SRI International) as a new approach for the design of artificial muscles. DEAs use electroactive polymers whose mechanical properties resemble those of natural muscles, thereby allowing them to work as light, soft, and energy-saving devices able to imitate the behavior of organisms. DEAs deform due to an electric field and provide comparatively large displacements and significant forces. They resemble a capacitor consisting of a thin silicone or acrylate elastomer placed between two flexible conducting electrodes. Upon the application of a voltage typically between 1 and 5 kV to the electrodes, an electrostatic pressure, known as Maxwell pressure, compresses the elastomer, which expands in the transverse direction. The Maxwell pressure is

$$p = \frac{\varepsilon_0 \varepsilon_r V^2}{d^2} \tag{1}$$

where ε_0 is the vacuum permittivity, ε_r is the relative permittivity of the elastomer, V is the applied voltage, and d is the thickness of the membrane [1]. Suitable elastomers sustain strains above 100% and lead to a high energy density. This is useful in soft robotics, prosthetics, and other fields requiring flexible and adaptive materials. DEAs also have multifunctional capabilities, being able to act simultaneously as actuators, sensors, information processors, and even energy harvesters–all within the same structure [2–8]. This multifunctionality can eliminate the need for rigid components and external control units in soft robotic systems, which simplifies their design and makes the robots more efficient.

Dielectric elastomer switches are a prominent example of multifunctionality. They were introduced as a concept by O'Brien [9] in 2010 to reduce the reliance on traditional external circuitry and enable fully soft, biologically inspired intelligent systems. A DES functions as a strain sensor, utilizing changes in resistance due to external mechanical stimuli, a phenomenon known as piezoresistive behavior. The sensitivity is described by the Gauge Factor (GF) defined as the ratio of the relative change in electrical resistance R to the mechanical strain ϵ [10]:

M. De Lorenzo · U. Marschner · E.-F. M. Vorrath
Institute of Semiconductors and Microsystems, TU Dresden, Dresden, Germany
e-mail: mario.de_lorenzo@tu-dresden.de

U. Marschner
e-mail: uwe.marschner@tu-dresden.de

E.-F. M. Vorrath
e-mail: markus.vorrath@tu-dresden.de

A. Richter
Institute of Semiconductors and Microsystems, Chair of Microsystems, Dresden, Germany
e-mail: andreas.richter7@tu-dresden.de

$$GF = \frac{\Delta R/R_0}{\Delta\epsilon} \tag{2}$$

The key advantage of the DES is its exceptional sensitivity to deformation-induced resistance changes, allowing it to achieve "ON-OFF" state conversion with a resistance shift spanning approximately three orders of magnitude. Material improvements and designs in the DES are promising for achieve bio-inspired electromechanical control in robotic structures.

So far, carbon-based liquid materials have been used as conductive components in DEA and DES, as discussed below. These materials have limitations related to reproducibility, stability, and performance. To overcome these limitations, we investigate the potential of new conductive polymer composites with high conductivity or piezoresistivity. For DEA applications, high conductivities have to be combined with low Young's modulus, to enable good actuation performance. DES requires large piezoresistivities with resistance changes of approximately three orders of magnitude over many cycles at limited stiffness. This project studies both materials and designs of DEA-DES combinations.

2 Material Development for Dielectric Elastomer Actuators and Dielectric Elastomer Switches

2.1 Conductive Silver-Based Composites for DEAs

Conductive Polymer Composites (CPCs) based on dielectric elastomers possess tunable resistivity properties and are suitable to create highly deformable electromechanical structures for actuators, particularly in soft robotics, where high actuation performance and low operating voltages are required. Highly conductive metal fillers, such as silver particles and flakes, are suitable fillers with high conductivity. Their comparatively high cost is compensated by conductivities that surpass those of graphite and carbon black by up to a few orders of magnitude [11]. However, the use of metal-filled dielectric elastomers in soft robotic structures presents challenges due to their high mechanical stiffness, with a Young's modulus in the range of 10^4 MPa to 10^6 MPa [12, 13]. Here, we investigate composites with silver particle fillers and small fractions of carbon black. The goal of this hybrid approach was to create a soft conductive composite that remains conductive after deformation for dielectric elastomer actuator material for soft robotics.

A conductive paste was prepared by dispersing silver particles (AgP) with an average size of 1 μm to 3 μm (Thermo scientific, USA), into the silicone liquid precursor Ecoflex 00-30 (SMOOTH-ON Inc., USA) at 24% volume fraction. Additionally, 2%

E.-F. M. Vorrath
Biomimetics Lab, Auckland Bioengineering Institute, The University of Auckland, Auckland, New Zealand

volume fraction of carbon black (CB) (Alfa Aesar, Germany) with an average particle size of 800 nm was added. The prepared paste was then screen-printed onto a 200 μm thick silicone film (Wacker, Germany), forming a conductive layer with an thickness around 20 μm. The printed films were cured at 80 °C for 3 h to remove the solvent and cross-link the silicone. Films were then immersed into 0.5 M salt water for 1 h and subjected to post-curing at 160 °C for 30 min. Electromechanical properties and durability were evaluated using uniaxial tensile testing with *in situ* two-probe resistance monitoring.

Figure 1a and b illustrate the relative resistance change of the AgP-Ecoflex 00-30 composites (without Carbon Black) and AgP-CB-Ecoflex 00-30 composites in response to loading of 10% during 4000 cycles. For AgP-Ecoflex 00-30, the initial resistance changed from 0.5 Ω to 1.6 Ω, while AgP-CB-Ecoflex 00-30 changed from 1.1 Ω to 3.6 Ω. The CB reduced conductivity, which implies that it is present between silver particles. We evaluated its effects on the durability of the composite. Figure 1c, clearly indicates that the overall cycling stability was enhanced by CB. We conclude that the incorporated carbon particles improved the connectivity of the conductive networks under cyclic loading, resulting in enhanced stability. This results in stable piezoresistive behavior in composites with CB already during the first 20 cycles, where the CB-free composite changes (Fig. 1d). The piezoresistive response of the CB-containing composite remained more stable than that of the purely metal-based composites after many cycles, too.

The fluctuation in relative resistance during early loading/unloading cycles that are particularly prominent for the CB-free composites are likely caused by crack formation and residual elongation of the matrix under large deformation. Previous work [14] reports that the resistance at rest of composites based on silver flakes and Ecoflex 00-30 changed from approximately 5 Ω to 50 Ω during 1000 cycles under 100% strain. The CB-based composites introduced here compare favorably, consistent with reports on the role of CB in conductive networks under strain [15].

2.2 *Piezoresistive Carbon-Based Thin Films for DES*

The original DES were often fabricated by smearing or stamping carbon grease or the inkjet-printing of carbon black-based inks [2, 8, 9, 16]. The highly viscous liquids required for such processes provide acceptable conductivity and softness, but they are prone to degradation due to Joule heating induced viscosity changes and poor fabrication repeatability. The resulting short lifespan and signal distortion highlight the need for the development of solid-state electrode materials to improve durability and ease of fabrication in DES. In solid strain sensor research, a primary goal is to enhance sensitivity, which is highly relevant to applications in human motion detection, soft robotics, and related fields. It is a challenge to retain sufficient piezoresistive response, with resistance changes spanning up to three orders of magnitude under external loading.

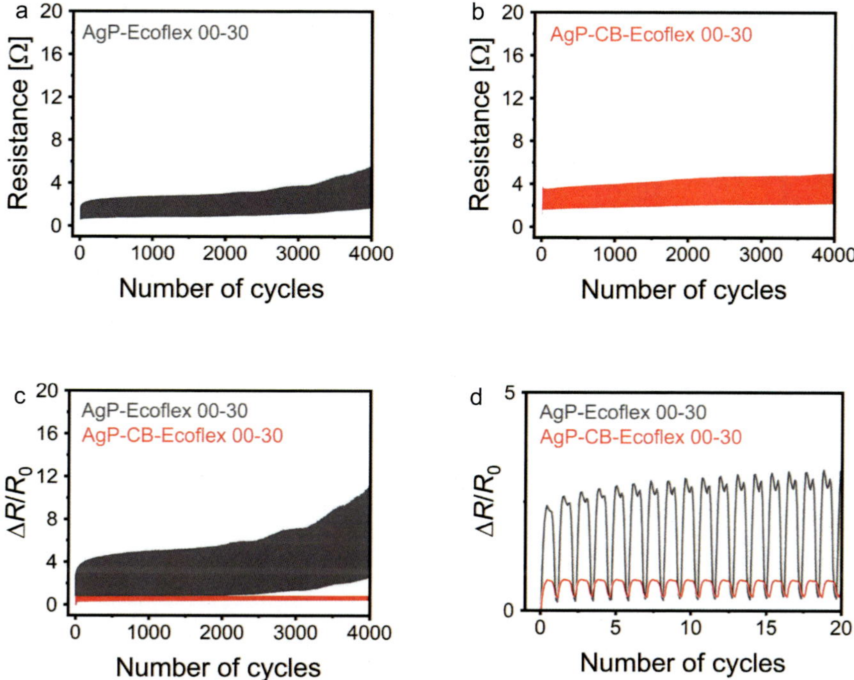

Fig. 1 Electromechanical characterization of printed composite electrodes for DEAs. **a** Relative resistance change of a AgP-Ecoflex 00-30 composite at 10% strain during 4000 cycles. **b** Relative resistance change of a AgP-CB-Ecoflex 00-30 composite at 10% strain during 4000 cycles. **c** Piezoresistivity of both composites during 4000 cycles. **d** Piezoresistivity during the first 20 cycles

In CPCs, conductive fillers lend insulating elastomers electrical conductivity by forming interconnected pathways. The transition from an insulating to conductive behavior is caused by a microstructural transition referred to as the percolation threshold [17]. At this critical filling ratio, random filler networks become space-filling, providing macroscopic electrical conductivity. According to standard percolation theory, the piezoresistive response of CPCs is particularly pronounced near the percolation threshold, where conductive pathways are highly sensitive to mechanical strain and can be easily disrupted [18].

We investigated whether strong and reliable piezoresistive responses can be achieved in thinner films. Conductive particles were confined within thin composite layers, which limits surface-normal motion of the fillers during deformation. Poisson's effect causes the contraction of the film perpendicular to the stretching direction. This rearranges conductive pathways, making them more susceptible to disruption under strain. Spatial confinement in thin films potentially reduces the effect and enhances the film's sensitivity to resistance changes under mechanical deformation.

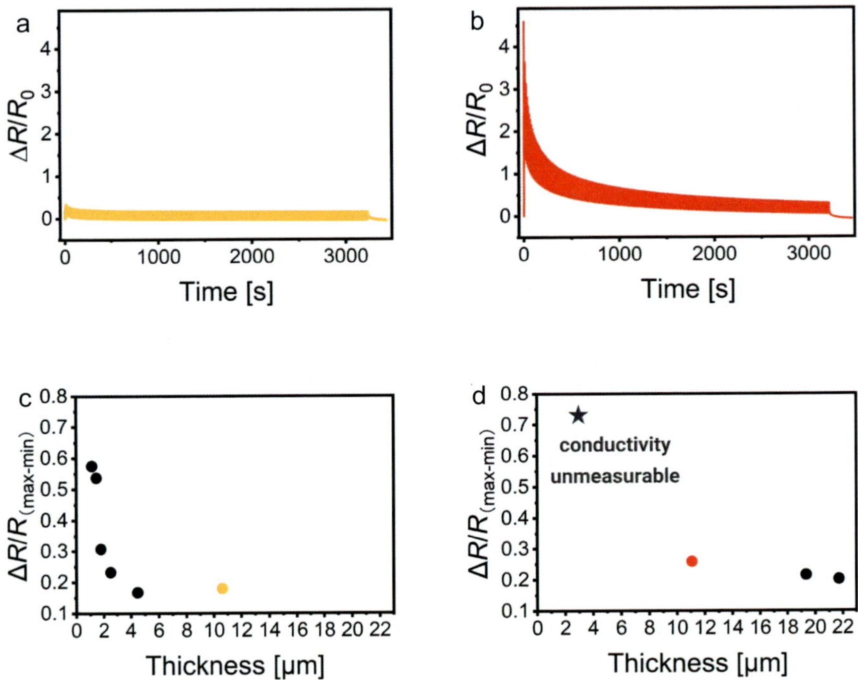

Fig. 2 Piezoresistivity of **a** 8 vol% CB/PDMS and **b** 8 vol% GF/PDMS composite films with approximately 10 μm thickness under 10% strain during 300 cycles. **c**, **d** Piezoresistance in the plateau region as a function of film thicknesses for **c** CB/PDMS and **d** GF/PDMS

We compared two different piezoresistive elastomer composites, one containing of CB (Alfa Aesar, Germany), the other of graphite flakes (GF) (Enerage Inc., Taiwan). Both were dispersed in the silicone elastomer Sylgard 184 (Dow Corning, USA), respectively. The composite paste was spin-coated onto 200 μm-thick silicone film substrates (Wacker, Germany) to create layers with thicknesses between 1 to 22 μm and cured at 80 °C. The electromechanical properties and durabilities of the films were evaluated using uniaxial tensile testing with in situ two-probe resistance monitoring.

To assess electromechanical properties, the films were subjected to a maximum strain of 10% for 300 cycles, and the relative resistance changes were recorded (Fig. 2a and b). Durability curves of CB/PDMS and GF/PDMS films with thicknesses of approximately 10 μm were measured. The GF/PDMS films exhibited an initial rapid increase in resistivity compared to CB/PDMS, after which the piezoresistive responses of both films plateaued at higher cycle counts.

Figure 2c and d show that the piezoresistive responses varied significantly for both CB/PDMS and GF/PDMS composites under the same strain conditions during the plateau phase. GF/PDMS films exhibited a gauge factor of 2.5, which exceeded that of the CB/PDMS film $GF = 1.8$. Thinner GF/PDMS films with a thickness of approximately 3 μm exhibited large piezoresistive responses: their resistance increased from 100 kΩ at rest to several hundred MΩ at a maximum strain of 20%, exceeding the limits of the measurement instrumentation (marked with a star in Fig. 2d). This corresponds to a dynamic resistance change spanning more than three orders of magnitude. These results indicate that ultrathin GF/PDMS films exhibit highly sensitive and tunable piezoresistive responses, making them promising candidates for application in dielectric elastomer switches (DESs).

3 Manufacturing and Modelling of Dielectric Elastomer Actuators and Switches for Soft Biomimetic Autonomous Soft Robots

3.1 Dielectric Elastomer Transistor for Smart Soft Robotics

One of the most notable achievements is the development of the dielectric elastomer transistor (DET) that works within a soft flexible material system in line with traditional transistors. DETs include a dielectric elastomer actuator (DEA) and a dielectric elastomer switch (DES), which work together to create an electrically controlled resistor. This innovation has enabled the manufacturing of complex logic circuits from a completely soft material which is an important step towards a fully integrated soft robotic system. We have successfully demonstrated the manufacturing of all basic logic gates (AND, OR, NOT, NAND, NOR, XOR, XNOR). These gates display the expected logic behavior with minimal errors, despite working on higher voltage (1–5 kV) and longer switching times than traditional electronics. This achievement is particularly important because it allows logic operations directly within DE material, which eliminates the requirement of external rigid electronic components in a soft robotic system [6, 7] (Fig. 3).

Building upon these basic logic gates, more complex circuits have been realized. A notable example is the development of a DE multiplexer (MUX) using pass transistor logic. This design significantly reduces the number of components needed compared to traditional circuit designs, showcasing the potential for efficient and compact DE-based circuits. The successful implementation of sequential logic circuits, such as flip-flops, further expands the capabilities of DE electronics, enabling memory and state retention in soft systems [2, 6, 8, 16, 19].

3.2 Simulink and ABAQUS Simulation

A major advance in the field is the creation of a comprehensive Simulink model to simulate circuit networks of DEs. This model integrates various aspects of behavior,

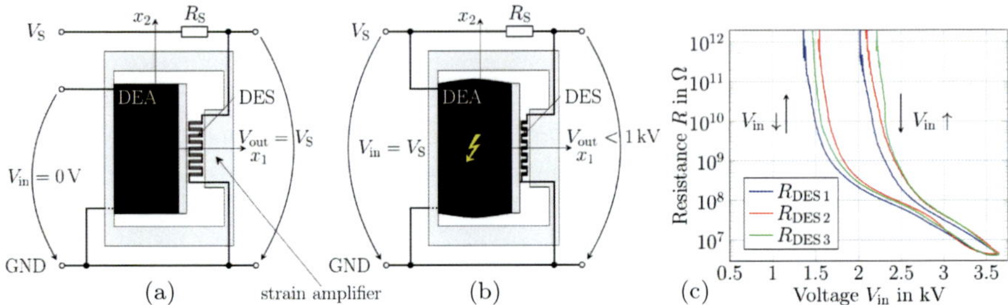

Fig. 3 DE inverter: **a** DE inverter with no input voltage V_{in}: R_{DES} is high and the output voltage V_{out} is high. **b** V_{in} is high, R_{DES} is low and V_{out} is low either. **c** Typical transition functions $R_{DES}(V_{in})$. Figure **c** by authors [4]. The figure is licensed under a creative commons attribution 3.0 international license. You should have received a copy of the license along with this work. If not, see https://creativecommons.org/licenses/by/4.0/

including viscoelasticity, leakage current, and in-plane actuation, providing a powerful tool to predict complex device performance. The precision of the model was validated through experimental results, with a maximum error of only 3.7% in the frequency prediction for digital dielectric elastomer oscillators (DEOs) [7].

This modeling capacity significantly accelerates the design and optimization process for DE devices, reducing the need for lengthy and expensive physical prototypes. The development of DEOs represents another important advance. These devices, inspired by standard central generators (CPGs) in biological systems, can generate rhythmic signs to control and synchronize processes in soft robotic systems. Successful modeling and implementation of DEOs open new possibilities for creating autonomous and inspired robots with distributed control mechanisms. The research also highlighted the multifunctionality of the DES, demonstrating its ability to serve as actuators, sensors and energy harvesters simultaneously. This versatility allows the creation of highly integrated systems that can perform various functions in a single material structure, closely imitating the efficiency and adaptability of biological organisms [2, 4, 7, 8].

Although block diagrams were used primarily for modeling the electromechanical behavior of DEOs in Simulink, Finite Element (FE) analysis were employed to analyze the mechanical properties of DEs with ABAQUS. We developed a comprehensive mathematical model implemented in Simulink to simulate the functionality of DEOs, which can autonomously generate oscillating signals from a DC voltage input. This model represents each component of the DEO as an electrical or electromechanical element, allowing a detailed analysis of their interactions. The integration of such models with ABAQUS simulations enhances the overall understanding of DE behavior in soft robotic systems [4] (Figs. 4 and 5).

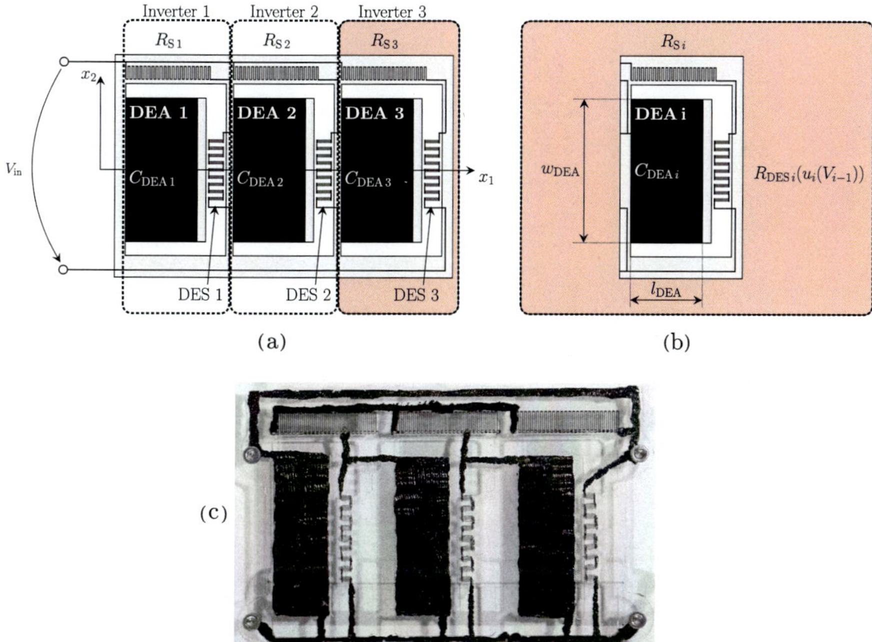

Fig. 4 Schematic set-up of a DEO: **a** DEO possessing three inverters, **b** Single inverter possessing a serial resistor R_S, a DEA of width w_{DEA} and length l_{DEA} and a DES, **c** Photograph of an experimental DEO demonstrator. Figure **c** by authors [4]. The figure is licensed under a creative commons attribution 3.0 international license. You should have received a copy of the license along with this work. If not, see https://creativecommons.org/licenses/by/4.0/

3.3 2D Bio-inspired DEA Gripper with Integrated Sensors

The application of these advances to biomimetic systems is exemplified by the development of a Venus flytrap-inspired gripper [8]. This device integrates sensors, actuators and logic circuits to create an autonomous system that mimics the reflexive movement of a Venus flytrap. The gripper can sense, process, and respond on environmental stimuli without the need for external control systems, operating only on a constant power supply. This achievement demonstrates the potential of DEs to create bio-inspired robots with embedded intelligence and autonomy (Fig. 6).

Although these advances are significant, researchers recognize challenges and ongoing challenges and areas for improvement [4, 6–8, 16, 20]. The viscoelasticity of current materials, particularly the VHB4905 acrylic tape, results in delays and limits switching speeds. Future work aims to explore alternative materials with reduced viscoelasticity to improve performance. In addition, manual manufacturing processes currently in use lead to variability in device performance, emphasizing the need for automated fabrication techniques to improve consistency and scalability.

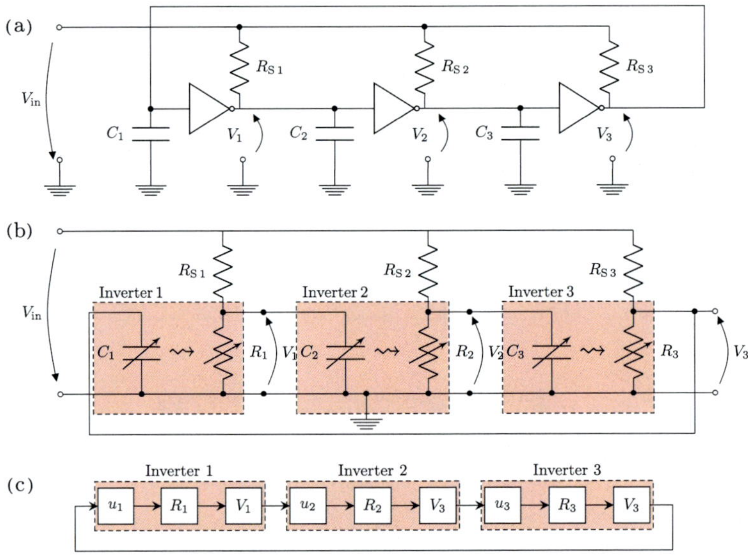

Fig. 5 Equivalent network of a DEO depicted in Fig. 4, with DEA capacitors C_i, variable DESs resistors R_i and serial resistors , generating three oscillating voltage signals from one single input voltage V_{in}. **a** Simplified circuit–DESs represented by an inverter module. **b** More realistic DEO model, representing the DESs as voltage-controlled variable resistors R_i. **c** Functional chain of the entire DEO. Figure **c** by authors [4]. The figure is licensed under a creative commons attribution 3.0 international license. You should have received a copy of the license along with this work. If not, see https://creativecommons.org/licenses/by/4.0/

3.4 3D Bio-inspired Autonomous Soft Robots with Skeletal Structures

Novel soft robotic structures driven by DEAs were made possible with bioinspired skeletal and muscular reinforcement. ABAQUS CAE was used for three-dimensional FE analysis to predict the initial shape and deformation of the chosen robotic structures. The simulation process involved iterative improvements to the silicone skeleton's geometry and variations in membrane pre-stretch to achieve desired deformation. ABAQUS was instrumental in modeling the complex interaction between the DE membrane and the silicone skeleton [3, 5] (Fig. 7).

This approach allowed for the optimization of the structure's anisotropic bending stiffness and actuation performance. We emphasize the critical role of ABAQUS in the design process, using it for three-dimensional finite element analysis to iteratively improve the silicone skeleton geometry and optimize membrane pre-strain. The ABAQUS simulations were crucial in predicting the initial shape and bending behavior of the structures under various loading conditions. The simulation results, which showed good agreement with experimental data, were used to finalize the design of the silicone skeleton and determine the optimal pre-strain level for the DE

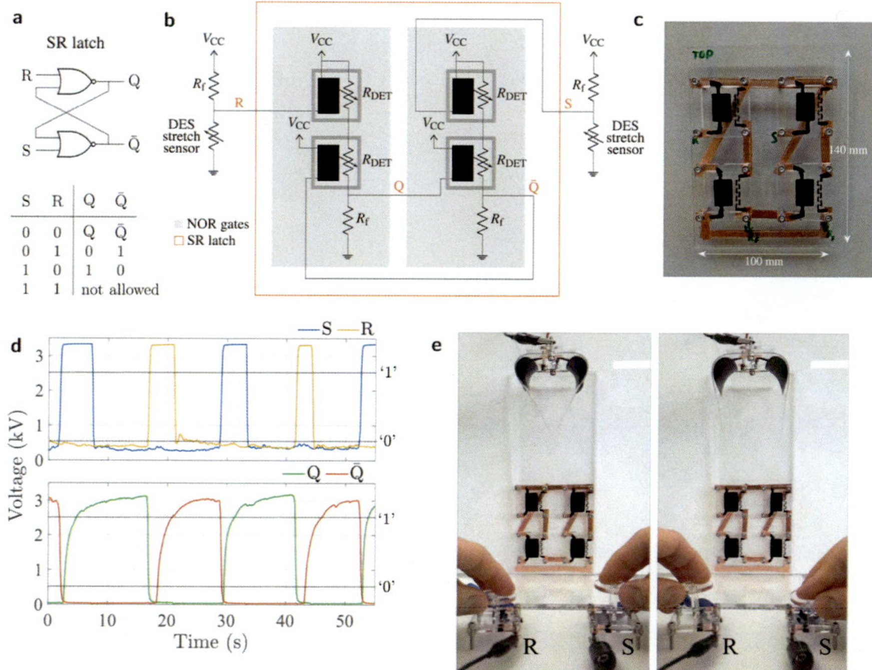

Fig. 6 **a** Schematic of a Set-Reset (RS) latch and its truth table. **b** Schematic of the circuit composed of two sensing systems and a set-reset latch made with DETs. **c** Photograph of the DE latch. **d** Measurements of the signals S, R, Q, and $\bar{Q}$. **e** The gripper is controlled by the DE latch. When the button for reset (R) is pushed, the gripper closes and maintains its state. When the button for set (S) is pushed, the gripper opens and maintains its state. Figure **c** by authors [8]. The figure is licensed under a creative commons attribution 4.0 international license. You should have received a copy of the license along with this work. If not, see https://creativecommons.org/licenses/by/4.0/

membrane. This approach highlights the effectiveness of ABAQUS in bridging the gap between theoretical design and practical implementation in soft robotics.

3.5 Combined Simulation of the Dynamics of Soft Robots with Skeletal Structures

In order to precalculate the dynamic behavior of the soft robotic finger, we applied the method of Combined Simulation [21]. It combines Finite Element methods and network methods, with preference to networks methods to simulate efficiently as well as exploring the dynamics. Figure 9 shows the mechanical network model of the soft robotic finger, where the low frequency model of the finger is used. It couples the DEA unimorph segments with stiff elements, where the latter are modeled only by their mass m in the first approach. The unimorph segments are described as finite

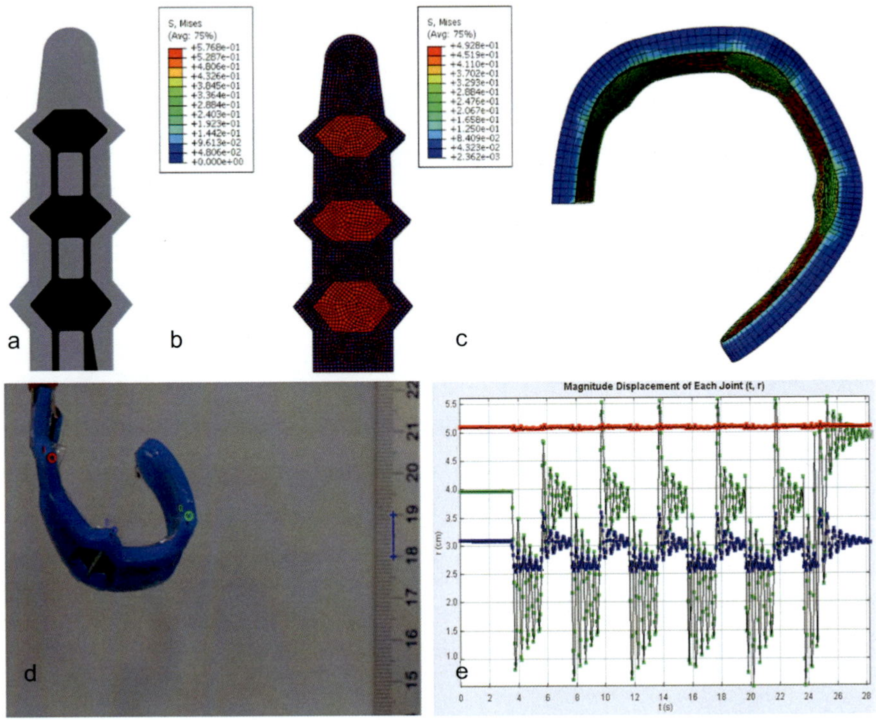

Fig. 7 **a** CAD design of a soft biomimetic finger with silicone frame and multilayer DEAs. **b** ABAQUS simulation of a 3 mm frame with a 40% pre-stretch of the silicone membrane. **c** Bending of the finger caused by the stress of the pre-stretched membrane. **d** Testing of a finger. **e** Displacement of the three joints of the finger measured with an open-source software tracker

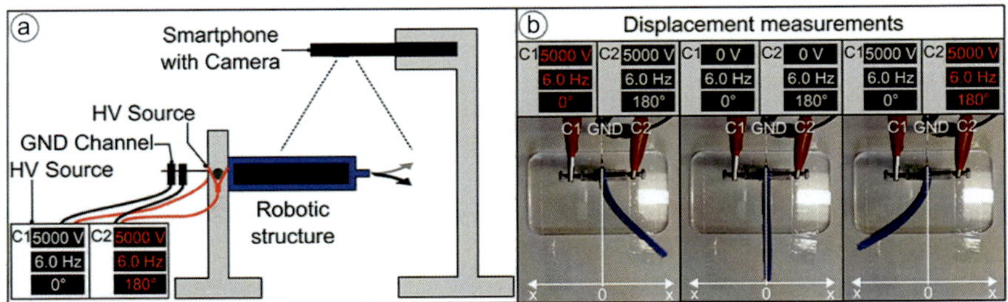

Fig. 8 Measurement of the displacement while running in static or dynamic mode. **a** Scheme of the experimental setup. **b** Video frames showing the left and right displacement x at maximum in dynamic mode depending on the voltage and frequency. By authors [5].

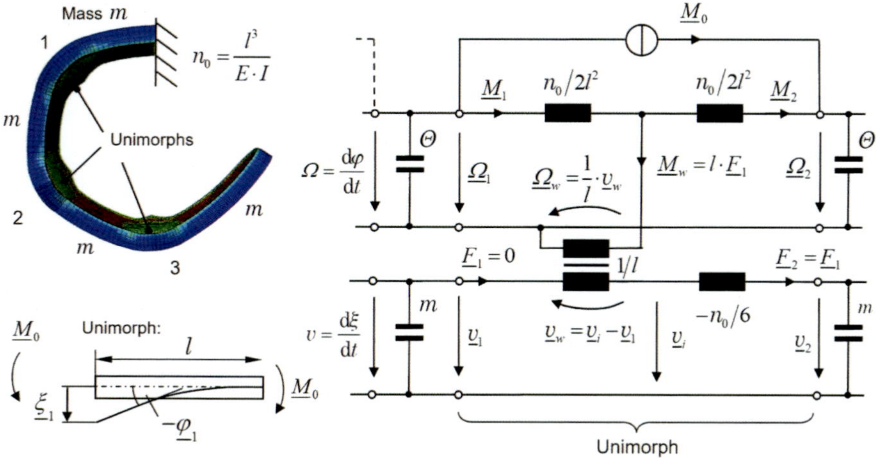

Fig. 9 Network model of the soft robotic finger with silicone frame and multilayer DEAs

network elements of ideal bending beams as derived in [22]. They are bent by the internal moment M_0 which is caused by the DEA element. The moment acts on the bending compliancies $n_0/2l^2$ with the length l of the segment, Young's modulus E and moment of inerta I in the rotational domain. In the translational domain these compliancies interact with the masses and constitute a resonating structure. Thus the network model explains the measured vibrations in Fig. 8 and gives a graphical overview of the dynamic elements. The Combined Simulation allows an efficient determination of the network parameters by evaluating the FE simulations.

4 Conclusion

In this study, we developed promising functional conductive silicone composites for use in DEAs and DESs with good reproducibility and electromechanical properties. Silver-carbon polymer composite films for DEAs exhibited stable conductivities under strain, high durability, and low piezoresistivities during 4000 strain cycles. Carbon-filled silicone elastomer thin films for DES had enhanced piezoresistive responses due to the confinement in thin films. The piezoresistive response increased with decreasing film thickness. This was particularly pronounced for flake-shaped carbon fillers. The resistance of the thinnest flake/elastomer films increased from 100 kΩ at rest to several hundred MΩ at a strain of 20 %, showing their potential for strain-sensitive applications.

The mathematical models relating to DEAs and circuit networks were effectively modeled and verified using finite element analysis with ABAQUS and dynamic modeling using Simulink as well as LTSPICE. These simulations were crucial in gaining insight into the electro-mechanical behavior of dielectric elastomer components and in optimizing soft robot design. One of the highlights was the development of a bioinspired gripper resembling a flytrap, which employs DES and exhibits autonomous functionality in grasping. This gripper is an example of the possibilities of fully soft dielectric elastomer robots with completely integrated actuation, sensing, and control within a single compliant structure.

In addition, our investigation of silicone frames for soft robotics highlighted their flexibility and versatility across a wide range of applications. The project also contributed to the development of DE-based digital electronics such as logic gates and multiplexers, thus allowing the creation of more complex soft robotic systems with integrated control systems. Future work will involve improving soft robots using silicone elastomers for both actuators and switches. Silicone elastomers offer a number of advantages over conventional VHB materials such as lower viscoelasticity, faster response times, better long-term stability, and improved mechanical robustness. These properties make silicone elastomers highly suitable for creating switches in the context of soft robots, which can result in more reliable, robust, and faster soft robotic systems. In addition, the use of silicone elastomers could allow simplified manufacturing procedures and better integration of components, furthering the creation of fully soft, autonomous robotic systems.

Acknowledgements This research was funded by Deutsche Forschungsgemeinschaft (DFG, German Research Foundation) under grant no. 498165449.

References

1. Pelrine, R., Kornbluh, R., Pei, Q., Joseph, J.: High-speed electrically actuated elastomers with strain greater than 100%. Sens. Actuators, A **64**(1), 77–85 (1998). https://doi.org/10.1126/science.287.5454.83
2. Henke, E.-F.M., Schlatter, S., Anderson, I.A.: Soft dielectric elastomer oscillators driving bioinspired robots. Soft Rob. **4**(4), 353–366 (2017). https://doi.org/10.1089/soro.2017.0022. Dec
3. Henke, E.-F.M., Wilson, K.E., Anderson, I.A.: Entirely soft dielectric elastomer robots, in Electroactive Polymer Actuators and Devices (EAPAD) 2017, vol. 10163, p. 101631N (2017). https://doi.org/10.1117/12.2260361
4. Henke, E.-F.M., Wilson, K.E., Anderson, I.A.: Modeling of dielectric elastomer oscillators for soft biomimetic applications. Bioinspiration Biomimetics **13**, 046009 (2018). https://doi.org/10.1088/1748-3190/aac911
5. Franke, M., Ehrenhofer, A., Lahiri, S., Henke, E.-F.M., Wallmersperger, T., Richter, A.: Dielectric elastomer actuator driven soft robotic structures with bioinspired skeletal and muscular reinforcement. Front. Robot. AI **7**, 510757 (2020). https://doi.org/10.3389/frobt.2020.510757
6. Ciarella, L., Richter, A., Henke, E.-F.M.: Digital electronics using dielectric elastomer structures as transistors. Appl. Phys. Lett. **119**, 261901 (2021). https://doi.org/10.1063/5.0074821
7. Ciarella, L., Wilson, K.E., Richter, A., Anderson, I.A., Henke, E.-F.M.: Modelling dielectric elastomer circuit networks for soft biomimetics. Bioinspiration Biomimetics **16**, 065006 (2021). https://doi.org/10.1088/1748-3190/ac2786

8. Ciarella, L., Richter, A., Henke, E.-F.M.: Integrated logic for dielectric elastomers: replicating the reflex of the venus flytrap. Adv. Mater. Technol. **8**(12), (2023). https://doi.org/10.1002/admt.202202000
9. O'Brien, B.M., Calius, E.P., Inamura, T., Xie, S.Q., Anderson, I.A.: Dielectric elastomer switches for smart artificial muscles. Appl. Phys. A **100**(2), 385–389 (2010). https://doi.org/10.1007/s00339-010-5857-z. Aug
10. Del Vecchio, R.M., Meiksin, Z.H.: A quantitative formulation of resistance and strain gauge factor expressions for discontinuous films. Thin Solid Films **61**, 65–71 (1979). https://doi.org/10.1016/0040-6090(79)90501-7
11. Leong, C.-K., Chung, D.D.L.: Improving the electrical and mechanical behavior of electrically conductive paint by partial replacement of silver by carbon black. J. Electron. Mater. **35**, 118–122 (2006). https://doi.org/10.1007/s11664-006-0193-y
12. Rich, S.I., Wood, R.J., Majidi, C.: Untethered soft robotics. Nat. Electron. **1**(2), 102–112 (2018). https://doi.org/10.1038/s41928-018-0024-1. Feb
13. Deignan, G., Goldthorpe, I.A.: The dependence of silver nanowire stability on network composition and processing parameters. RSC Adv. **7**(57), 35590–35597 (2017). https://doi.org/10.1039/C7RA06791K
14. Yoon, I.S., Kim, S.H., Oh, Y., Ju, B.K., Hong, J.M.: Ag flake/silicone rubber composite with high stability and stretching speed insensitive resistance via conductive bridge formation. Sci. Rep. **10**(1), (2020). https://doi.org/10.1038/s41598-020-61752-2
15. Mu, Q., Hu, T., Tian, X., Li, T., Kuang, X.: The effect of filler dimensionality and content on resistive viscoelasticity of conductive polymer composites for soft strain sensors. Polymers **15**(16), (2023). https://doi.org/10.3390/polym15163379
16. Yi, J., Ciarella, L., Rosset, S., Wilson, K., Anderson, I., Richter, A., Vorrath, E.F.M.: A piezoresistive dielectric elastomer switch consisting of inkjet-printed carbon black. Chem. Eng. J. **500**, 156718 (2024). https://doi.org/10.1016/j.cej.2024.156718. Nov
17. Foulger, S.H.: Electrical properties of composites in the vicinity of the percolation threshold. J. Appl. Polym. Sci. **72**(12), 1573–1582 (1999). https://doi.org/10.1002/(SICI)1097-4628(19990620)72:12<1573::AID-APP10>3.0.CO;2-6
18. Liu, A., Ni, Z., Chen, J., Huang, Y.: Highly sensitive graphene/polydimethylsiloxane composite films near the threshold concentration with biaxial stretching. Polymers **12**(1), 71 (2020). https://doi.org/10.3390/polym12010071
19. Yi, J., Babick, F., Strobel, C., Rosset, S., Ciarella, L., Borin, D., Wilson, K., Anderson, I., Richter, A., Henke, E.-F.M.: Characterizations and inkjet printing of carbon black electrodes for dielectric elastomer actuators. ACS Appl. Mater. Interfaces. **15**, 41992–42003 (2023). https://doi.org/10.1021/acsami.3c05444
20. Ciarella, L., Wilson, K.E., Richter, A., Anderson, I.A., Henke, E.-F.M.: A model for dielectric elastomer based electronics. IFAC-PapersOnLine **55**(20), 588–593 (2022). https://doi.org/10.1016/j.ifacol.2022.09.159
21. Starke, E., Marschner, U., Pfeifer, G., Fischer, W.-J., Flatau, A.B.: Combining network models and FE-models for the simulation of electromechanical systems, in Mehrdad, N., Ghasemi-Nejhad (eds.) Proceedings Active and Passive Smart Structures and Integrated Systems 2011. SPIE International Society for Optics and Photonics, vol. 7977, (2011). https://doi.org/10.1117/12.885633
22. Marschner, U., Gerlach, G., Starke, E., Lenk, A.: Equivalent circuit models of two-layer flexure beams with excitation by temperature, humidity, pressure, piezoelectric or piezomagnetic interactions. J. Sens. Sens. Syst. **3**, 187–211 (2014). https://doi.org/10.5194/jsss-3-187-2014

BY NC ND

Design of Soft Robots

Trust the Hand: Lessons from 15 Years of Applied Co-Design for Soft Manipulation

Adrian Sieler, Alexander Koenig, and Oliver Brock

Abstract A robot's behavior is governed by its body and the controllers that actuate it, making robotics a co-design problem. Until the emergence of soft robotics, the control side dominated robotics. In particular, soft robotic hands shifted the responsibility from complex control strategies to simpler, morphology-driven solutions. Our RBO Lab accumulated 15 years of experience building and applying soft hands for contact-rich manipulation tasks such as grasping and in-hand manipulation. By sharing the lessons we learned along the way, we hope to convince future roboticists that co-design is the right approach for advancing robotics.

1 Introduction

Soft robotics is causing a paradigm shift in robotics [1]. In this shift, the role of the robot's body is changing from a dutiful executor of control policies to an active contributor to robot behavior. The consequences are substantial. Much of our understanding of how to build robotic systems and our intuition about generating behavior has become incomplete.

To advance robotics following this paradigm shift, we must understand how to orchestrate the role of the body and control in generating robot behavior. We must learn how to build robot bodies that complement the capabilities of robot control and vice versa. This is not only new scientific territory; this paradigm shift will require us to overcome some of the well-established wisdom of robotics engineering and science.

Adrian Sieler and Alexander Koenig Equal contributions.

A. Sieler · A. Koenig · O. Brock (✉)
Robotics and Biology Laboratory, Technische Universität Berlin, Berlin, Germany
e-mail: oliver.brock@tu-berlin.de

Robotics Institute Germany, Munich, Germany

O. Brock
Science of Intelligence, Research Cluster of Excellence, Berlin, Germany

A. Raatz et al. (eds.), *Soft Material Robotic Systems*,
https://doi.org/10.1007/978-3-032-22453-8_6

Over the past 15 years, our laboratory has grappled with the consequences of this paradigm shift in the context of soft robot manipulation. During this time, we questioned and changed our intuitions time and time again. We had to realize that traditional wisdom was sometimes hindering, not helping. We had to learn many new lessons, some the hard way. In this paper, we have collected the most important of these lessons. We present each lesson as a story. This format is unusual, but we transmit our transformative experiences most effectively through these stories. We feel justified in taking this unusual approach to transmitting research results by the perspective of Peschl and Fundneider about how drastic innovation takes place in science [2]. In their Emergent Innovation approach, they argue that radically new knowledge emerges via (1) a deep understanding of the innovation object and (2) the letting go of deeply ingrained assumptions. In accordance with this view, our stories explain the insights from co-designing a soft manipulation system and highlight which parts of traditional robotics wisdom we needed "to let go of."

2 Lessons from Co-Designing Soft Manipulation

We continuously and iteratively designed and built a soft, anthropomorphic, pneumatically actuated robot hand (Fig. 1). Each step of this 15-year-long evolution included changes to the hardware and the way we control the hardware. We designed and evaluated the entire system, control and hand together, keeping in mind that improvements could be achieved in either or both [3–5]. We, therefore, lived through 15 years of practical, comprehensive, applied co-design. By sharing our most insightful stories, we want to provide the reader with a "crash course" in co-design, a fast way of benefiting from our experience. We hope this work lays the foundation for a new generation of roboticists for whom co-design is a natural choice and engineering practice for advancing soft robotics.

2.1 *Trust the Hand*

In 2019, a new generation of PhD students joined our effort to advance soft in-hand manipulation. The students came from a traditional robotics background and overlooked the role of the hand's embodiment. During their first year, they experimented with standard approaches like simulating the hand, reinforcement learning, and probabilistic movement primitives (against Oliver's advice). However, none of these methods worked. Why? Probably because the underlying assumptions of well-defined kinematics, dynamics, and states simply do not provide the most appropriate framing for soft systems and need to be let go of, as predicted by the Emergent Innovation paradigm [2].

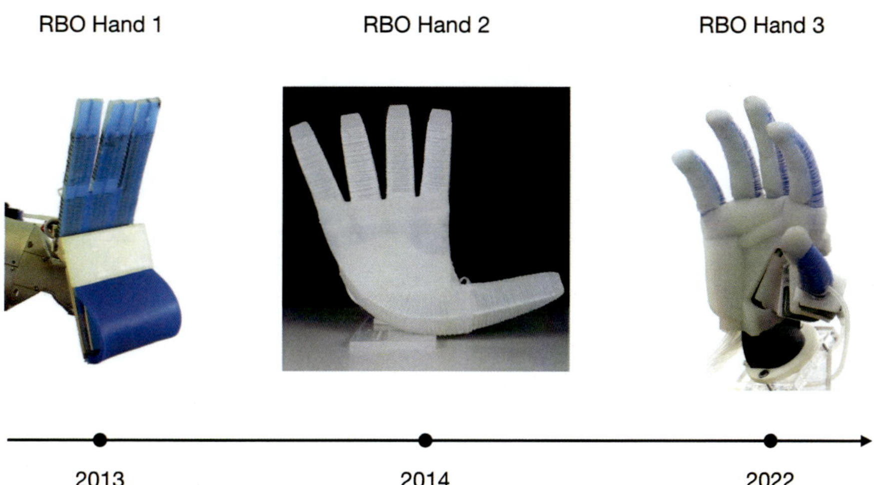

Fig. 1 Three generations of RBO Hands: RBO Hand 1 (photo reprinted from [6] with permission, © 2013 IEEE), RBO Hand 2 (photo reprinted from [7] with permission of the authors), and RBO Hand 3 (photo reprinted from [8], licensed under CC BY 4.0). Anthropomorphism increased over time, driven by improvements in thumb dexterity

As frustration grew, they began manually exploring the system (and the advice they received). They mapped each actuator to a manual slider and programmed simple open-loop sequences. One of these sequences should rotate a cube. To their surprise, it worked. Not only did it work, it generalized across different object shapes, poses, and even to 80-fold speed-ups despite the hand being sensorless [9].[1] They learned that one can *trust the hand* in taking over substantial control responsibilities.

But which responsibilities can the hand's body take over and why? The soft body of compliant hands can store and release energy via material deformation. This property enables the hand to passively conform to the object's shape and to reject disturbances automatically [10, 11].[2] This self-stabilizing effect [12] simplifies control because the hand's soft body handles large parts of disturbance rejection [13, 14], which typically requires active control [15]. In summary, our soft hand has desirable *morphological computation* abilities, which can be accessed and "programmed" via appropriate digital computation of controllers by changing the overall shape of the hand [4, 16]. Distributing sub-tasks and their solutions across these different types of computation (here: morphological and digital) is at the core of co-design.

---

[1] View the surprising robustness of open-loop soft manipulation in this playlist: https://youtube.com/playlist?list=PLb-CNILz7vmt6Ae_yD9i15TrCw0S8bKCn.

[2] The RBO hand passively rejects disturbances: https://youtu.be/U6KgnqitfvY&t=100.

2.2 Help the Hand Help You

How can one access the helpful morphological computation of soft hands? In the early days of our work on robotic grasping, we focused heavily on perception. The idea was simple: sense the object accurately, plan exact contact points, and optimize for performance metrics like force closure. In this view, any additional contact with the environment is treated as a problem, not an opportunity. It was necessary and productive to let go of this view, again confirming the ideas expressed in Emergent Innovation [2]. Instead, we now consider extensive physical contact essential to competent grasping.

We reached a turning point when we compared robotic grasping strategies to how humans grasp and manipulate objects [17]. We realized that grasping is not a single moment of perfect contact. Rather, it is a process full of interaction and adjustment. With bewilderment, we watched a chef cutting potatoes[3]: almost none of the hand/environment interactions, except for the grasp on the knife, could be explained by classical grasping concepts. Instead, the chef used her fingers, the counter, and the knife to guide the cutting motion of the potato. We observed that humans deliberately leverage environmental constraints (ECs), such as table surfaces, which enable simple and robust grasping [17, 18]. The same principle of leveraging ECs applies to in-hand manipulation. The entire hand, not just the fingertips, provides ECs to constrain an object [19]. ECs like gravity and inertia can also be leveraged for simplified manipulation [20, 21].

Our soft actuator's compliant properties allow us to safely drive them "into" the environment without risking damage. As shown in Fig. 2, the hand passively conforms to ECs such as table surfaces, enabling guided finger motion without the need for accurate control or perception [17, 22]. We realized that the hand's helpful morphological computation abilities are accessed by bringing the hand into contact and leveraging these ECs. In essence, we *make contact* instead of avoiding contact, and by making contact, we *help the hand help us*. The ability to exploit ECs can be considered an inductive bias for co-design.

2.3 Do Not Always Trust the Hand

When we designed the RBO Hand 1 [6] and 2 [7], our main goal was to maximize compliance, as we were enthusiastic about our early results on grasping.

However, we quickly observed that not all compliance is helpful [23]. In the RBO Hand 2, the actuators were sometimes too soft. They struggled to lift heavy objects and often buckled when coming into contact with the environment. Their compliance made their behavior unpredictable and limited what the hand could do, as shown in Fig. 3. To fix this, we made the actuators of the RBO Hand 3 stiffer, giving them more strength and significantly reducing the risk of buckling.

[3] Julia Child, "The Potato Show", see: https://youtu.be/Vjq5P24AkwM.

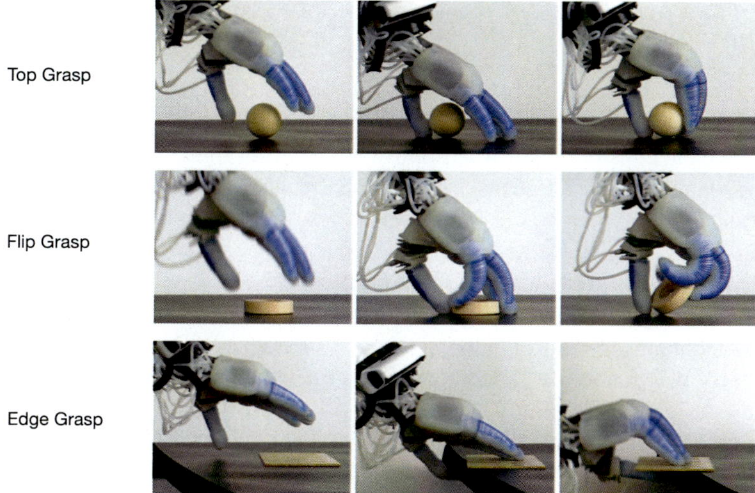

Fig. 2 Our soft actuators can safely make contact with the environment. Therefore, our hand can replicate human grasping strategies that exploit the environment, such as table surfaces, without risking damage. These robust grasping strategies leverage the hand's morphological computation, reducing the need for accurate sensing and control [17] (figure adapted from [8], licensed under CC BY 4.0). View our videos of the top grasp https://youtu.be/ENbrUOmDsSI, the edge grasp https://youtu.be/0LnvVSEINH4, and more EC-exploiting grasps https://youtu.be/WuSbj64F4N0

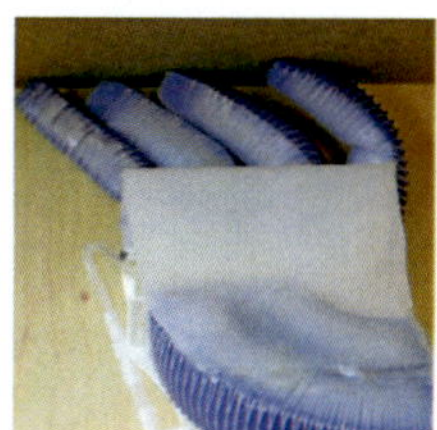

Fig. 3 Examples of bad morphological computation. Left: soft fingers yield while pulling on a heavy door handle. Middle: fingers buckle while sliding along a wall. Right: fingers buckle while lifting a heavy bottle (photos reprinted from [24] with permission, © 2017 IEEE)

Realizing that not all compliance is good highlights the importance of adjusting the hand's compliance for a given task. Our actuators' compliance is tunable by changing their inflation level: high inflation means lower compliance. Our RBO Hand 3 has more degrees of actuation per finger than its predecessors and can, therefore, more accurately tune its compliant response [8]. Furthermore, variable stiffness actuators can improve the versatility of hands by selectively stiffening their body [25, 26]. In summary, the degree of material compliance and the actuation abilities of a soft hand are key factors that must be considered in a co-design process.

2.4 The Human Hand Inspires

After successfully using soft hands for grasping through environmental constraint exploitation, where the hand simply closes upon force feedback triggered by contact, we began considering how to control the hand for more fine-grained tasks, like in-hand manipulation. These tasks posed a challenge because common control, planning, and learning methods [27–32], which rely on precise models, are unsuitable for soft hands. Again, we had to question our deeply ingrained assumptions from traditional robotics approaches to these problems [2].

Motivated by a programming error in a class exercise, we experimented with random Jacobians to control an object's position tracked using an AprilTag marker. Each entry was randomly set to 1 or -1, and if we could not reduce the error, we flipped the signs. Surprisingly, this naive approach worked far better than expected. This revealed that the hand's natural compliance and self-stabilizing properties could compensate for the lack of precise models. Around this time, we came across a paper on human motor control, suggesting that forceful tasks could be performed using "sloppy" models rather than precise ones [33], which aligned with our findings. Inspired, we developed a robust feedback control algorithm using coarse, inaccurate models [34]. These models were not inaccurate for the sake of inaccuracy. But making them more accurate was very costly and did not lead to any improvements in performance—something that deeply seems to contradict standard engineering intuitions.

Our soft, anthropomorphic hands look like human hands and behave surprisingly similarly. Their shape and compliance naturally aid in guiding and stabilizing interactions, even without complex control. This insight allowed us to investigate bio-inspired control strategies for grasping and in-hand manipulation [17, 33, 35–37]. As the human hand is already the product of co-design through evolution, human control strategies can serve as inductive biases for control in our co-design efforts.

Interestingly, anthropomorphic design also impacts how the community perceives research. While the grasping performance of the RBO Hand 1 was impressive, it did not manage to capture the community's substantial attention. However, by making the RBO Hand 2 more anthropomorphic (among other things, of course), we were able to win a best student paper award at RSS 2014 [7]. The imbalance in attention is also reflected in the citation counts as of May 2025: the RBO Hand 1 paper [6] has 431 citations while the later RBO Hand 2 paper [7] has 1364 citations.

2.5 Think Beyond the Human Hand

Figure 1 shows the three generations of hands developed in our lab to make the hand as anthropomorphic as possible. While developing in-hand manipulation skills with the RBO Hand 3, we encountered a surprisingly human problem: clamping an object between two fingers to let the thumb gait along the object. People do this

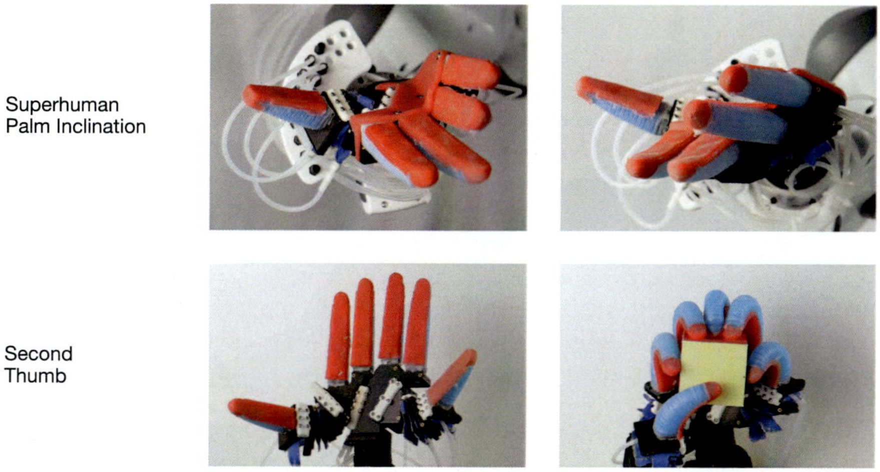

Superhuman
Palm Inclination

Second
Thumb

Fig. 4 Bioinspiration is only a starting point. Top row: we equipped the hand with a superhuman palm inclination of up to 80°, increasing the range and dexterity of the ring and little fingers. Bottom row: A second thumb can secure an object from one more side and increases the hand's overall dexterity. View the second thumb in action: https://www.youtube.com/playlist?list=PLb-CNILz7vmuvzliy-YiXME-ksJyFBwOd

instinctively, but with their high lateral compliance, our soft robotic fingers required precise contact placement, which was far from trivial.

Guided by lesson 2 (Sect. 2.2) encouraging us to exploit contact, we avoided solving this problem with control for the current hand. Instead, we went to hand design again (Fig. 4). We increased the bending range of the actuated palm from 30° to 80° to enable a firmer grip between the fingers. Then, we made an even larger change by adding a second thumb. This extra thumb offered an additional contact patch, helping to stabilize the grasp while gaiting the other thumb.

Addressing this problem solely at the control level would have required intricate control. But by co-changing the hand's design, we could explore capabilities beyond the human form factor. This iterative co-design aims to identify the fundamental properties required for a general manipulation platform rather than simply copying the human hand. The second thumb simplified the original control problem and enhanced the hand's co-designed abilities.

2.6 *Treat Friction Like an Actuator*

New challenges emerged when we started to exploit contact with the environment for grasping. While using the RBO Hand 2 to pick objects from cluttered boxes in the *Soft Manipulation (SOMA)* [38] project, we kept running into a frustrating issue.

As the hand slid across the bottom of the box, its soft silicone fingers, made from high-friction *Dragon Skin 10*, would stick, bend awkwardly, and buckle under load.

Then came an unexpectedly simple fix. One PhD student, watching the hand struggle with the surface, grabbed a bottle of baby powder and lightly dusted the back of the actuators. Suddenly, the fingers stopped catching. They slid cleanly, and the buckling problem disappeared. It worked so well that we started using baby powder even during in-hand manipulation, as the actuators had been designed for grasping and had too much friction.

The key lesson here is that selecting the appropriate level of friction is crucial for performance. To address this, we designed a special pulp layer that we attach to the soft actuators, allowing us to tune both the stiffness and friction at the contact. Eventually, we turned away from using baby powder. Now, we adjust the stiffness using different silicones. We also modify friction by adding a *Slide STD/1 Surface Tension Diffuser* into the silicone during manufacturing. Specific compliance structures in the pulp can further enhance grasping performance [39]. In the future, we hope to build soft actuators that can adjust their friction through control, allowing even greater adaptability [40]. Similar to compliance, friction is an important variable in a co-design process.

2.7 *Material Compliance Simplifies Exhibiting Compliance*

In 2020, our PhD students set out to better understand the unique contributions of soft bodies in robotic manipulation. In a simple but revealing experiment (Fig. 5), they clamped a wooden annulus between the index and the thumb of a soft robotic hand. Then, they actuated only the thumb for just half a second.

What happened next was unexpected. The annulus kept rotating in a coordinated motion with the thumb, even though the thumb was not actuated at all. The students realized that this behavior is not just acceptable—it is desirable! The body itself is manipulating, reducing the burden on control. It is a feature, not a bug! And again, this violated our intuitions.

What makes this passive, continued motion possible? The answer lies in the omnidirectional compliance of soft actuators (Fig. 6). Thanks to its compliance, the index finger automatically stays opposed to the thumb. The thumb essentially actuates the *passive workspace* of the index. The thumb's brief actuation initiated a cascade of energy storage and release, naturally moving the system into a new energy minimum—without extra control. Rigid hands would require intricate joints and complex control schemes to achieve similar results. Meanwhile, the soft hand achieves it effortlessly. Enabled by the soft body, we could let go of the typical assumption that fine manipulation, such as rotating the annulus, requires accurate sensing and control, further supporting the Emergent Innovation paradigm [2].

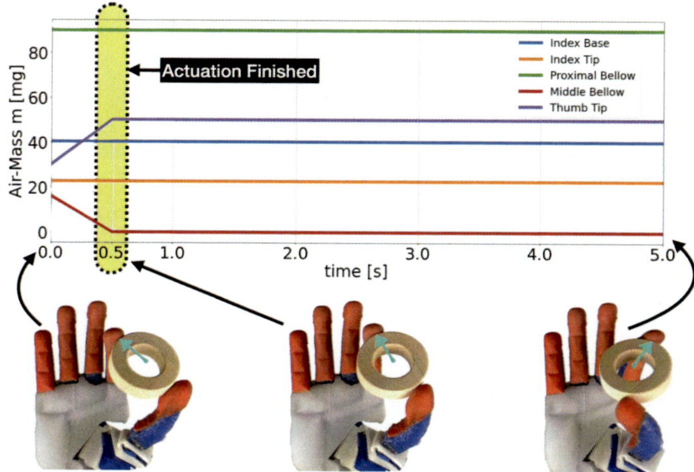

Fig. 5 Embodied intelligence at work: a brief thumb actuation introduced energy into the system, rotating the annulus. After the actuation stops at 0.5 s, the hand continues rotating the annulus as it progresses into a new energy minimum. The turquoise arrow indicates annulus rotation. Watch the video here: https://youtu.be/Nor2QEtM4W8 (figure adapted from [9] with permission of the authors)

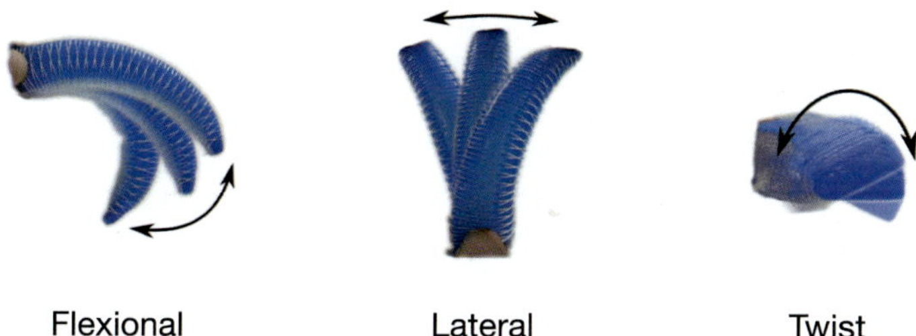

Fig. 6 The omnidirectional compliance of our fingers is a key enabler for its morphological computation, as demonstrated in Fig. 5. Exhibiting a similar level of compliance on a rigid hand would require complex articulation and control. A hand that is compliant in more directions is more self-stabilizing because its body can compensate for disturbances in more directions (photos reprinted from [24] with permission, © 2017 IEEE)

2.8 *The Body Is the Sensor*

Although soft hands enable robust manipulation in a sensorless, i.e. open-loop, fashion [9, 10, 21], we need sensorization to avoid, detect, and recover from failures, as well as for learning manipulation behaviors. With this in mind, we began building sensors specifically tailored to soft hands, i.e., sensors that preserve the compliance of the hand [41]. Oliver voiced an unconventional idea—embedding a speaker and a microphone into the squishy fingers to listen to their deformations. Everyone was skeptical. It sounded too strange, too different. Nobody wanted to take on such a

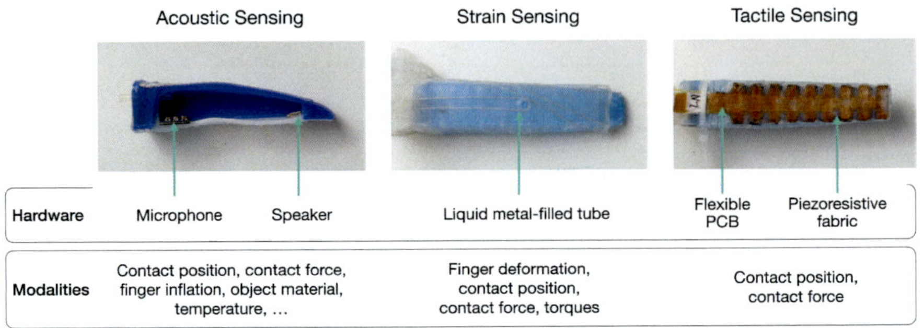

Fig. 7 Our three sensor types for soft fingers: The sensors preserve compliance because they are either flexible or their rigid components are placed in a way that preserves actuator compliance. The acoustic sensor is placed inside the finger [42]. Flexible strain sensors are glued onto the finger [24]. Stretchable tactile arrays are mounted on the non-stretchable inside of the fingers [45]. The table lists the physical properties that can be sensed by the respective sensor

risky project. However, Gabriel Zöller, a Master's student at the time, decided (in alternative accounts: was coerced) to give it a try during a project course we were offering at the time.

To everyone's—even Oliver's—surprise, the soft fingers had much to say. The internal sounds, captured by the microphone, revealed rich information like contact position, force, and even temperature, all decoded using simple machine learning models. The results were so unexpectedly accurate and general that Gabriel did not initially trust them. Something must have gone wrong. But he could not find the mistake. He wiped the slate clean, rebuilt the entire setup from scratch, and collected an entirely new dataset. But again, the soft fingers told the same astonishing story.

Acoustic sensing [42], together with other sensing technologies we explored (Fig. 7), gave rise to the idea of *morphological sensing* [43]: The entire morphology of the body serves as a powerful sensor. Figure 8 illustrates this approach. Morphological sensing is based on the observation that all interactions between hand and environment manifest themselves in changes to the *morphology*. Sound propagating through these changed morphologies is modulated to capture these changes. The desired sensor information can then be reconstructed from the recorded, modulated sound signal via computation [44], reconstructing the relevant *measurement* from the sound. Computation can use the morphological sensor to emulate several different "classical" sensor types (Fig. 7). Therefore, it is essential to recognize that sensing naturally becomes part of the co-design process and should not be considered a separate problem.

2.9 Do Not Control Compliance "Away"

In the early days of designing our soft hands, we needed a way to inflate and deflate the actuators. The obvious choice was to control the internal pressure since pressure

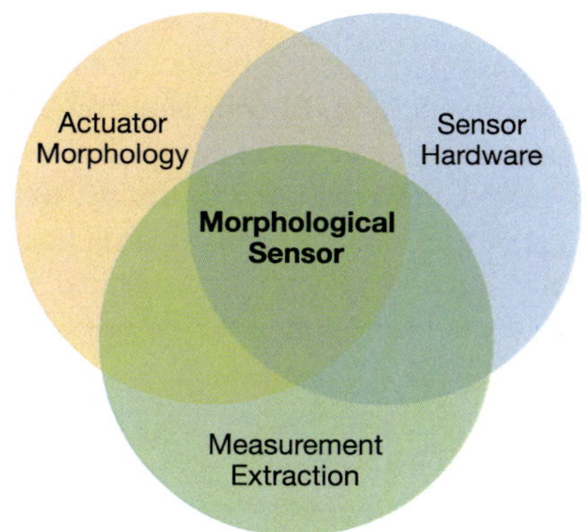

Fig. 8 The morphological sensing paradigm: the interactions between hand and environment influence the actuator morphology, a sensor captures a raw signal from this morphology, and a digital process extracts the measurement that caused the change in actuator morphology (figure reproduced from [43], licensed under CC BY 4.0)

sensors are inexpensive and reliable. So, we followed the conventional approach and built a pneumatic controller based on pressure.

As we worked with this setup, problems started to emerge. The issue was that soft actuators change shape when they contact something, altering their internal volume. As a result, the controller kept trying to regulate pressure even during contact. When the hand grasped an object, the pressure increased, and the controller would deflate it, almost as if the hand was "scared" of the contact, controlling its natural compliance "away". To circumvent this behavior, we tried something different. We opened the valve for a fixed amount of time to let air in, then shut it. That was it. With no controller constantly adjusting the pressure, the hand was free to respond naturally to contact, just as we intended.

Closing the valve locks in a fixed amount of air mass, which sets the shape the actuator returns to when no external forces are applied—like a stretched spring returning to its original position [9]. This makes air mass a more suitable control variable than pressure [46]. Air mass control effectively turns soft pneumatic actuators into programmable springs, helpful in sensing actuator deformation [34] and modeling the hand's free-motion kinematics, which is useful for in-hand manipulation tasks [11]. This perspective on control lets us disambiguate the contributions of software and hardware. Knowing these contributions can inform the design of an objective function of a co-design algorithm [23].

2.10 Pneumatics and Silicone Are Good for Prototyping

Building a robotic hand is traditionally a complex task. Therefore, only a few are built, making it costly to break one. As a result, contact—something a hand is meant to handle—becomes something to avoid, which is counterintuitive.

In 2011, we discovered the universal jamming gripper [47] and the starfish gripper [48]. Suddenly, a whole new way of thinking about hands opened up. Pneumatic actuation was a game-changer. It allows us to decouple the control system (valves and compressors) from the actuators, enabling quick replacement of broken actuators. In addition, silicone actuators are easy to manufacture. All we do is 3D-print a new mold, and we can change its design at will.

With this newfound flexibility, we began building. Over the past 15 years, we've created around 70 hands, sharing them with research groups and showcasing them at exhibitions. Each time we built a new hand, the hardware continuously improved. Every student joining the lab gets their own hand (and most often builds it), which has shifted our mindset. We no longer fear breaking the hand. Instead, we embrace hardware failure as an opportunity to improve the system. In the language of Emergent Innovation [2]: By building pneumatically actuated hands out of silicone, we can quickly prototype and, therefore, efficiently obtain a deeper understanding of our soft manipulation system.

2.11 Dare to Change Everything

When the new PhD cohort joined the RBO lab in 2019, the outgoing group, which had been working on soft manipulation, was beginning to hand over their projects. At first, they assumed they would simply inherit the existing version of the robotic hand and develop software solutions for manipulation tasks. They were afraid to change the hand design because they were unfamiliar with hardware development and were trained in machine learning and control.

But that assumption did not last long. As the students started working with the hand, it became clear that real progress could not come from treating the hardware as fixed. They quickly found themselves getting hands-on with every part of the stack—improving the low-level pneumatic control, making the manufacturing of actuators more reproducible, enhancing the design of the hand for easy and quick assembly, and even changing the kinematics of the hand. The hand is far from perfect, and there was and still is much room for improvement.

Daring to change everything is essential for a successful co-design project, where every part of the system must be open to change. While this might seem daunting initially, it pays off in the long run by providing the deep understanding of every aspect of the hand needed to truly engage in an Emergent Innovation process [2]. Otherwise, you risk being limited by artificially imposed, unnecessary constraints, like those of traditional robotics that focus narrowly on the control aspect and disregard the body's capabilities.

3 Conclusion

We have shared our experiences with orchestrating the role of the body and control through a real-world co-design process in the context of robotic in-hand manipulation and grasping. Our findings demonstrate that simple control can lead to remarkably general behaviors when making appropriate hardware choices, which starkly contrasts with how current deep learning approaches achieve generalization [27, 28, 31]. This new paradigm calls for us to be generalists. While this perspective is not new—robotics was interdisciplinary in its early days [49]—it had become more compartmentalized over time. We encourage future roboticists to embrace co-design, thoughtfully distributing responsibilities between a robot's body and its controls to unlock simple, general, and robust robots.

Acknowledgements Over the years, many people have contributed to the efforts described here. We thank Raphael Deimel, Steffen Puhlmann, Vincent Wall, Aditya Bhatt, Gabriel Zöller, Jason Harris, Veranika Pavlova, Lion Weber, Martin Splettstößer, Reyk Carstens, Anton Ohler, Tessa Pannen, Caroline Duncan, Juan Gomez Daza, Nikolas Thelenberg, Hutomo Saleh, Marius Hebecker, Callum Waters, Henning Ray, Jenny Berger, Fabian Heinemann, Samuel Glombitza and Tim Helge Faß.

This work has been partially funded by the Federal Ministry of Research, Technology and Space (BMFTR) within the Robotics Institute Germany, grant No. 16ME1000, the Deutsche Forschungsgemeinschaft (DFG, German Research Foundation) under Germany's Excellence Strategy—EXC 2002/1 "Science of Intelligence"—project No. 390523135, and the Deutsche Forschungsgemeinschaft (DFG, German Research Foundation) under the project "Soft Material Robotic Systems"—project No. 405033880.

References

1. Freyberg, S., Hauser, H.: The morphological paradigm in robotics. Stud. Hist. Philos. Sci. **100**, 1–11 (2023)
2. Peschl, M., Fundneider, T.: Emergent innovation and sustainable knowledge co-creation. A socio-epistemological approach to innovation from within. In: Lytras, M., Carroll, J., Damiani, E., et al. (eds.) The Open Knowledge Society: A Computer Science and Information Systems Manifesto, vol. CCIS 19, pp. 101–108. Springer (2008)
3. Gilday, K., Hughes, J., Iida, F.: Sensing, actuating, and interacting through passive body dynamics: a framework for soft robotic hand design. Soft Rob. **10**(1), 159–173 (2023)
4. Hauser, H., Hughes, J.: Morphological computation—past, present and future. Device **2**(9), 100439 (2024)
5. Milana, E., Santina, C.D., Gorissen, B., Rothemund, P.: Physical control: a new avenue to achieve intelligence in soft robotics. Sci. Robot. **10**(102), eadw7660 (2025)
6. Deimel, R., Brock, O.: A compliant hand based on a novel pneumatic actuator. In: IEEE International Conference on Robotics and Automation, pp. 2047–2053 (2013)
7. Deimel, R., Brock, O.: A novel type of compliant, underactuated robotic hand for dexterous grasping. In: Proceedings of Robotics: Science and Systems (RSS) (2014)
8. Puhlmann, S., Harris, J., Brock, O.: RBO Hand 3: a platform for soft dexterous manipulation. IEEE Trans. Robot. (T-RO) **38**(6), 3434–3449 (2022)
9. Bhatt, A., Sieler, A., Puhlmann, S., Brock, O.: Surprisingly robust in-hand manipulation: an empirical study. In: Proceedings of Robotics: Science and Systems (RSS) (2021)
10. Abondance, S., Teeple, C.B., Wood, R.J.: A dexterous soft robotic hand for delicate in-hand manipulation. IEEE Robot. Autom. Lett. (RA-L) **5**(4), 5502–5509 (2020)
11. Sieler, A., Koenig, A., Brock, O.: What is the key to dexterous manipulation: learning or compliance? In: German Robotics Conference (GRC) (2025)

12. Iida, F., Pfeifer, R.: Cheap rapid locomotion of a quadruped robot: self-stabilization of bounding gait. Intell. Auton. Syst. **8** (2004)
13. Mason, M.T.: The mechanics of manipulation. In: IEEE International Conference on Robotics and Automation (ICRA), vol. 2, pp. 544–548 (1985)
14. Shapiro, A., Rimon, E., Shoval, S.: On the passive force closure set of planar grasps and fixtures. Int. J. Robot. Res. (IJRR) **29**(11), 1435–1454 (2010)
15. Khadivar, F., Billard, A.: Adaptive fingers coordination for robust grasp and in-hand manipulation under disturbances and unknown dynamics. IEEE Trans. Robot. (T-RO) **39**(5), 3350–3367 (2023)
16. Pfeifer, R., Bongard, J.C.: How the Body Shapes the Way We Think: A New View of Intelligence (Bradford Books). The MIT Press (2006)
17. Eppner, C., Deimel, R., Alvarez-Ruiz, J., Maertens, M., Brock, O.: Exploitation of environmental constraints in human and robotic grasping. Int. J. Robot. Res. (IJRR) **34**(7), 1021–1038 (2015)
18. Nakamura, Y.C., Troniak, D.M., Rodriguez, A., Mason, M.T., Pollard, N.S.: The complexities of grasping in the wild. In: IEEE-RAS International Conference on Humanoid Robotics (Humanoids), pp. 233–240 (2017)
19. Brahmbhatt, S., Ham, C., Kemp, C.C., Hays, J.: ContactDB: analyzing and predicting grasp contact via thermal imaging. In: The IEEE Conference on Computer Vision and Pattern Recognition (CVPR) (2019)
20. Dafle, N.C., Rodriguez, A., Paolini, R., Tang, B., Srinivasa, S.S., Erdmann, M., Mason, M.T., Lundberg, I., Staab, H., Fuhlbrigge, T.: Extrinsic dexterity: in-hand manipulation with external forces. In: IEEE International Conference on Robotics and Automation (ICRA), pp. 1578–1585 (2014)
21. Patidar, S., Sieler, A., Brock, O.: In-hand cube reconfiguration: simplified. In: IEEE/RSJ International Conference on Intelligent Robots and Systems (IROS), pp. 8751–8756 (2023)
22. Bonilla, M., Farnioli, E., Piazza, C., Catalano, M., Grioli, G., Garabini, M., Gabiccini, M., Bicchi, A.: Grasping with soft hands. In: IEEE-RAS International Conference on Humanoid Robots (Humanoids), pp. 581–587 (2014)
23. Ghazi-Zahedi, K., Deimel, R., Montúfar, G., Wall, V., Brock, O.: Morphological computation: the good, the bad, and the ugly. In: IEEE/RSJ International Conference on Intelligent Robots and Systems (IROS), pp. 464–469 (2017)
24. Wall, V., Zöller, G., Brock, O.: A method for sensorizing soft actuators and its application to the RBO Hand 2. In: IEEE International Conference on Robotics and Automation (ICRA), pp. 4965–4970 (2017)
25. Manti, M., Cacucciolo, V., Cianchetti, M.: Stiffening in soft robotics: a review of the state of the art. IEEE Robot. Autom. Mag. (RAM) **23**(3), 93–106 (2016)
26. Wall, V., Deimel, R., Brock, O.: Selective stiffening of soft actuators based on jamming. In: IEEE International Conference on Robotics and Automation (ICRA), pp. 252–257 (2015)
27. Andrychowicz, O.M., Baker, B., Chociej, M., Jozefowicz, R., McGrew, B., Pachocki, J., Petron, A., Plappert, M., Powell, G., Ray, A., et al.: Learning dexterous in-hand manipulation. Int. J. Robot. Res. (IJRR) **39**(1), 3–20 (2020)
28. Chen, T., Tippur, M., Wu, S., Kumar, V., Adelson, E., Agrawal, P.: Visual dexterity: in-hand dexterous manipulation from depth. In: ICML Workshop on New Frontiers in Learning, Control, and Dynamical Systems (2023)
29. Jin, W., Posa, M.: Task-driven hybrid model reduction for dexterous manipulation. IEEE Trans. Robot. (T-RO) **40**, 1774–1794 (2024)
30. Pfanne, M., Chalon, M., Stulp, F., Ritter, H., Albu-Schäffer, A.: Object-level impedance control for dexterous in-hand manipulation. IEEE Robot. Autom. Lett. (RA-L) **5**(2), 2987–2994 (2020)
31. Pitz, J., Röstel, L., Sievers, L., Burschka, D., Bäuml, B.: Learning a shape-conditioned agent for purely tactile in-hand manipulation of various objects. In: IEEE/RSJ International Conference on Intelligent Robots and Systems (IROS), pp. 13112–13119 (2024)
32. Suh, H.J.T., Pang, T., Zhao, T., Tedrake, R.: Dexterous contact-rich manipulation via the contact trust region (2025). https://arxiv.org/abs/2505.02291

33. Akulin, V.M., Carlier, F., Solnik, S., Latash, M.L.: Sloppy, but acceptable, control of biological movement: algorithm-based stabilization of subspaces in abundant spaces. J. Hum. Kinet. **67**(1), 49–72 (2019)
34. Sieler, A., Brock, O.: Dexterous soft hands linearize feedback-control for in-hand manipulation. In: IEEE/RSJ International Conference on Intelligent Robots and Systems (IROS), pp. 8757–8764 (2023)
35. Elliott, J.M., Connolly, K.J.: A classification of manipulative hand movements. Dev. Med. Child Neurol. **26**(3), 283–296 (1984)
36. Feldman, A., Levin, M.: The equilibrium-point hypothesis—past, present and future. Adv. Exp. Med. Biol. **629**, 699–726 (2009)
37. Iberall, T., Bingham, G., Arbib, M.A.: Opposition space as a structuring concept for the analysis of skilled hand movements. In: Generation and Modulation of Action Patterns, pp. 158–173. Springer Berlin Heidelberg (1986)
38. Páll, E.: Motion Generation with Contact-Based Environmental Constraints. Dissertation, Technische Universität Berlin (2023)
39. Puhlmann, S., Weber, L.C., Höppner, H.: Programming passive fingertip deformation for improved grasping and manipulation. In: IEEE/RSJ International Conference on Intelligent Robots and Systems (IROS), pp. 9175–9181 (2024)
40. Spiers, A.J., Calli, B., Dollar, A.M.: Variable-friction finger surfaces to enable within-hand manipulation via gripping and sliding. IEEE Robot. Autom. Lett. (RA-L) **3**(4), 4116–4123 (2018)
41. Hegde, C., Su, J., Tan, J.M.R., He, K., Chen, X., Magdassi, S.: Sensing in soft robotics. ACS Nano **17**(16), 15277–15307 (2023)
42. Wall, V., Zöller, G., Brock, O.: Passive and active acoustic sensing for soft pneumatic actuators. Int. J. Robot. Res. (IJRR) **42**(3), 108–122 (2023)
43. Wall, V.: Morphological sensing for soft pneumatic actuators based on acoustics and strain. Dissertation, Technische Universität Berlin (2024)
44. der Spiegel, J.V.: Computational sensors: the basis for truly intelligent machines. In: Intelligent Sensors, *Handbook of Sensors and Actuators*, vol. 3, pp. 19–37. Elsevier Science B.V. (1996)
45. Pannen, T.J., Puhlmann, S., Brock, O.: A low-cost, easy-to-manufacture, flexible, multi-taxel tactile sensor and its application to in-hand object recognition. In: 2022 International Conference on Robotics and Automation (ICRA), pp. 10939–10944 (2022)
46. Deimel, R., Radke, M., Brock, O.: Mass control of pneumatic soft continuum actuators with commodity components. In: IEEE/RSJ International Conference on Intelligent Robots and Systems (IROS), pp. 774–779 (2016)
47. Brown, E., Rodenberg, N., Amend, J., Mozeika, A., Steltz, E., Zakin, M.R., Lipson, H., Jaeger, H.M.: Universal robotic gripper based on the jamming of granular material. Proc. Natl. Acad. Sci. **107**(44), 18809–18814 (2010)
48. Ilievski, F., Mazzeo, A.D., Shepherd, R.F., Chen, X., Whitesides, G.M.: Soft robotics for chemists. Angew. Chem. Int. Ed. **50**(8), 1890–1895 (2011)
49. Brooks, R.: Artificial life and real robots. In: Varela, F.J., Bourgine, P. (eds.) Toward a Practice of Autonomous Systems: Proceedings of the First European Conference on Artificial Life, Complex Adaptive Systems, pp. 3–10. The MIT Press, Cambridge, Mass (1991)

BY NC ND

Automated Design and Fabrication of Pneumatic Logic Circuits to Control Soft Robots

Marco Pontin and Perla Maiolino

Abstract A key obstacle for pneumatic soft robots to operate in the real world is represented by the bulky and heavy hardware (pumps and valves) needed for their control. Recent advancements have made it possible to create fully pneumatic, 3D printed robots that integrate all control logic onboard, using readily available 3D printing devices. Despite this, critical breakthroughs are needed to automate and optimize the design process of such onboard controllers. In this chapter, we present an end-to-end framework covering the journey from single logic component design to an automated compiler that turns user-input truth tables into an optimized, ready-for-printing CAD file of the corresponding logic circuit for soft robot control.

Keywords Soft robotics · Fluidic logic compiler · Monolithic fabrication · Embedded control · Design automation · Soft systems design · Soft valves

1 Introduction

Soft robotics has nowadays established itself as a key field of robotics research [1]. Over the years, hundreds of actuator and sensor designs have been explored [2–6], with applications ranging from medical devices (millimeter scale) [7–9] to space and environment exploration [10–14] or the construction industry (meter scale) [15]. The intrinsic compliance of soft robots makes them the optimal choice for highly unstructured environments and dynamic interactions where safety is a concern, but this comes at the cost of ease of control. Their high degrees-of-freedom bodies require complex sensing networks and control algorithms to be able to achieve the desired performance.

M. Pontin · P. Maiolino (✉)
Department of Engineering Science, University of Oxford, Oxford, UK
e-mail: perla.maiolino@eng.ox.ac.uk

M. Pontin
e-mail: marco.pontin@eng.ox.ac.uk
URL: https://ori.ox.ac.uk/labs/srl/research-soft-robotics/

A. Raatz et al. (eds.), *Soft Material Robotic Systems*,
https://doi.org/10.1007/978-3-032-22453-8_7

To mitigate this challenge, soft robotics has strongly tied itself to the concept of embodied intelligence, meaning achieving desired behaviors out of the robot design itself, rather than the control logic alone [16]. This is particularly clear in the case of pneumatically actuated soft actuators, which represent one, if not the most, commonly used type of actuator in the field. To achieve the desired shape of the actuator upon inflation, constraints are added to specific parts of the actuator's body, making them expand less, while others are left free to inflate. The result is devices which can achieve complex trajectories completely open-loop, the only control signal being the actuating pressure. In addition, breakthroughs in additive manufacturing techniques have transformed what used to require highly skilled researchers and multiple steps of manual work into semi to fully automated processes where a CAD file for the actuator is generated and printed, being ready for use after minimal post-processing [17].

Despite the recent advancements, for soft robots to see widespread adoption, a lot of work still needs to be done on the aspects of control and 3D printable logic circuits. So far, most robots rely on rigid pumps, valves and digital microcontroller units (MCUs) and circuits, which, due to their weight and size, are not onboard the robot itself. This constitutes a critical limitation for untethered operation [18]. Researchers have explored the possibility of using custom-design soft valves and advanced additive manufacturing technologies and monopropellant fluids to manufacture soft robots that include onboard computation in the form of a fluidic controller and storage for the actuation fluid [19]. The cost of such technologies, though, renders their widespread adoption unfeasible. Several state-of-the-art works have explored the possibility of using Fused Deposition Modeling (FDM) 3D printing as a viable, low-cost solution for monolithically printed soft robots, including onboard control logic [17, 20, 21]; the key contribution being the use of commonly available 3D printing machines. To strengthen the success of these approaches, some limitations still need to be addressed. Specifically, particular techniques need to be adopted to be able to print air-tight chambers and actuators [20], which are a crucial requirement in pneumatic circuits. Designs need to avoid any supports, or more complex multi-material setups are needed. In addition, the resolution of FDM printing limits the minimum size of channels and details that can be printed. For these reasons, the design of such systems still requires expert engineers and design iterations to be able to achieve robots that are functional directly out of the printer. One of the key challenges is the lack of automated procedures to turn desired custom logic functions into optimized circuits using libraries of standardized 3D printable logic elements. In [22], the authors make a first attempt at developing an online compiler using pneumatic component designs available in the literature. The approach limits itself to taking a user-input truth table and displaying the necessary piping and connections between logic components, leaving to the user the labor-intensive and error-prone fabrication steps. A more all-encompassing approach is therefore needed to speed up progress in this field.

Throughout this chapter, we present a design framework for monolithically 3D printed control circuits for pneumatic soft robots, starting from an example of a bistable mechanism for valve control [23], to more complex modular designs featuring multiple inlets and outlets [24], culminating in an attempt to fully automate the pneumatic circuit design pipeline end-to-end [25]. The first two sections demonstrate how a systematic library of fully 3D printable components can be built, allowing any user-specified function to be replicated. Components should ideally be printed in place, including all necessary connections, without any further assembly required. Such logic elements can then be used to power compiler algorithms such as the one presented in this chapter, to generate space-optimized pneumatic circuits for onboard control of soft robots.

2 3D Printing Pneumatic Valves and Controllers

As briefly mentioned, the state-of-the-art includes countless soft valve designs, not just for computation, but also for more advanced scopes like fault detection and isolation [26]. For the sake of conciseness, this section will only focus on the several examples of 3D printed valves for soft robots. Arguably, some of the other designs might still be 3D printable, with some adaptations, but what we want to focus on is the specific function provided by the components rather than their specific design.

MultiJet Modelling (MJM), a process in which different materials are deposited in droplets layer-by-layer, was used in [27] to create fluidic components equivalent to capacitors, diodes and gain-tunable transistors. In [28], Hubbard et al. present designs for 3D printed fluidic diodes as well as normally open and normally closed transistors. These components are printed using a high-precision PolyJet method, similar to MJM. As a result, all components of the valves can be printed in place with varied mechanical properties, from the rigid body to the deformable membranes and seals. The authors then use the components to create an oscillator circuit for the onboard control of a 3D printed soft robotic turtle. Bartlett et al. [29] use an analogous technology to manufacture a robot powered by combustion. The capability of mixing materials of different mechanical properties is exploited to create a shell with graded elastic modulus, ranging from 1 MPa to 1 GPa, to focus the explosive energy towards a specific part of the membrane, while providing a strong and reliable interface between its soft, deformable part and the rigid structure of the robot.

More recently, researchers have focused on inexpensive and readily available Fused Deposition Modeling (FDM) printers, to lower the barrier to entry to the field. One of the drawbacks of FDM is the limited precision of the process compared to PolyJet printing. In addition, even though multi-material setups are now available, printing designs with graded mechanical properties remains challenging. Most importantly, though, due to the extrusion-based process, printing with very soft materials, below 70A Shore hardness, is very complex and time-consuming. Despite these limitations, Zhai et al. demonstrated techniques for printing air-tight actuators and valves using Thermoplastic Polyurethane (TPU) without supports using FDM printing [20]. These have been built upon in [17] to create a highly integrated monolithically printed walking robot featuring a ring oscillator circuit to power the actuators moving the six legs the robot uses to crawl. A similar result was achieved by Conrad et al. [21] with their valve design that can operate as a NOT, AND or OR gate, depending on the way it is connected into the circuit.

All examples so far have dealt with the printing of the logic components and actuators, but the source of actuation itself was provided after the fact either through a tether or using small canisters. In [19], Wehner et al. present the use of EMB3D printing using fugitive and catalytic inks to print a fully soft octopus robot. A microfluidic oscillator is used to control the robot motion, but, contrary to previous examples, the actuation power comes from the decomposition of a monopropellant fuel stored in onboard reservoirs.

3 3D Printed Multi-channel Bistable Valves for Pneumatically-Driven Soft Robots

This section presents the method for monolithic fabrication through PolyJet 3D printing of modular multi-channel valves for complex pneumatic circuit creation developed in [23]. First, we investigate the design and characterization of the bistable switching mechanism for a simple ON-OFF valve. Then, designs for 3/2-way, 4/2-way, and 5/2-way valves are proposed [24]. The valves occupy a volume of $50 \times 18 \times 18$ mm and can operate at pressures up to 187 kPa. Furthermore, thanks to their multiple inlets and outlets, a single valve can be used to control multi-chamber soft actuators in a space-efficient manner, responding to key needs of untethered operation.

3.1 Design and Fabrication of the Bistable Valve

The internal structure of the valve, visible in Fig. 1, consists of the two conical membranes connected by a rigid cylinder. The membranes actuate two small pistons which can open or block the airflow in the air channels they are coupled to. Thanks to its configuration, the system has two stable states, activated by pressurizing chambers P1 and P2 alternatively. As the system moves from one state to the other, it goes through a sudden transition state, characterized by a snap-through behavior of the membranes. These undergo compression, until a critical load is reached, after which buckling occurs, moving the system from one stable state to the other. The bistable design allows the valve to maintain its state even in the absence of the control signal. In addition, small changes to the control pressure do not cause the valve to switch, making it robust to perturbations in the supply pressure level and its operation more reliable.

Thanks to its design, the valve can be monolithically printed on a multi-material PolyJet printer (a Stratasys J735 was used in the study) with soluble supports. The printer allows controlling the mechanical properties of the different parts of the

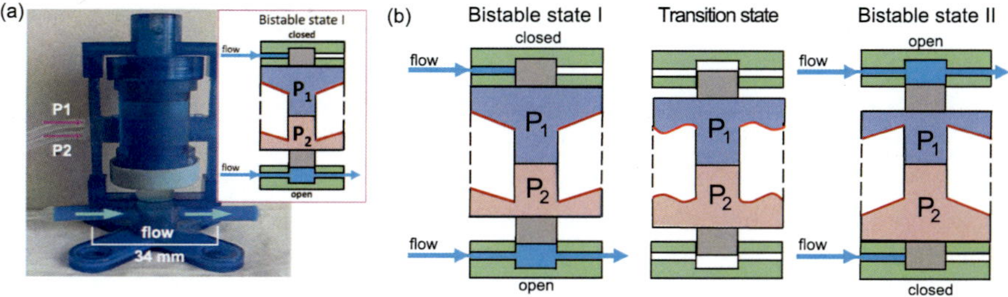

Fig. 1 The bistable valve and its working principle. **a** The valve is monolithically printed on a MultiJet multi-material printer using soluble supports. **b** The valve features two internal membranes that enable the bistable behavior. When pressurizing either chambers P_1 or P_2, the membranes deform until a critical condition is reached when snap-through is observed

design, going from soft Agilus30 (rubber-like material with tensile strength of 2.16–2.6 MPa) to VeroWhite, rigid and plastic-like, with a tensile strength of 50–65 MPa. After printing, the support material can be removed either through mechanical action, using a pressurized water jet, or by chemical bathing in a solution of 0.02 kg/l of Sodium Hydroxide and 0.01 kg/l of Sodium Metasilicate in water.

The behavior of the valve can be modeled and controlled, assuming the membranes are hinged at the edges, knowing the geometrical parameters of the design and applying the Rayleigh-Ritz method for buckling [30]. The details of the procedure are presented in [23], while Fig. 2 shows the results of the characterization on the valves. As expected, increasing the membrane material's Young's modulus increases the pressure difference needed for the snap-through behavior to occur and makes for faster responding valves. A similar trend is observed for the shell thickness, but the response times don't display a clear monotonic trend.

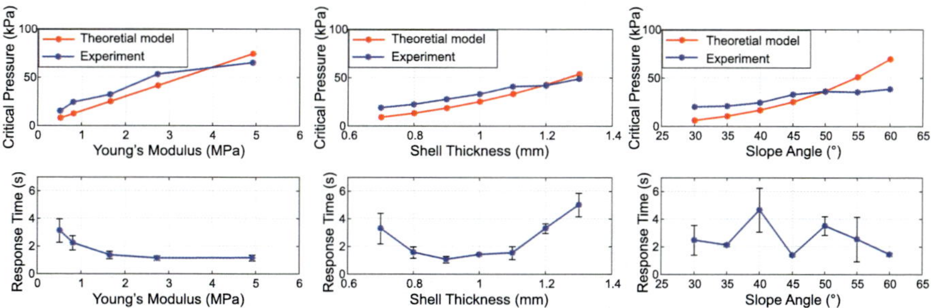

Fig. 2 Characterization results. By tuning the geometrical and mechanical parameters of the membranes such as the material's Young's modulus or their thickness, it is possible to tune the snap-through behavior to suit different supply pressure ranges

3.2 Modular Design of the Multi-channel Bistable Valves

By pairing the bistable actuation mechanism with a modular spool design for the flow control portion of the valve, it is possible to create more complex directional control valves, featuring multiple inlets and outlets [24]. These components are key in creating more complex logic functions and compact systems for controlling multi-chamber actuators and robots. Fig. 3a summarizes the design of the 3/2 4/2 and 5/2 valves: the first number representing the number of functional ports, the second referring to the number of stable states of the valve. Compared to the previous example, the piston is replaced by a longer spool, featuring soft Agilus30 seals. These slide against the surface of the rigid housing, and separate pairs of inlet and outlet channels from each other. By actuating the bistable membranes, different inlet-outlet configurations can be achieved. Furthermore, the design allows for a pressure gain of almost 3 to be achieved, meaning a control pressure of 60 kPa can be used to switch inlet pressures of up to 180 kPa (the maximum pressure the valve can withstand without internal leakage). This allows for effective separation between logic and power signals within the pneumatic circuit, limiting driving fluid consumption in untethered applications.

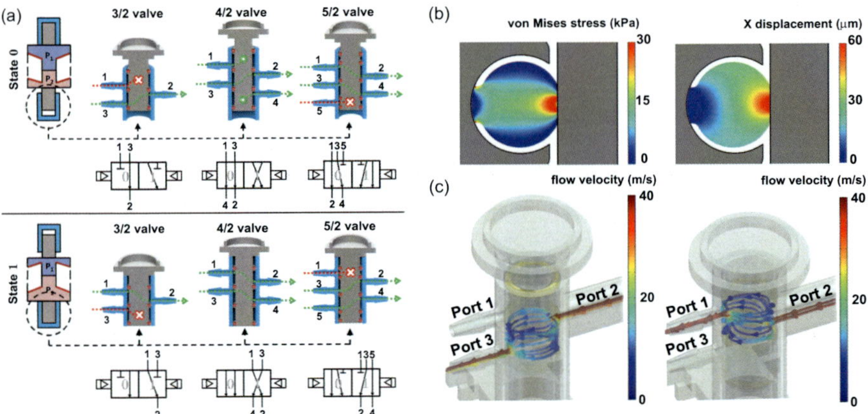

Fig. 3 Design of the 3D printed multi-channel valves. **a** All multi-channel valves include the same bistable modular unit in the middle. A spool with soft seals (shown in gray) is attached to each end of the central bistable structure and is placed in a cylinder (shown in blue), controlling the opening and closing of multiple air channels based on the state of the bistable structure. **b** FEA and CFD results of the multi-channel valve unit. The stress distribution and deformation of the spool seal during operation are displayed as well as the airflow in the valve in both its operating states

3.3 *Characterization of the Multi-channel Valves*

The mechanical action of the seals and flow through can be simulated using FEA and CFD methods, to aid during the design process. In [24], COMSOL Multiphysics was used to verify that the seals would be able to withstand the mechanical stresses caused by the interaction with the housing's walls and to visualize the flow through the valves in the various configurations (Fig. 3b).

Thanks to their modular design, the valves can be used to control a bi-directional soft actuator (Fig. 4). The bi-directional motion is obtained by pressurizing two antagonistic bellow actuators. By regulating the pressure value in the chambers, different positions of the end effector can be achieved. In the example of Fig. 4, two different supply pressure levels, 20 kPa and 40 kPa are used to achieve four different configurations. In particular, only four valves are needed to achieve the control sequence, compared to a total of six 2/2 valves that would otherwise be required.

4 A Design Automation Toolbox for 3D Printed Pneumatic Logic Circuits

A crucial need for the design of soft robots with onboard fluidic control is being able to create controllers that achieve the desired functionality while respecting physical space constraints. Recently, researchers have tried automating this process, usually

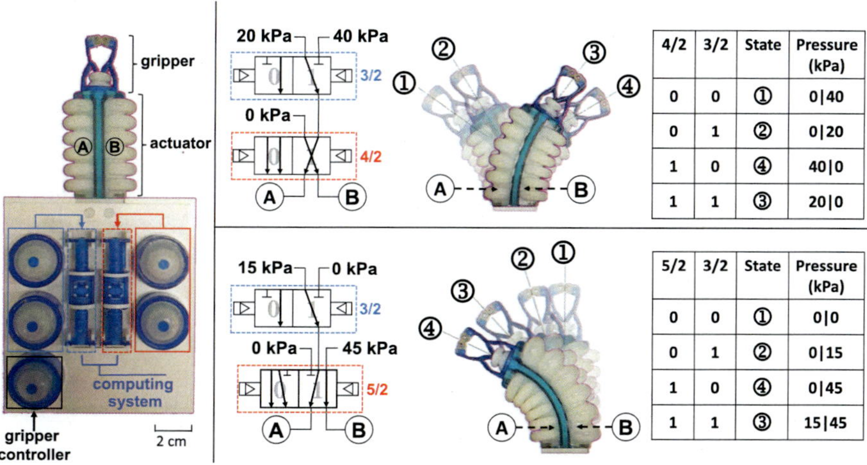

Fig. 4 Control of a multi-chamber bi-directional soft actuator. Two valves are needed, each with two control signals, and two different supply pressure values are used to control the actuator in four distinct states. The 3/2 valve highlighted in green manages the supply pressure selection, while the 4/2 in the red box selects which chamber to connect to the supply, while the other is depressurized

requiring expert engineers and multiple design iterations, starting from a library of components achieving base logic functions and combining them together in an optimal manner to obtain more complex functionalities. An example of this is represented by the toolbox developed by Wang et al. [25], which we analyze in detail in the following.

4.1 Design Automation of the Pneumatic Logic Circuit

Figure 5a presents a synoptic view of the design pipeline, from user input in terms of truth tables to the final .stl file ready for 3D printing. Just three base modular logic gates—NOT, AND, and OR—derived from optimized versions of previously established 3D printed pneumatic transistors (Fig. 5b) are used in the process. These components are adapted for compact integration, routing all inlets and outlets to the base. This makes it possible to easily combine them together in sequences to achieve any desired logic function.

The toolbox uses MATLAB to interpret truth tables input by the user into logical expressions via sum-of-products (SOP) algorithms. The Symbolic Toolbox is then used to simplify the expressions to minimize the number of logic gates needed. The simplified expressions are then parsed into circuit schematics to generate the final layout. For this, the user is asked to define the available space by coloring in an area in a graphical user interface. The planning algorithm then uses backtracking to identify all possible gate placements within the user-specified 2D workspace.

Air channels connecting the gates are routed next in a dedicated manifold layer of the controller. Total layer count is minimized by using the Bentley-Ottmann algorithm to detect intersected channels with the lowest complexity and a greedy algorithm to arrange the channels in an optimal way so that no intersecting channels are in the same layer. Final layouts are assessed for spatial feasibility, layer count, and total channel length to optimize performance. The toolbox then uses COMSOL, interfaced through MATLAB, to generate ready-to-print 3D CAD models of the hub layer with integrated ports and channels. The logic gates are printed separately and directly plugged in without requiring any tools.

4.2 Validation of the Framework

Thanks to the generality of the design automation toolbox complex logic functions and configurations can be achieved. In the study, the creation of XOR and 1:4 demultiplexer circuits constrained by differing spatial layouts is demonstrated. All circuits were fabricated using polyjet 3D printing and tested for response time, and durability. Results confirmed correct logic operation across configurations, while response times varied based on layout efficiency, with less constrained designs performing

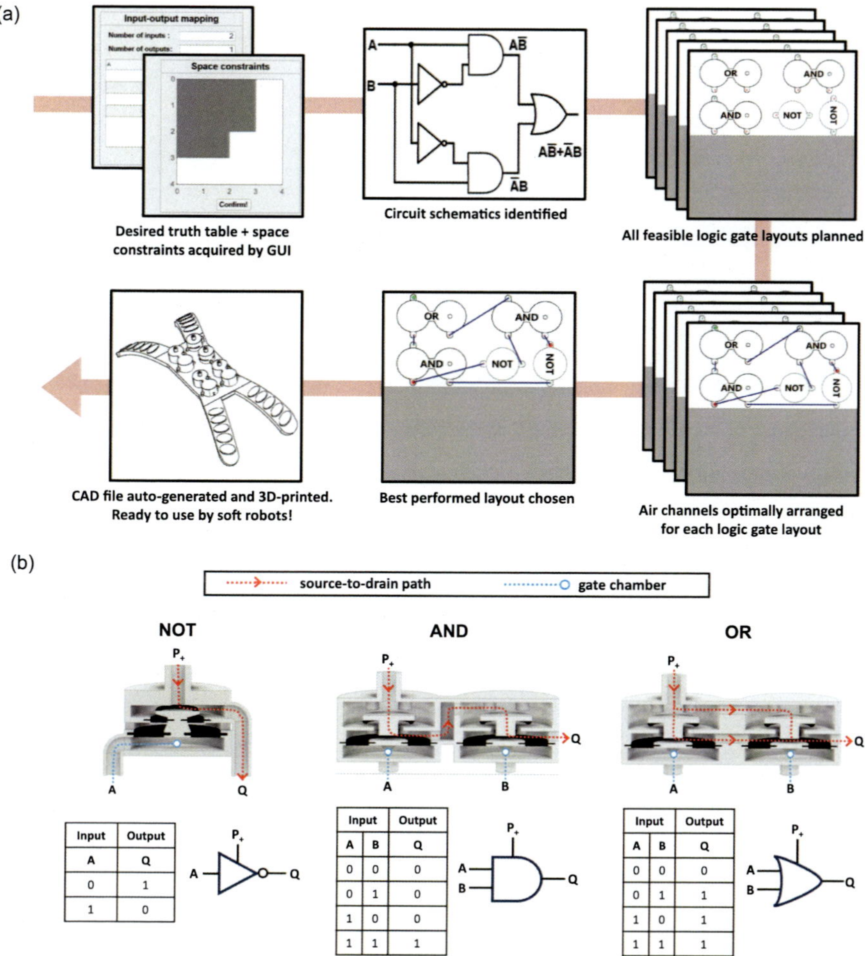

Fig. 5 The automated controller design pipeline. **a** Synoptic view of the framework displaying its key steps. The user input is converted into logic equations which are then simplified. The minimum number of basic logic gates is then arranged according to the user-provided space constraints and connected together. Finally, the *.stl* file for printing is generated. **b** Only three basic gates—NOT, AND and OR—are designed for printing and used to generate any user-specified logic function. Thanks to the multi-material printing capability of the PolyJet process used, rigid (white in the figure) and soft (black) parts can be printed together, without requiring further assembly

faster due to reduced channel lengths and fewer hub layers. A limitation of the proposed approach is linked to the slow response times of the gates compared to their electronic counterparts: propagation of signals in the circuit is asynchronous, and this can cause glitches in specific configurations due to channel lengths. Long-term

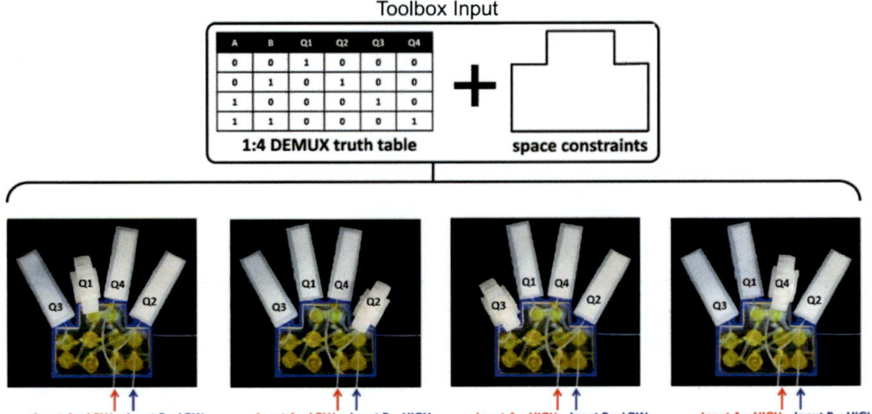

Fig. 6 Validation of the automatic design toolbox. A 1:4 demux is fit inside the constraints of the palm of a 4-finger hand. PneuNet actuators serve as fingers and only two inputs are needed, to control each finger individually

fatigue tests revealed that circuits remained accurate over 10,000 cycles at low frequency, though performance degraded at higher frequencies due to viscoelastic aging in the soft materials.

A soft hand is then used to showcase a possible use case for the toolbox. A 1:4 demultiplexer is fully contained in the palm of the hand (Fig. 6), while four PneuNet actuators take the place of the fingers. Thanks to the embedded logic, the opening and closing of each finger is controlled using only 2 inputs to the system, underscoring the capability of the toolbox for scalable automated design of highly integrated, electronics-free pneumatic control systems.

5 Conclusions

Thanks to the framework presented in the previous sections, it is possible to almost-completely automate the design and fabrication of space-optimized pneumatic controllers for soft robots. This achieves critical advancements on a number of fronts, from faster prototyping, to higher control and more repeatability in the manufactured components. As highlighted, a limitation of the algorithm is that it tries to minimize air channel lengths, but a more nuanced approach focused on matching lengths of specific channel pairs, or carefully controlling the resistance of some of the channels, might be needed to avoid conditions in which wrong logic configurations are reached due to the non-negligible response times of the 3D printed valves. Furthermore, the process used for logic function simplification doesn't make use of advanced techniques like Karnaugh maps for eliminating race conditions in the final circuits.

The manufacturing technique adopted in the study allows for complex valve designs to be printed in place without assembly or specific considerations about printing orientation, leaving a lot of freedom to the designer and simplifying the overall process. The high precision achievable allows for small details to be repeatably replicated, while the possibility to freely mix materials to get the desired mechanical properties for each part of the design further enhances the capabilities of the printed components. PolyJet multi-materials 3D printers, though, are still expensive and not readily available in research labs. Having in the library of components specific ones that can be printed using more common FDM processes would be a beneficial addition, despite the aforementioned limitations of such systems.

In terms of components, the selection of logic components in the library makes it possible to replicate any logic function, but further improvements could be achieved by expanding the library to include multi-channel valves like the one presented in Sect. 3.2. Furthermore, only logic functions have been so far developed, but for more complex autonomous operation of soft robots, clock sources are needed. A few examples exist in the literature that exploit bistable behaviors and some examples of fully 3D printed ones have also been demonstrated. These components would further enhance the compiler functionalities, leading to more nuanced autonomous behaviors of the final robots. Nevertheless, the results presented in this chapter constitute a solid foundation for a scalable approach to the autonomous creation of logic circuits for onboard control of pneumatic soft robots.

References

1. Jumet, B., Bell, M.D., Sanchez, V., Preston, D.J.: A data-driven review of soft robotics. Adv. Intell. Syst. **4**(4), 2100163 (2022)
2. Li, M., Pal, A., Aghakhani, A., Pena-Francesch, A., Sitti, M.: Soft actuators for real-world applications. Nat. Rev. Mater. **7**(3), 235–249 (2022)
3. El-Atab, N., Mishra, R.B., Al-Modaf, F., Joharji, L., Alsharif, A.A., Alamoudi, H., Diaz, M., Qaiser, N., Hussain, M.M.: Soft actuators for soft robotic applications: a review. Adv. Intell. Syst. **2**(10), 2000128 (2020)
4. Jung, Y., Kwon, K., Lee, J., Ko, S.H.: Untethered soft actuators for soft standalone robotics. Nat. Commun. **15**(1), 3510 (2024)
5. Polygerinos, P., Correll, N., Morin, S.A., Mosadegh, B., Onal, C.D., Petersen, K., Cianchetti, M., Tolley, M.T., Shepherd, R.F.: Soft robotics: review of fluid-driven intrinsically soft devices; manufacturing, sensing, control, and applications in human-robot interaction. Adv. Eng. Mater. **19**(12), 1700016 (2017)
6. Wang, H., Totaro, M., Beccai, L.: Toward perceptive soft robots: progress and challenges. Adv. Sci. **5**(9), 1800541 (2018)
7. Roche, E.T., Horvath, M.A., Wamala, I., Alazmani, A., Song, S.E., Whyte, W., Machaidze, Z., Payne, C.J., Weaver, J.C., Fishbein, G, et al.: Soft robotic sleeve supports heart function. Sci. Transl. Med. **9**(373), eaaf3925 (2017)
8. Cianchetti, M., Ranzani, T., Gerboni, G., Nanayakkara, T., Althoefer, K., Dasgupta, P., Menciassi, A.: Soft robotics technologies to address shortcomings in today's minimally invasive surgery: the stiff-flop approach. Soft Rob. **1**(2), 122–131 (2014)
9. Cianchetti, M., Laschi, C., Menciassi, A., Dario, P.: Biomedical applications of soft robotics. Nat. Rev. Mater. **3**(6), 143–153 (2018)

10. Chelsea Shan Xian Ng and Guo Zhan Lum: Untethered soft robots for future planetary explorations? Adv. Intell. Syst. **5**(3), 2100106 (2023)
11. Zhang, Y., Li, P., Quan, J., Li, L., Zhang, G., Zhou, D.: Progress, challenges, and prospects of soft robotics for space applications. Adv. Intell. Syst. **5**(3), 2200071 (2023)
12. Hawkes, E.W., Blumenschein, L.H., Greer, J.D., Okamura, A.M.: A soft robot that navigates its environment through growth. Sci. Robot. **2**(8), eaan3028 (2017)
13. Mazzolai, B., Mondini, A., Del Dottore, E., Sadeghi, A.: Self-growing adaptable soft robots. Mech. Responsive Mater. Soft Robot. 363–394 (2020)
14. Del Dottore, E., Mondini, A., Rowe, N., Mazzolai, B.: A growing soft robot with climbing plant—inspired adaptive behaviors for navigation in unstructured environments. Sci. Robot. **9**(86), eadi5908 (2024)
15. Soana, V., Minooee Sabery, S., Bosi, F., Wurdemann, H.: Elastic robotic structures: a multidisciplinary framework for the design and control of shape-morphing elastic system for architectural and design applications. Constr. Robot. **9**(1), 3 (2025)
16. Pfeifer, R., Bongard, J.: How the Body Shapes the Way We Think: A New View of Intelligence. MIT Press (2006)
17. Zhai, Y., Yan, J., De Boer, A., Faber, M., Gupta, R., Tolley, M.T.: Monolithic desktop digital fabrication of autonomous walking robots. Adv. Intell. Syst. 2400876 (2025)
18. Drotman, D., Jadhav, S., Sharp, D., Chan, C., Tolley, M.T.: Electronics-free pneumatic circuits for controlling soft-legged robots. Sci. Rob. **6**(51), eaay2627 (2021)
19. Wehner, M., Truby, R.L., Fitzgerald, D.J., Mosadegh, B., Whitesides, G.M., Lewis, J.A., Wood, R.J.: An integrated design and fabrication strategy for entirely soft, autonomous robots. Nature **536**(7617), 451–455 (2016)
20. Zhai, Y., De Boer, A., Yan, J., Shih, B., Faber, M., Speros, J., Gupta, R., Tolley, M.T.: Desktop fabrication of monolithic soft robotic devices with embedded fluidic control circuits. Sci. Rob. **8**(79), eadg3792 (2023)
21. Conrad, S., Teichmann, J., Auth, P., Knorr, N., Ulrich, K., Bellin, D., Speck, T., Tauber, F.J.: 3D-printed digital pneumatic logic for the control of soft robotic actuators. Sci. Rob. **9**(86), eadh4060 (2024)
22. Kendre, S.V., Whiteside, L., Fan, T.Y., Tracz, J.A., Teran, G.T., Underwood, T.C., Sayed, M.E., Jiang, H.J., Stokes, A.A., Preston, D.J., et al.: The soft compiler: a web-based tool for the design of modular pneumatic circuits for soft robots. IEEE Robot. Autom. Lett. **7**(3), 6060–6066 (2022)
23. Wang, S., He, L., Maiolino, P.: Design and characterization of a 3d-printed pneumatically-driven bistable valve with tunable characteristics. IEEE Robot. Autom. Lett. **7**(1), 112–119 (2021)
24. Wang, S., He, L., Maiolino, P.: A modular approach to design multi-channel bistable valves for integrated pneumatically-driven soft robots via 3D-printing. IEEE Robot. Autom. Lett. **7**(2), 3412–3418 (2022)
25. Wang, S., He, L., Yao, Y., Liu, C., Maiolino, P.: A toolbox for designing 3D-printing-ready pneumatic circuits for controlling soft robots. Adv. Intell. Syst. **5**(12), 2300394 (2023)
26. Pontin, M., Damian, D.D.: Multimodal soft valve enables physical responsiveness for preemptive resilience of soft robots. Sci. Rob. **9**(92), eadk9978 (2024)
27. Sochol, R.D., Sweet, E., Glick, C.C., Venkatesh, S., Avetisyan, A., Ekman, K.F., Raulinaitis, A., Tsai, A., Wienkers, A., Korner, K., et al.: 3D printed microfluidic circuitry via multijet-based additive manufacturing. Lab Chip **16**(4), 668–678 (2016)
28. Hubbard, J.D., Acevedo, R., Edwards, K.M., Alsharhan, A.T., Wen, Z., Landry, J., Wang, K., Schaffer, S., Sochol, R.D.: Fully 3D-printed soft robots with integrated fluidic circuitry. Sci. Adv. **7**, eabe5257 (2021)
29. Bartlett, N.W., Tolley, M.T., Overvelde, J.T., Weaver, J.C., Mosadegh, B., Bertoldi, K., Whitesides, G.M., Wood, R.J.: A 3D-printed, functionally graded soft robot powered by combustion. Science **349**(6244), 161–165 (2015)
30. Henry,T., Megson, G.: Structural and Stress Analysis. Butterworth-Heinemann (2019)

Engineering Soft Robots—Towards Automated Frameworks to Accelerate Design Exploration

Kristin M. de Payrebrune, Daniel Müller, Noah Tillmann, Dominik Sturm, Alfred Jose Puthoor, and Philipp Tchistiakov

Abstract Designing soft robotic systems is challenging due to the nonlinear behavior and high compliance of soft materials, making traditional trial-and-error methods inefficient and difficult to reproduce. To overcome these limitations, we developed a structured, simulation-guided methodology for the systematic design and prototyping of soft robots. We first investigated how different design aspects affect actuator performance, highlighting complex dependencies and the need for an automated approach to efficiently evaluate variations. In response, we implemented two simulation pipelines: one for a pneumatically actuated bending actuator using parametric CAD and finite element analysis, and another for optimizing a pneumatic artificial muscle in the SOFA simulation environment. These frameworks allow for rapid exploration of design parameters, reducing the reliance on physical testing. To validate the methodology, we developed three prototypes: an autonomous bending actuator, a muscle for skeletal movement, and a peristaltic actuator for fluid transport. Targeted simulations supported the development process and helped identify potential design issues early. Our results show that combining automated simulation with application-driven prototyping enables a scalable and reproducible approach to soft robot design. This work demonstrates how new soft robotic systems can be systematically invented using integrated, simulation-based strategies.

1 Introduction

The development of bioinspired actuation systems has received increasing attention in recent years, particularly in the fields of soft robotics and wearable assistive devices. Various soft actuator concepts, such as McKibben muscles [1, 2], PneuNet actuators [3, 4], and tendon-driven structures [5, 6], offer compelling alternatives

K. M. de Payrebrune (✉) · D. Müller · N. Tillmann · D. Sturm · A. J. Puthoor · P. Tchistiakov
Institute for Computational Physics in Engineering, RPTU Kaiserslautern-Landau, Kaiserslautern, Germany
e-mail: kristin.payrebrune@mv.rptu.de

A. J. Puthoor
e-mail: alfred.puthoor@mv.rptu.de

A. Raatz et al. (eds.), *Soft Material Robotic Systems*,
https://doi.org/10.1007/978-3-032-22453-8_8

to traditional rigid actuators due to their inherent compliance, flexibility, and safety during interaction with humans. These characteristics make soft actuators especially suitable for applications that demand safe human–machine interaction, including prosthetics, exoskeletons, and soft robotic manipulators [7, 8].

Despite these advantages, designing reliable systems based on soft actuators remains a complex challenge [9]. Their nonlinear mechanical behavior, sensitivity to initial conditions, and strong dependence on attachment geometry and boundary conditions complicate the prediction of system-level performance [10]. This creates a growing demand for structured design guidelines and computational frameworks that can support actuator placement, loading strategies, and overall system configuration. Although several design tools and models have been proposed in the literature, practical implementation often requires extensive experimental validation and task-specific adaptation.

This work contributes to the development of systematic design strategies through the use of numerical modeling and application-driven analysis. Using a soft bending actuator (Sect. 2.1) and a reinforcing structure (Sect. 2.2) as representative examples, we explore how different design parameters influence mechanical behavior and identify generalizable guidelines for future development. Building on this foundation, we present two example frameworks that enable automated simulation-based exploration of design specifications: one based on the commercial finite element platform Abaqus (Sect. 3.1) and another using the open-source simulation toolkit SOFA (Sect. 3.2).

Finally, we apply these methods to three case studies representing different soft robotic applications, demonstrating how simulation insights can guide the design process through to functional prototype realization (Sect. 4). The resulting findings are discussed with respect to design robustness, scalability, and their potential integration into real-world assistive and robotic systems.

Through this investigation, we aim to bridge the gap between theoretical modeling and practical design. By developing a transferable methodology that integrates simulation, validation, and design feedback, we provide a reproducible framework to support engineers and researchers in translating soft actuator concepts into effective biomechanical and robotic solutions.

2 Structured Design Methodology

The design of soft robotic systems presents unique challenges due to their highly compliant structures and nonlinear material behavior. Traditionally, the development of such systems has relied heavily on iterative trial-and-error processes, where physical prototypes are manually adjusted and tested until desired performance is achieved. While this approach can lead to functional designs, it is often time-consuming, resource-intensive, and lacks systematic reproducibility.

In this section, we present a structured design methodology, illustrated using a pneumatically driven bending actuator and a 3D printable reinforcement structure as representative examples, to improve efficiency and guide the design process in a more targeted manner. Finite element simulations are employed to systematically analyze and identify optimal design solutions.

2.1 *Derivation of Design Guidelines of a Soft Bending Actuator*

Soft robots feature a wide range of actuation strategies and designs. A key aspect in the development of such systems is the choice of the actuation mechanism, whether pneumatic, tendon-driven, or alternative methods. In addition, the layout of the soft structure and the underlying deformation principle must be carefully defined.

In pneumatically driven soft robots, deformation can generally be categorized based on whether it arises from the geometry of a single chamber or from the coordinated interaction of multiple chambers. Likewise, actuation may result from axial lengthening or shortening of simple chamber structures [11], or from direct bending induced by geometric asymmetries [4]. This section explores design principles for a versatile actuator enabling controlled bending and elongation.

We identified six design criteria, which form the basis for a systematic investigation into how structural parameters influence bending and elongation behavior. To ensure consistency across simulations, all actuator modules were standardized to a length of 100 mm and an outer diameter of approximately 30 mm. Dragon Skin 10 was selected as the chamber material and modeled as a neo-Hookean solid with material parameters $C_{10} = 0.0425$ MPa and $D_1 = 0$ [12, 13]. Finite element simulations were conducted in Abaqus 2019 using the C3D10H hybrid element type.

Study of design criteria

First, the influence of the **number of air chambers** is investigated. The FE simulations are based on the sector-shaped chambers proposed by Suzumori et al., featuring a soft robot with three, four, or five air chambers molded into a single silicone core [14]. The outer diameter is kept equal, while the wall thickness between the chambers is adjusted to keep the total cross-sectional area of all chambers constant. The performance is evaluated by examining the bending under varying numbers of pressurized chambers, the elongation under pressurization of all chambers, and the module's XY range of motion, cf. Fig. 1.

Among the configurations tested, the design with three air chambers was found to be the most efficient in terms of both bending and elongation. To verify its effectiveness not only in extreme positions but also during transitions, the XY-range of motion was analyzed. For this, a complete actuation cycle was simulated, in which each chamber is pressurized and depressurized in sequence, and the XY coordinates of the top surface center point were recorded and plotted. The five-chamber variant (red) offers a more uniform motion pattern, but its range is entirely covered by the three-chamber design (blue), which also achieves greater bending.

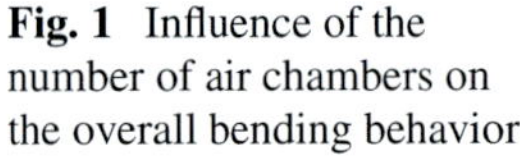

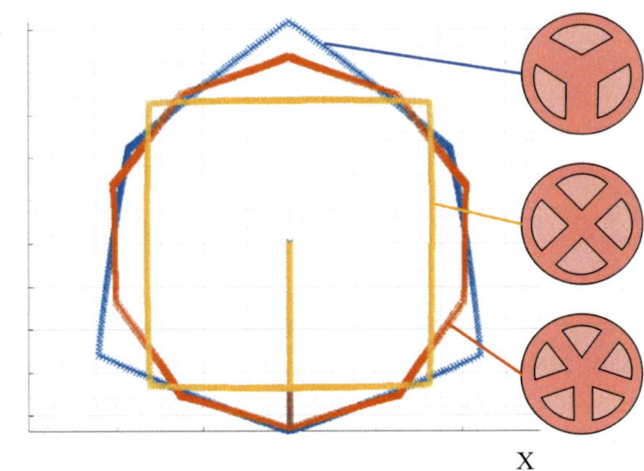

Fig. 1 Influence of the number of air chambers on the overall bending behavior

Therefore, the three-chamber layout is considered the most efficient solution overall.

The **air chamber geometry** has been extensively studied in literature. Suzumori et al. optimized sector-shaped chambers using finite element simulations and topology optimization, resulting in more rounded and efficient profiles [14]. Elsayed et al. compared circular, semicircular, ring, and sector-shaped geometries, concluding that sector-shaped chambers require the least pressure for a 90° bend, but also exhibit the highest radial expansion [15]. Polygerinos et al. showed that circular and semicircular cross-sections perform similarly under fiber reinforcement, whereas rectangular shapes suffer from stress concentrations and require higher pressures [16].

Building on these studies, we simulated and evaluated sector-shaped, circular, semicircular, and inverse semicircular chamber geometries under consistent conditions. All chamber variants had identical cross-sectional areas. The chambers were connected using multiple discrete ABS elements evenly distributed along the actuator length (see blue parts in Fig. 2). To simplify simulation, adhesive bonding between chambers and ABS elements was assumed. The bending performance of the soft robot was assessed under 55 kPa pressure, applied to either one or two chambers.

Among the tested shapes, the circular cross-section achieved the highest bending angles. Additional tests with inverted semicircular chambers—placing the rounded side inward—showed slightly improved bending when two chambers were actuated. However, they underperformed under single-chamber actuation and exhibited greater radial expansion, which may lead to stress peaks despite fiber reinforcement.

Overall, the circular geometry demonstrated the best performance, combining high bending efficiency with good manufacturability.

We investigated various methods for **connecting the air chambers** of a soft robot to assess their influence on bending performance and manufacturing feasibility. The primary approach used discrete connectors to join individually cast circular air chambers, each connector formed by three interlinked ring-like structures, see Fig. 3 (connectors). To evaluate the influence of stiffness, three connector materials were tested: rigid ABS, and two soft silicones (Elastosil and Dragon Skin 10). Each

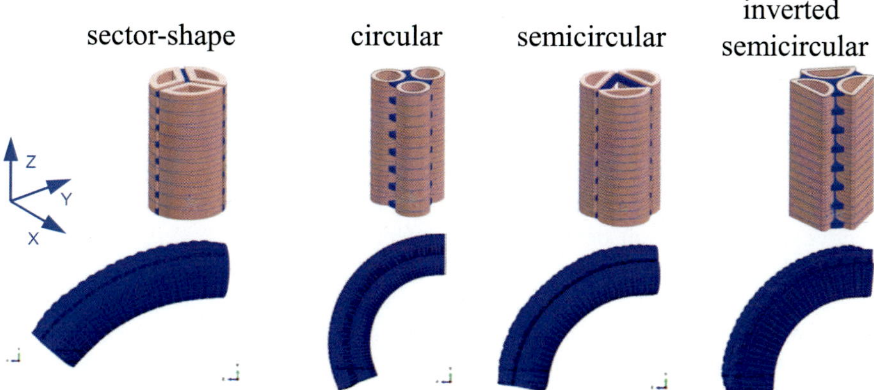

Fig. 2 Examined geometries of the air chambers with connectors made of ABS (blue) and simulated bending when one air chamber is pressurized with 55 kPa

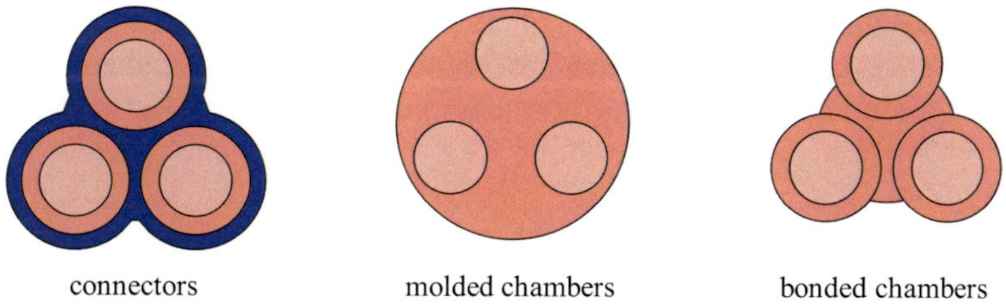

Fig. 3 Different methods to connect individual chambers or include the chambers in one core

actuator module incorporated 15 connectors, evenly spaced along its length, with individual connectors measuring 2 mm in thickness and placed 7 mm apart to provide a continuous and stable connection between adjacent chambers. This modular approach allows flexibility in material choice and ease of assembly. The investigation show that the choice of connector material has only a minor influence on the bending behavior.

Beyond discrete connectors, alternative methods included directly casting the air chambers in a single mold (molded chambers) or manufacturing the chambers separately and then bonding them together afterward (bonded chambers). This post-casting approach allows for a more targeted use of material, maintaining thin wall thicknesses while adding structural mass only where necessary for connection.

The numerical tests show that connector elements achieve the largest bending angles, followed by bonding the chambers in a subsequent step. Casting the chambers in one piece shows the worst performance due to the extra material, which hinders bending.

In summary, connectors offer a good trade-off between ease of manufacturing and performance and were used for the final design.

Various **reinforcement strategies** were investigated to control radial expansion and improve the performance of pneumatically driven soft actuators. Therefore, fiber-based reinforcements provide effective radial constraint with relatively simple implementation, cf. Fig. 4.

Symmetrical Cross-Fiber Reinforcement (e.g., opposing Kevlar windings) effectively limits radial expansion while preserving strong bending performance. Fiber angle has little impact on bending but larger angles increase expansion. A balance between reinforcement density and manufacturing ease is needed, with a recommended pitch of 3° and spacing under 5 mm.

Single Fiber Windings with a shallow pitch reduce manufacturing effort but introduce unwanted torsion, especially when multiple chambers with the same winding direction are actuated simultaneously. This can negatively impact the controllability of the robot.

Integrated Reinforcements (e.g., molded lamellae inside the soft body) are more complex to realize during casting and show limited effectiveness. Simulations indicate significant radial expansion even at moderate pressures (45 kPa), reducing their practical benefit.

Reinforcement Rings offer a low-cost and easy-to-implement alternative. When spaced appropriately, they achieve bending angles comparable to fiber-reinforced actuators while maintaining the soft character of the robot and reducing manufacturing effort. However, their effectiveness is limited by the spacing and material properties of the rings, as elastic rings also tend to expand radially.

In summary, cross-fiber reinforcement offers the best balance between performance and control, while rings represent a viable alternative for simplified fabrication.

To investigate the influence of **wall thickness** on bending performance, the chambers were designed with identical internal diameters, ensuring the same air volume across all variants. The outer diameter of each air chamber was offset 2 mm from the center of the module, while the thickness of the wall varied between 2 and 4 mm.

The results in Fig. 5 show a clear correlation between wall thickness and bending capability under a constant pressure of 55 kPa. As the wall thickness increased, the bending angle decreased over-proportionally.

As a conclusion, thicker walls are more suitable when higher mechanical loads need to be supported, while thinner walls are advantageous for achieving high bending angles. However, very thin walls can pose manufacturing challenges.

In a separate study, the influence of the **inner diameter** of the air chambers was examined. In this case, wall thickness and chamber positioning were held constant while only the inner diameter was varied. The results revealed a nearly linear relationship between the inner diameter and the bending performance. However, the effect was relatively small, reducing the diameter from 12 to 8 mm led to only a 10 % reduction in bending.

This indicates that the inner diameter is a secondary factor compared to the wall thickness when optimizing for bending behavior.

Final design of a soft bending actuator

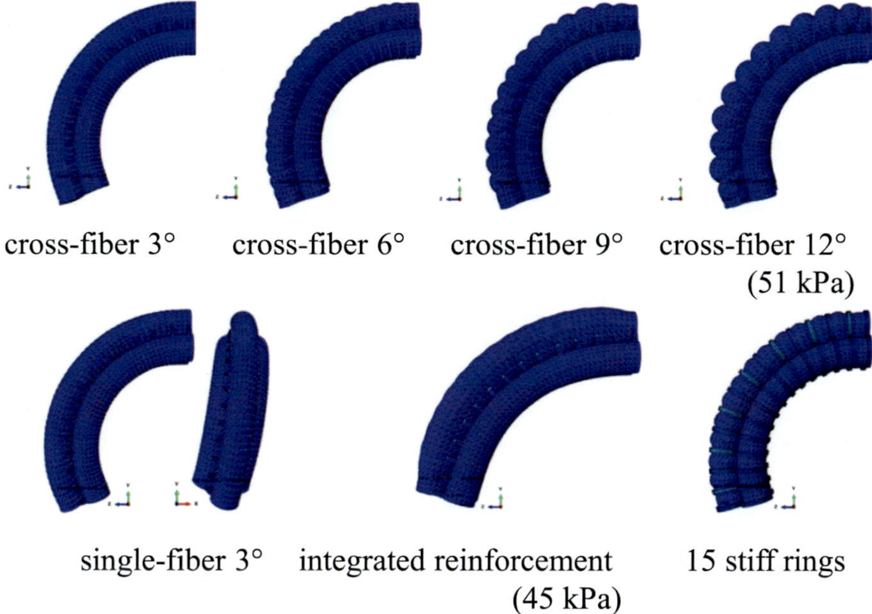

Fig. 4 Examined reinforcements and simulated bending when one air chamber is pressurized with 55 kPa (if not indicated otherwise)

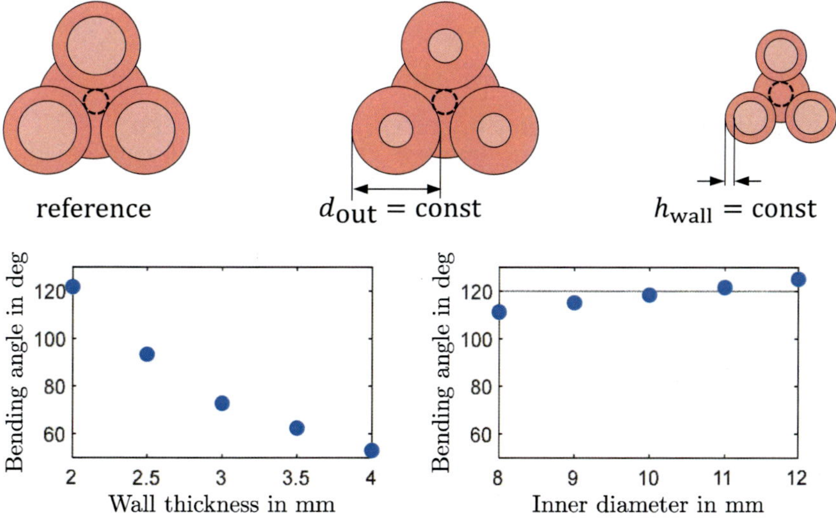

Fig. 5 Simulated bending angle for a varying wall thickness and inner diameter when one air chamber is pressurized with 55 kPa

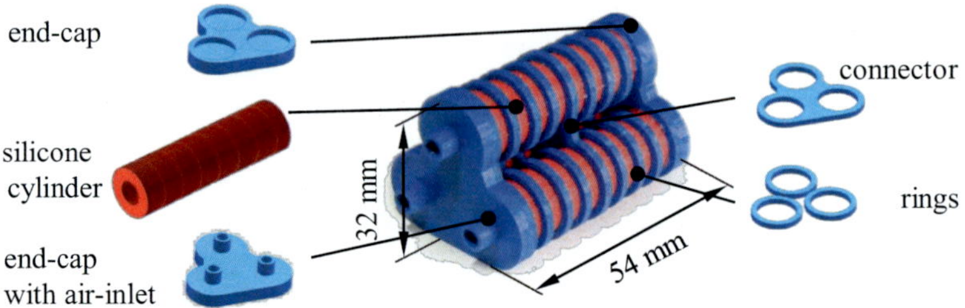

Fig. 6 Final design of the soft bending actuator

Based on all previously discussed investigations and design considerations, the final design of the soft bending actuator has taken shape as follows [17]: it features a modular and efficient structure composed of three cylindrical chambers, each reinforced with rings and connected via specialized end-caps and connectors, cf. Fig. 6. This configuration is optimized for both performance and adaptability. The use of three chambers simplifies the control system while maintaining precise movement. A cylindrical base geometry helps to minimize material stress, and the efficient interconnection of chambers facilitates both modeling and manufacturing. The actuator dimensions fall within standard ranges, and each component is independently adjustable, allowing for high customization. Furthermore, the actuator length can be scaled by serially connecting multiple modules, enhancing its versatility for a wide range of soft robotic applications.

2.2 Derivation Guidelines of a Reinforcement Design

The developed soft robot performs reliably under typical operating conditions. However, at higher applied pressures, the reinforcement rings may shift, causing ballooning between them and reducing actuator performance. To overcome this limitation, a novel 3D-printable reinforcement structure was explored as an alternative to the ring-based design. A key objective was to limit radial expansion while allowing unrestricted axial elongation. To this end, we investigated and numerically compared metamaterials with anisotropic stiffness. The aim was to improve bending performance while retaining a simple and robust manufacturing process. The final design was fabricated using additive manufacturing, enabling precise realization of the complex geometry.

Design study of metamaterials
The investigated reinforcement structure is a thin, 3D-printable metamaterial developed for use in soft robotic actuators. It provides anisotropic stiffness by combining rigid beams arranged in the circumferential direction with flexible segments in between. This structure ensures high radial stiffness while allowing significant axial stretch up to 80% within the elastic range. To facilitate design and maintain compa-

Table 1 Displayed are the unit cells of each tested design, along with an exemplary simulation illustrating both the initial and the deformed state under an applied elongation of 40 mm

Design No	1	2	3	4	5	6	7	8	9	10
CAD										
Δl in mm	24	12.2	12.8	11.2	14.8	13.6	9	8.6	8	8.5
F in N	0.006	0.031	0.034	0.11	1.67	4.71	5.64	13.07	20.27	23.72
note	ad	pad			ad	ad				

ad = asymmetric deformation; pad = partly asymmetric deformation

rability, all prototypes are created as 2D sketches within a $50 \times 50\,\text{mm}^2$. The design thickness is limited to 2.5 mm during simulation. The manufactured and tested design was scaled to the circumferential area of the silicone chamber with a thickness of 2 mm.

Ten different reinforcement structure designs were selected and modeled as 2D sketches using Siemens NX CAD software (version 1888). To evaluate their mechanical behavior, finite element simulations for each design were created in Abaqus CAE (versions 2018 and 2020). The CAD models were imported as STEP files into Abaqus, where a linear elastic material model with Young's modulus $E = 22$ MPa and Poisson ration $\nu = 0.45$ were assigned. The boundary conditions were established by fixing the structure on the left side, while a prescribed axial elongation of 80% (40 mm) was applied on the right side. The geometry was discretized using eight-node linear hexahedral elements with reduced integration (C3D8R). A mesh convergence study determined a global mesh size of 0.1 as optimal for detailed simulations, balancing accuracy and computational effort. The designs were evaluated according to the reaction forces that occur after deformation. Table 1 shows one unit cell of the investigated designs along with their maximal available deformable length Δl and the overall calculated reaction forces F. Furthermore, Fig. 7 shows the simulation setup and the deformed state of Design 1. Due to the geometric asymmetry in a unit cell, an unwanted asymmetric overall deformation occurs.

From the evaluation of the simulation results, several clear **design guidelines for the reinforcement structure** can be derived. The most critical factor influencing the resulting reaction forces F is the available deformable length Δl of the intermediate structure, which should be maximized to reduce the required force. Additionally, the intermediate structures should start and end on a horizontal axis to avoid directional dependencies. The restoring force generated by deformation should act in parallel to the direction of elongation to prevent unwanted torsional forces. Symmetry in the intermediate structure is also essential to minimize directional sensitivity. Lastly, the number of sharp corners in the intermediate structures should be kept to a minimum, as they lead to increased stress concentrations and higher required forces.

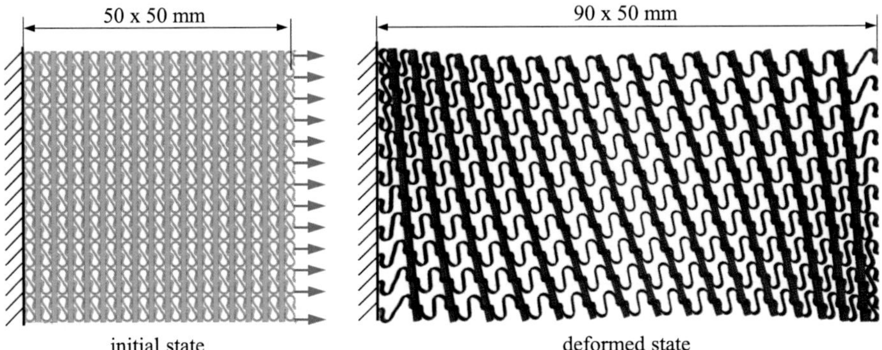

Fig. 7 Simulation setup of the metamaterial with its initial configuration (left) and the deformed state (right) showing an unsymmetrical overall deformation

Experimental evaluation of reinforcement structure

A modified and combined version of Design 2 and Design 4 was used to fabricate a reinforcement for one soft robot cylinder. The dimensions were adapted to match the actuator geometry, reducing the original $50 \times 50\,\text{mm}^2$ to a $50 \times 15\pi$ mm rectangle. This resulted in a denser arrangement of elementary cells and a reduced total cell count, potentially increasing the axial stiffness due to the decreased elastic displacement capacity.

The main challenge was building cylindrical prototypes. Two design approaches were tested: (1) a flat structure with self-locking hooks for wrapping around the actuator, and (2) a modular design with separate rings and a flexible mesh, assembled afterward. Due to material constraints, the second approach was tested on the soft actuator. The modular design, as shown in Fig. 8 allows the rings to be made from a stiffer material, improving radial stiffness while keeping the reinforcement structure flexible in the axial direction. Figure 8 illustrates the CAD models of the modular components along with the assembled actuators, featuring reinforcement and rings made of Filaflex 82A (top) and white rings made of Polyflex 95-HF (bottom).

The experiments were carried out with a stereo camera set-up to study expansion behavior while the soft actuator is pressurized in 10 kPa steps up to 100 kPa and the deformation was automatically analyzed. Two soft actuator configurations were tested: with rings but without reinforcement and with rings and reinforcement, cf. Fig. 9. Without the reinforcement, significant radial expansion was observed, reaching up to 1.45 times the original size at 60 kPa.

With the reinforcement, radial expansion was greatly reduced. The axial expansion improved slightly in both reinforcement configurations, by about 6.1 % with rings and reinforcement structure made from Filaflex and 5.9 % with reinforcement structure made from Filalfex and rings made from Polyflex. Despite the added reinforcement, the mounting rings exhibited structural weakness due to the grooves required to integrate the reinforcement structure. Greater axial expansion was observed in the

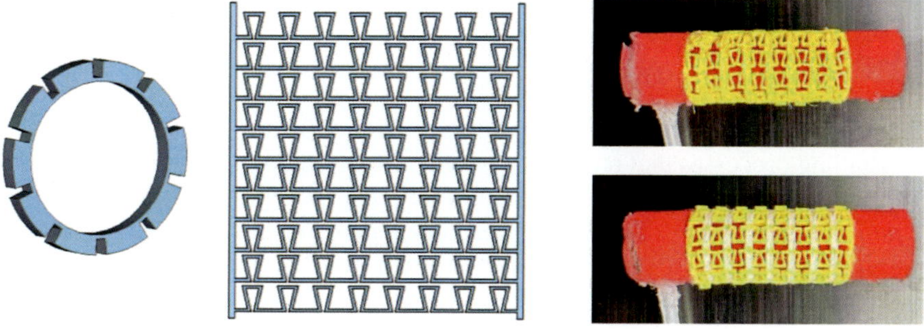

Fig. 8 CAD model of ring and final reinforcement structure (left) and assembly with reinforcement and structure made from Filafrlex 82A (top right) and rings made of Polyflex 95-HF (bottom right)

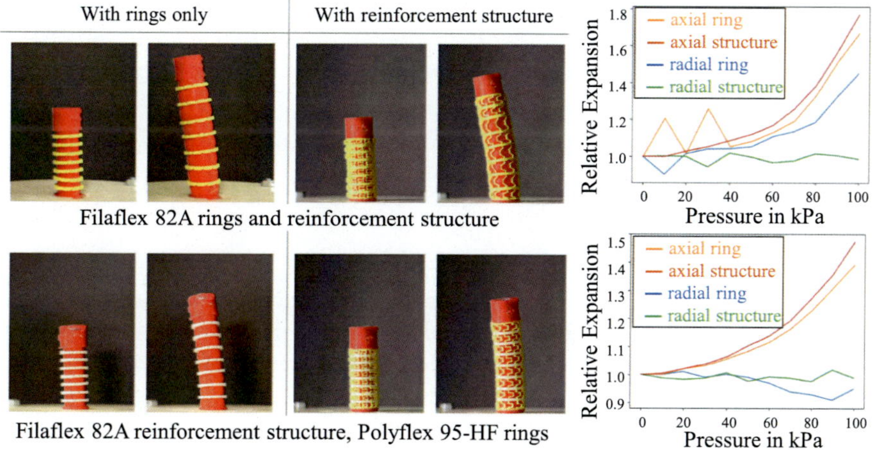

Fig. 9 Initial state at 0 kPa and deformed state at 100 kPa for both configurations (left) and results for radial and axial expansion without and with reinforcement structure (right). The slight bending of the pressurized actuator results from inaccuracies in manufacturing

configuration with lower radial stiffness, likely because the softer rings deformed more easily, allowing the actuator to inflate more uniformly.

Concluding, the simulation-based design of the reinforcement structures was successful and met the intended performance targets. This was further confirmed through experimental testing, demonstrating the reliability and effectiveness of the developed approach.

3 Automated Simulation Framework for Exploring Design Modifications

The development of soft robots is often hindered by application-specific, time-consuming processes that rely heavily on repeated simulations and physical testing. Accurate modeling of the non-linear deformation behavior of soft materials adds additional complexity. To address these challenges, in this section, we show two simulation pipelines to automatically analyze design aspects of different soft robot systems.

The first system (Sect. 3.1) focuses on a soft bending actuator, for which automated dataset generation using parametric CAD modeling and finite element analysis (FEA) enables systematic variation of design parameters and boundary conditions. This approach reduces the need for extensive experimental testing and accelerates the development process.

The second simulation framework focuses on optimizing a single soft chamber within the SOFA simulation environment for use as an artificial muscle (Sect. 3.2).

3.1 Parametric Finite Element Modeling for Design Evaluation

The developed simulation framework enables numerical analysis of the mechanical behavior of a soft bending actuator under varying design parameters. To ensure system consistency and seamless integration, a main script shall control the entire process, starting from the creation of parametric CAD models, through geometry meshing, to the final finite element analysis. To ensure a flexible, efficient, and scalable simulation framework, **a proper selection of software tools** is essential. Although finite element analysis was performed using the commercial software Abaqus, we preferred (i) open-source solutions that offer full compatibility with Python, as the main control script is written in Python. To guarantee smooth data exchange between pipeline stages, such as CAD modeling, meshing, and FE simulation, we selected tools that (ii) support standard file formats like STEP or STL. Software with (iii) comprehensive and well-structured documentation was preferred for simple on-boarding and accelerating troubleshooting. To speed up testing and reduce setup time, we focus on (iv) tools with user-friendly interfaces or intuitive command-line operation. Finally, (v) modular and extensible software is preferred to meet specific project requirements and to ensure long-term adaptability of the framework.

For the **CAD modeling step**, we chose CADQuery, a powerful Python-based library for parametric 3D modeling. CADQuery enables the automated generation of CAD models based on external design parameters, which we stored in a structured JSON file. This setup allows for quick and reproducible design changes without manual intervention, supporting a seamless integration into the simulation pipeline.

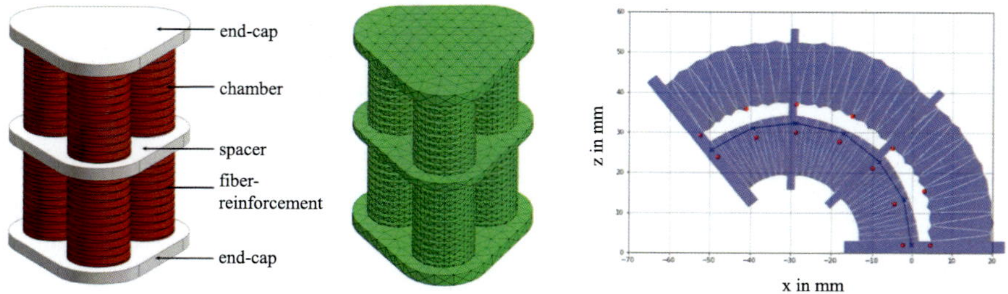

Fig. 10 The soft bending actuators with fiber reinforcement (right), the automatically generated mesh in GMSH (middle), and the simulation results in Abaqus when one chamber is pressurized with 60 kPa

CADQuery also supports exporting models in the STEP format, ensuring compatibility with downstream tools like GMSH for meshing.

The modeling process of a soft bending actuator with fiber reinforcement (Fig. 10) begins by loading geometric parameters such as outer diameter, wall thickness, fiber pitch, and chamber spacing from the JSON file. These parameters are used to generate individual parts of the soft robot, which are three fiber-reinforced chambers, two end-caps, and one connector plate. This parametric modeling approach ensures that any modifications in the design parameters are immediately reflected in the generated geometry, significantly simplifying and accelerating the overall simulation workflow. Finally, the model is exported as a STEP file.

To ensure a **flexible meshing process** within the simulation framework, GMSH was selected, since it supports the import of widely used geometry formats, such as STEP and IGES, and enables the direct export of mesh data in the Abaqus input file format (.inp). This compatibility ensures efficient integration with the Abaqus simulation environment and reduces the need for manual conversion or preprocessing steps.

Each CAD part is imported into GMSH in STEP format and meshed individually using Python scripts. A Delaunay algorithm generates high-quality tetrahedral elements, with mesh sizes adapted to the geometry—coarser for end-caps and finer for chambers (Fig. 10 middle). Since GMSH does not provide all Abaqus-specific keywords (e.g., *Part, *End Part, *Surface), a custom Python script modifies the exported .inp file by inserting missing definitions, converting C3D10 elements into C3D10H for hyperelastic materials, and transforming quadratic line elements into B32 beam elements for fiber reinforcements. A high-order optimization step further improves mesh quality. This workflow enables the seamless integration of all part meshes into a unified Abaqus model with minimal manual effort.

Abaqus was selected as the **simulation software** due to its advanced finite element capabilities and the team's prior experience, which allowed for efficient and confident setup. To ensure long-term flexibility, the simulation framework also supports future integration of open-source alternatives (see Sect. 3.2).

The simulation process begins by importing the .inp file generated in GMSH. Since all parts were already aligned during the CAD design, no additional positioning is needed. Material properties are then assigned to each component. Common materials used include Dragon Skin 10, Ecoflex, ABS, and Kevlar [18]. These are defined through sections—solid sections for structural parts and beam sections for fibers. Each section includes relevant material data such as Poisson's ratio and cross-sectional geometry. Nonlinear geometry (nlgeom=ON) is enabled to account for large deformations.

Contact interactions are defined using surfaces created in GMSH. Surface-to-surface tie constraints connect chambers to end-caps and connectors, ensuring they move as a single rigid body. Stiffer components with coarser meshes serve as master surfaces, while more flexible parts with finer meshes act as slave surfaces. Node-to-surface contact is also used to model the interaction between fibers and chambers.

Finally, boundary conditions and loads are applied. The base of the end-cap is fixed in all directions, and pressure is applied incrementally to selected chamber surfaces based on their x- and y-coordinates. After the simulation, nodal coordinates along the center of the actuator are extracted to reconstruct the backbone curve. This is done by computing cross-section centroids and fitting a curve through them. The bending angle is then calculated from this curve. This process is repeated as different design parameters—such as wall thickness, inner diameter, chamber spacing, number of connector plates, and fiber pitch—are systematically varied.

The developed simulation framework now enables fully automated execution of complex analyses, significantly streamlining the evaluation of specific design aspects in soft robotics. This automation not only accelerates the simulation process but also facilitates the generation of extensive datasets, which are essential for training and improving artificial intelligence models. By reducing manual intervention and minimizing potential errors, the framework supports efficient exploration of a wide range of design parameters, making it a powerful tool for both detailed design optimization and data-driven AI applications in soft robot development.

3.2 Systematic Evaluation of Pneumatic Artificial Muscle for Elbow Flexion

While the previous simulation framework provided a robust foundation for analyzing pneumatic actuators and generating large datasets, it was computationally intensive and often required long computational times. To address this limitation, the focus shifted towards leveraging the SOFA framework [19], which is specifically designed for finite element analysis of soft materials and real-time simulation of deformable bodies. SOFA's optimized solvers and modular architecture offer significant improvements in efficiency, making it well-suited for simulating the complex behavior of soft pneumatic actuators with greater speed and flexibility. To demonstrate the capabilities of the adapted simulation framework, parameter studies are conducted on a

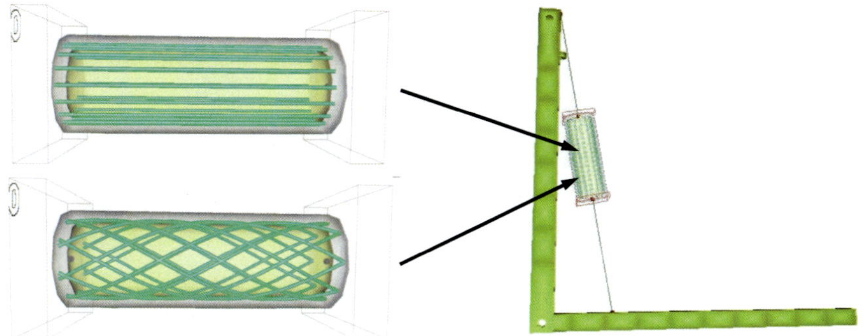

Fig. 11 Detailed view of the pneumatic artificial muscle with parallel or spiral fiber reinforcement (left), which are used as biceps at a simplified arm setup (right)

pneumatic artificial muscles (PAM), which functions as a biceps to lift the lower arm, as displayed in Fig. 11.

In contrast to the previous framework, which relied on CADQuery for part modeling, the current approach uses FreeCAD to design the components of the pneumatic actuator and robotic arm. After the parametric model generation in FreeCAD, the parts are imported into GMSH, and discretized using tetrahedral elements. Finally, the meshed parts are exported in VTK format and loaded into the SOFA framework, enabling detailed finite element analysis and dynamic simulation.

The simulation setup in SOFA is designed to model the behavior of a soft pneumatic actuator functioning as a biceps. To simulate the actuation process, a cavity component is integrated into the model and linked with a *SurfacePressureConstraint*, which dynamically applies internal pressure to the cavity surface. A Python controller regulates the pressure values over time, allowing realistic simulation of the actuator's contraction and expansion cycles. Reinforcements are added in the form of either parallel or spiral spring systems, which mimic the fiber structures. The fibers are defined by custom Python functions and implemented through *StiffSpringForceField* components, with adjustable stiffness and damping parameters. The reinforcements are mapped onto the deformable shell of the cavity component using *BarycentricMapping* to ensure synchronized deformation. The actuator is anchored at one end and connected via cable constraints to a simplified arm structure, allowing it to contract and generate movement of the lower arm. With this framework, extensive simulation studies are conducted to investigate the behavior of PAMs under various internal pressures and reinforcement strategies. The focus lies on evaluating two main reinforcement types, spiral and parallel fibers, by analyzing their effects on contraction behavior, structural integrity, and deformation characteristics.

Spiral reinforcements demonstrate a distinct nonlinear response that can be divided into three phases: low initial contraction due to high stiffness, followed by a near-linear increase as the spiral fibers reorient, and eventually reaching a stagnation point where further contraction is limited. The simulation also reveals lower braid-

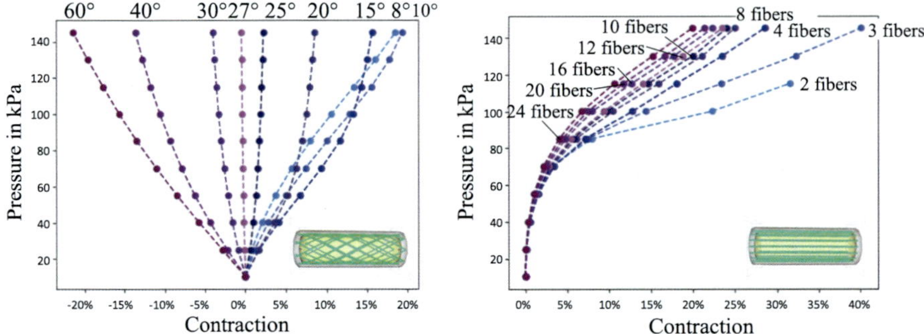

Fig. 12 Results of parameter studies for spiral fiber reinforcement with 8 pairs of fibers and different pitch angles (left), and with a parallel fiber reinforcement with different numbers of fibers (right)

ing angles (e.g., 10°, ref. Fig. 12 (left)) result in greater contraction up to 19–20%, whereas increasing the number of spiral pairs offers only marginal gains due to rising structural stiffness. In contrast, PAMs with parallel reinforcement show high initial resistance to contraction but reach up to 40 % contraction at high pressures when fewer fibers are used. However, this comes at the cost of structural stability, with lower fiber counts leading to rupture at elevated pressures, see Fig. 12 (right).

The framework thus facilitates a comprehensive comparison of the trade-offs between contraction efficiency and stability. Spiral reinforcements are better suited for applications that require robustness and pressure tolerance, while parallel reinforcements provide higher contraction but are more sensitive to failure. These findings guide the design and optimization of soft actuators for specific use cases, balancing flexibility, efficiency, and durability.

4 Application-Driven Prototyping of Soft Robots

In addition to the simulation-based analysis, this section presents the development and evaluation of prototypes for three very different applications, (i) the modification of the soft bending actuator such that it can run autonomously by carrying a pump, valves and a controller, (ii) the practical application of pneumatic artificial muscles to lift the arm of a skeleton, and (iii) finding a new soft actuator design to realize peristaltic motions of an artificial urea, which can be used in a phantom of a human body.

The goal is to investigate the basic behavior of the actuator concepts in a real-world setting and assess their feasibility for soft robotic applications. Although based on similar design principles, this section stands independently from the previous simulation framework. Instead of developing a comprehensive simulation environment, specific simulations were carried out to support the design and fabrication of the prototypes. These targeted simulations aimed to evaluate the basic functionality of each configuration and to identify potential design issues before production. The resulting prototypes enable hands-on exploration of design strategies, material behavior, and actuation performance under controlled conditions.

4.1 Structural Adaptation of a Soft Robot Bending Actuator for Autonomous Operation

Following the design optimization presented in Sect. 2.1, the objective now was to adapt the soft bending actuator for autonomous operation without modifying its underlying working principle. The soft bending actuator is retained in its functional concept, while its dimensions are scaled to allow the integration of all required system components, this includes an air pump, valves, and control units, which must be embedded within the module itself to enable standalone operation. The key question addressed here is whether the actuator can be adapted in size and layout to support complete system integration while maintaining its mechanical performance and reliability.

Based on the validated functionality of the original soft robotic design, the geometric structure of the actuator was maintained while its size was adapted to allow integration of commercially available components. Since these components are not available at the original scale, the **actuator was scaled up** using our developed virtual work model described in Lamping et al. [20]. This model allows error-free scaling as long as the ratio between the material volume and the internal chamber volume remains constant, which was maintained throughout this application. According to the required bending capability, the actuator length was increased to 200 mm. With an outer diameter of 60 mm of the final chamber geometry a direct scale-up of the original module geometry is confirmed and the operating pressure of approximately 100 kPa does not change. This provided the basis for selecting an appropriate pump capable of delivering the required pressure within the new dimensions.

The **construction of the soft robot** consists of three main components: two newly designed functional modules for pneumatic control and pressure generation, and the actual soft actuator, see Fig. 13. The pneumatic module is responsible for controlling and monitoring airflow to the soft actuator chambers, while the pump module manages pressure generation. Both are classified as active modules, as they contain powered components and require electrical input. Each module has a height of 73 mm and is fabricated using 3D printing with materials such as ABS or PLA.

The **pneumatic module** houses all components required for precise airflow control of the soft actuator. Each of the three chambers is individually actuated using a combination of one electrically controlled 3/2-way valve (Festo MHJ10) and one 2/2-way valve (Festo VOVG-LK), together replicating the function of a 3/3-way valve, as compact 3/3-way valves in the required size are not available. This setup enables each chamber to be filled, vented, or held in place as needed. In the blocking state, all ports are closed, preventing any unintentional pressure loss.

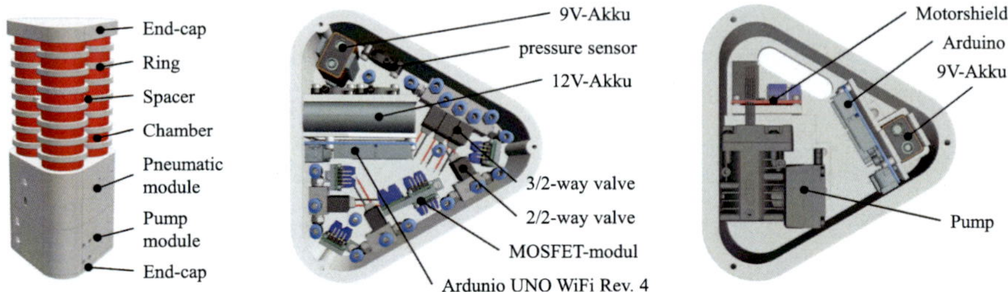

Fig. 13 Final design of the upscaled soft bending actuator (left), and detailed view of the pneumatic modules (middle) and the pump modules (right)

To control these valves, N-channel MOSFETs are used as switches. These are triggered by the Arduino and allow control of the 12 V valves using 5 V logic levels. Prefabricated MOSFET modules, specifically designed for microcontroller integration, are used instead of custom PCBs to simplify the assembly.

For closed-loop control, the module integrates three absolute pressure sensors (MPX4250AP), each monitoring one chamber. These sensors operate at 5 V and are connected to the analog inputs of the Arduino. Due to their 6 mm pneumatic connection, they require different tubing from the 4 mm used for the valves. An Arduino UNO WiFi Rev2 manages the sensors and valves. It is powered by a 9 V block battery (6LR61), while the valves draw power provided by a dedicated 12 V battery shared with the pump module. All USB and power ports of the Arduino remain accessible from the outside, facilitating programming.

The **pump module** integrates a diaphragm pump (KNF NMP03) for pressure generation, along with an L298N motor driver, an Arduino UNO Rev3, a 12 V battery, and a 9 V battery. The pump is driven via the L298N motor driver, which enables pulse width modulation control of motor speed and direction. Although the driver causes a voltage drop of approximately 2 V (reducing the pump input to around 10 V) this has negligible impact due to the pump's oversized specification. The Arduino Rev3 is powered separately by the 9 V battery to isolate logic circuitry from power electronics.

Two Arduinos, one in each module, coordinate operation. While only the Arduino UNO WiFi Rev2 has wireless connectivity, the boards communicate via I^2C using SDA and SCL pins and a shared ground. Electrical and pneumatic connections between modules run through dedicated cutouts in the 3D-printed enclosure and wiring guides, ensuring clean integration and modular scalability.

The actual **soft actuator module** consists of three inflatable silicone chambers, each with an outer radius of 30 mm, an inner radius of 16 mm, and a wall thickness of 14 mm. Each chamber is 200 mm long and is stabilized at the midpoint (100 mm) by a 3D-printed connector plate. Additional structural elements, such as rings and end-caps are also 3D-printed and serve to support the actuator's geometry while enabling secure integration into the modular system. These elements are designed for mechanical reliability and simple assembly.

Table 2 Overview of components and costs for the pump module

No.	Component description	Quantity	Unit cost (€)	Total cost (€)
1	Pump NMP850KPDC-BI4 12V	1	303.45	303.45
2	Motor shield L298N	1	2.60	2.60
3	12V Battery	1	16.16	16.16
4	9V Battery	1	5.65	5.65
5	ARDUINO UNO Rev3	1	20.40	20.40
Total				348.26

Table 3 Overview of components and costs for the pneumatic module

No.	Component description	Quantity	Unit cost (€)	Total cost (€)
1	3/2-Way valve N362.2	3	18.12	54.36
2	3/2-Way valve N372.2	3	18.12	54.36
3	Base plate for 3/2-way valve 395.03	1	24.77	24.77
4	Base plate for 2/2-way Valve 396.01	3	24.77	74.31
5	ARDUINO UNO WiFi REV2	1	46.70	46.70
6	Pneumatic tubing	1	0.50	0.50
7	Pressure sensor MPX4250AP	3	18.95	56.85
8	MOSFET module	6	1.00	6.00
9	12 V Battery	1	16.16	16.16
10	9 V Battery	1	5.65	5.65
Total				339.66

This construction ensures both functionality and scalability, and allows multiple modules to be aligned in series. The system architecture thus offers an efficient and flexible solution. The estimated cost of all purchased components for one bending actuator (with pneumatic module, pump module and soft actuator module) amounts to approximately 690€, based on market prices at the time of research. This excludes standard fasteners and in-house manufactured parts such as structural bodies, mounting brackets, and molding tools for the chambers. A detailed breakdown of all required components and their corresponding costs is provided in Table 2 (pump module) and Table 3 (pneumatic module).

The described modules and components represent the planned design and cost estimation; however, physical manufacturing and assembly have not yet been carried out. The design is based on a previously investigated and developed soft bending actuator, which demonstrated scalable characteristics, allowing the current concept to be extrapolated from that foundation. Future work will focus on fabrication, integration, and testing to validate the system's functionality and performance.

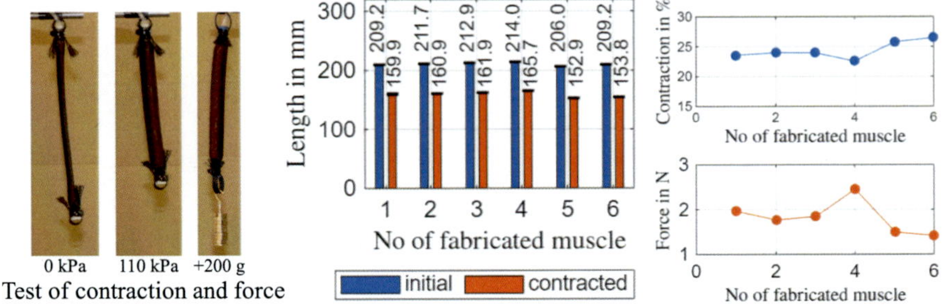

Fig. 14 Experimental characterization of fabricated PAMs shown in their initial state, pressurized at 110 kPa, and additionally loaded (left); measured actuator lengths in initial and pressurized states (middle); and corresponding percentage contraction alongside measured contraction forces (right)

4.2 Use of Pneumatic Artificial Muscles for an Arm Flexion Movement

As an example of soft robotic application, we used soft pneumatic artificial muscles (PAMs) that shorten in length to function as biceps for lifting an arm, similar to the numerical investigation in Sect. 3.2. The fabricated PAMs consist of a molded silicone cylinder with a length of 200 mm ($\approx$ 210 mm when assembled), an inner diameter of 6 mm, and a wall thickness of 1 mm. Unlike the fiber reinforcement of the simulated design, a braided sheath surrounds the cylinder and enables the characteristic ballooning and axial contraction during pressurization. The silicone cylinder is sealed at both ends with 3D-printed end-caps, which also secure the braided sheath.

First, six fabricated **PAMs were tested individually** to measure their contraction and contraction force at an internal pressure of 110 kPa. The contraction force was determined by gradually applying an increasing load to each actuator while it was fixed at one end, and measuring the load at which the actuator began to elongate, cf. Fig. 14. Despite manual fabrication, all actuators showed consistent contraction values within the expected range of 20–30 %. However, the measured contraction forces varied significantly due to irregular wall thickness, and friction between the silicone cylinder and braided sheath.

Following these tests, the **PAMs were integrated into a custom built test rig** simulating a human arm. The setup consisted of plastic tubes representing the upper and lower arm, connected by a 3D-printed fork joint that enabled planar rotation. The upper arm was fixed in place, while the lower arm was free to move and subjected to external loads. The PAMs were added to evaluate their combined ability to generate sufficient torque for lifting the lower arm through coordinated contraction. Tests were conducted with varying numbers of PAMs, ranging from two to six, which were mounted on 3D-printed brackets and secured with cable ties. The actuators were pressurized to 110 kPa, and loads from 20 to 250 g were incrementally added.

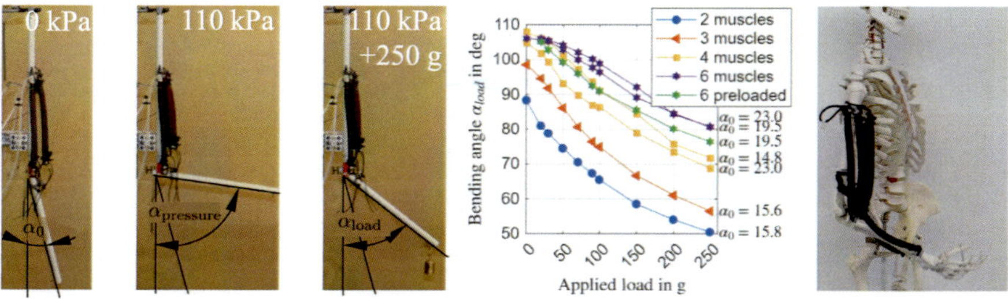

Fig. 15 Experimental setup of the arm test rig with 2 mounted PAMs shown in their initial state, pressurized at 110 kPa, and pressurized under additional load state (left); measured bending angle for varying added weights (middle); and 6 PAMs applied on a skeleton (right)

The flexion angles were measured using the image analysis software ImageJ by comparing the arm's position before and after actuation.

The results showed that increasing the number of actuators led to greater bending and higher load capacity, reaching up to approximately $\alpha_{\text{pressure}} = 106°$ with four PAMs, cf. Fig. 15. Adding two side-mounted actuators (Fig. 15 (right)) had only a minor effect on the bending angle α_{pressure} but did enhance the system's ability to support heavier loads. When the arm was loaded before the actuators were pressurized (preload condition), the resulting bending angle decreased. Furthermore, the initial bending angle α_0 appears to influence the overall bending behavior, as initial deflection promotes a larger absolute bending angle.

This investigation confirms consistent trends across tests. However, while simulations are valuable for guiding the design, discrepancies caused by manual fabrication and mounting tolerances significantly affect actual performance, highlighting the need for additional experimental validation.

4.3 Generating Peristaltic Ureter Movements with Soft Actuators in an Endourology Phantom Model

As a final application example, a soft robotic ring muscle was developed for integration into body-like phantoms, specifically the EndoUroPhantom [21], a urological organ model used for endoscopic surgery training. The aim is to replicate the peristaltic contraction waves of the human ureter to realistically simulate urine transport from the kidneys to the bladder.

To replicate realistic ureteral motion, the actuator must meet specific parameters: a wavelength of 20–30 mm, up to four contractions per minute (every 20–25 s), and a wave propagation speed of 20–60 mm/s. The design incorporates a central silicone tube with a lumen diameter of 5 mm, while limiting the actuator's outer diameter to 40 mm to maintain compactness and functionality.

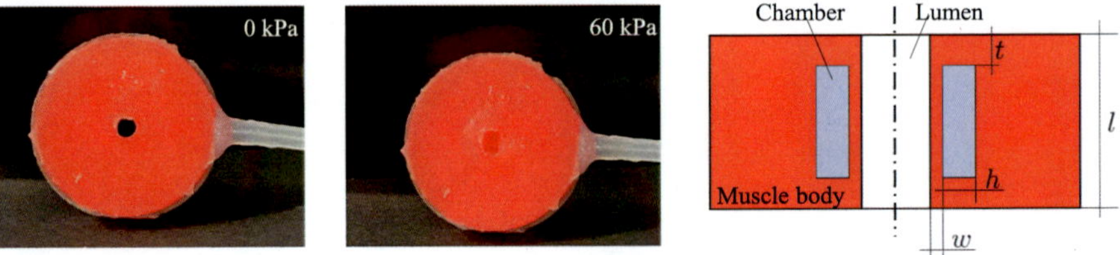

Fig. 16 Fabricated soft ring muscle from Dragon Skin 10 in its initial state (left), and in pressurized state (middle), and cross sectional view with numerically analyzed geometrical parameters (right)

A simulation-driven design approach using finite element analysis in Abaqus was employed to optimize geometric and material parameters such as chamber size and wall thickness. Multiple CAD models were generated and simulated under varying conditions to identify the geometry that maximized cross-sectional compression while ensuring manufacturability and functional reliability.

Hyperelastic materials, including Dragon Skin 10 and Eccoflex 00-30, were selected for their large deformation capabilities and modeled using the third-order Ogden model. Simulations applied static pressure loads ranging from 10 to 70 kPa to evaluate deformation behavior.

The final design was experimentally tested to validate the accuracy of the simulations and confirm the reliability of the modeling approach. Furthermore, a custom control system was implemented to allow adjustment of wavelength, frequency, wave speed, and contraction intensity, enabling analyses of the ureter's peristaltic motion.

Design study of geometrical parameters
The soft ring muscle is a circular actuator with a 40 mm outer diameter and a 5 mm central lumen for the silicone ureter tube. It features a pneumatic chamber encircling the lumen, with dimensions optimized through numerical analysis (see Fig. 16).

In the first simulation series, the **length of the muscle** l was varied (10 mm, 15 mm, 20 mm) to examine its influence on lumen compression. Shorter muscles required less force to deform and produced stronger contractions at lower pressures. Although longer muscles also showed significant deformation in simulations, shorter segments offer greater flexibility in shaping the contraction wave. Therefore, a length of 10 mm was chosen to enable modular and adjustable peristaltic motion.

Next, the **chamber's cross-sectional shape** was evaluated. Even though trapezoidal chambers offer asymmetric compression that may aid fluid transport, they caused excessive sidewall expansion and less effective lumen closure. Rectangular chambers provided more focused compression and were therefore chosen.

The **air chamber height** h was varied between 1 and 4 mm. Results indicated that higher chambers produced more pronounced deformations due to increased stretchability. Based on this, a 3 mm chamber height was selected to achieve a balance between mechanical integrity and sufficient deformation.

The **thickness of the inner wall** w between the air chamber and the central lumen was further investigated. Simulations with 1 mm and 2 mm wall thicknesses revealed

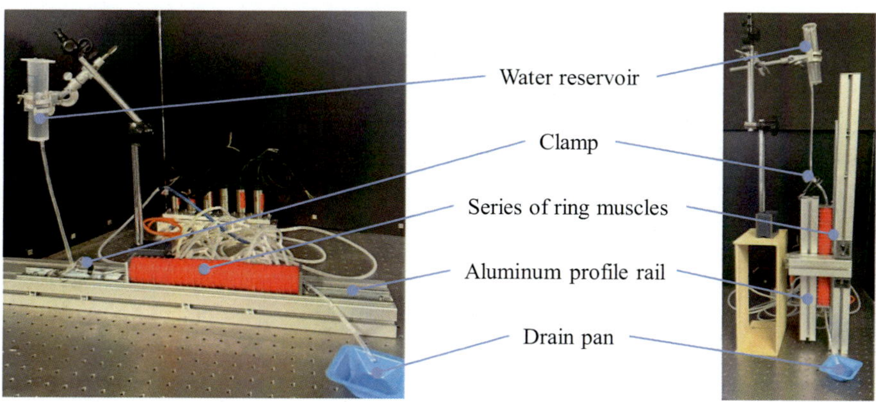

Fig. 17 Experimental setup with horizontal orientation (left), and vertical orientation (right) of the ring muscles

that thinner walls permitted greater compression due to lower material resistance and, hence, a 1 mm inner wall was chosen for optimal performance.

Additionally, the **outer wall thickness** t of the muscle was minimized to 1 mm to allow the largest possible contraction length while maintaining structural coherence during manufacturing.

Material studies initially suggested that Eccoflex 00-30, due to its lower stiffness, would allow greater deformation. However, experimental tests revealed that muscles made from Eccoflex were too soft, resulting in poor performance and insufficient lumen compression. As a result, Dragon Skin 10 was selected as the preferred material for the muscle body, offering better structural integrity and more effective actuation.

Overall, the systematic simulation-based approach allowed the identification of key design parameters, setting the foundation for manufacturing and testing a soft ring muscle capable of mimicking realistic ureteral peristalsis.

Experimental evaluation of the peristaltic motion of the soft ring muscles
The experimental setup consists of 25 fabricated ring muscles arranged in series and fixed on a profile rail. Each actuator is designed to circumferentially compress a soft silicone tube with an outer diameter of 5 mm and a wall thickness of 0.5 mm.

For actuation, a modular pneumatic terminal from Festo (Motion Terminal VTEM) is used, equipped with six electronically controlled valves and integrated pressure regulators. Due to the limited number of valves, the 25 soft muscles are grouped such that every fifth actuator is controlled simultaneously through shared silicone tubing, with each valve operating four to five actuators. The applied pressure is limited to 60 kPa to prevent mechanical damage. A CompactRIO (National Instruments) interface links the terminal to a custom LabVIEW-based control program on a PC, enabling adjustment of parameters such as contraction frequency (1–6 cycles/min), contraction speed (10–50 mm/s), and the number of simultaneously activated actuator groups (1–5).

The experimental procedure includes two test scenarios: (1) a horizontal setup of the soft muscles and (2) a vertical configuration to assess the effect of gravity (see Fig. 17). In both cases, the tube was pre-filled with water and initially clamped upstream to evaluate whether peristaltic activation of the ring muscles alone could drive active fluid transport. The peristaltic sequence was then triggered, and any displaced water was collected at the outlet for the tube. A precision scale measures the output, and each test is repeated five times to determine the mean transported volume and standard deviation for each set of parameters.

The study showed that increasing the contraction frequency raises the fluid volume transported, but the increase is limited by a vacuum forming inside the sealed tube, which reduces efficiency at lower frequencies. An optimal frequency of 6 Hz was identified. Increasing contraction speed improved flow up to 30 mm/s; beyond this, efficiency dropped because the actuators didn't have enough time to fully return to their original shape (actuator recovery), reducing their ability to compress the tube effectively. Longer contraction lengths, meaning more actuators are activated simultaneously, also improved flow, with five actuators performing best. Tests with the vertical setup were inconclusive due to air bubbles affecting the measurements. However, we could show that with the developed ring muscles a fluid transport is possible.

Building on the preceding numerical design analysis, the initial experimental tests were successful already at the first trial, confirming that the simulation-driven approach offered a robust foundation for achieving effective actuator performance and realistic peristaltic motion.

5 Conclusion

This work presents a structured and application-driven methodology for the design and prototyping of soft robotic systems, combining automated simulation frameworks with targeted experimental validation. By integrating finite element simulations, parametric modeling, and physical prototyping, we demonstrated efficient and systematic approaches to address the inherent challenges posed by the nonlinear behavior and high compliance of soft materials.

Key contributions include the development of simulation pipelines for two distinct soft robotic systems: a pneumatically actuated bending actuator and a pneumatically driven artificial muscle. These pipelines enable automated dataset generation, parameter variation, and performance evaluation, significantly reducing the need for manual iteration and physical testing. Furthermore, we demonstrated the real-world applicability of these methods through three application-driven prototypes—each addressing a different functional requirement, from autonomous actuation to skeletal movement and peristaltic transport.

The findings illustrate that combining simulation-driven design with focused prototyping enables a deeper understanding of soft actuator behavior and accelerates innovation in the field. This work serves as an example of how soft robots can be systematically invented by linking simulation tools with application-oriented development strategies. The proposed methodology provides a versatile foundation for future research and development of task-specific, functional soft robotic systems.

Acknowledgements This work was funded by the Deutsche Forschungsgemeinschaft (DFG, German Research Foundation)—Project 404986830—SPP2100 and Project 501861263—SPP2353.

Competing Interests The authors have no conflicts of interest to declare that are relevant to the content of this chapter.

References

1. G.K. Klute, J.M. Czerniecki, B. Hannaford.: McKibben artificial muscles: pneumatic actuators with biomechanical intelligence. In: 1999 IEEE/ASME International Conference on Advanced Intelligent Mechatronics (Cat. No.99TH8399), pp. 221–226 (1999). https://doi.org/10.1109/AIM.1999.803170
2. Antonelli, G., et al.: Mechanical design of McKibben muscles predicting developed force by artificial neural networks. Actuators **14**, 153 (2025). https://doi.org/10.3390/act14030153
3. Liu, T., Wang, X.: Modeling and analysis of oblique-chamber and symmetric oblique-chamber Pneu-Net soft actuators. IEEE Robot. Autom. Lett. **910**, 8682–8689 (2024). https://doi.org/10.1109/LRA.2024.3451392
4. Awada, Z., Haddab, Y., Gouttefarde, M.: PneuNet actuators design: trade-offs between deformation, force, and resistance to buckling. Sens. Actuators A: Phys. **386**, 116307 (2025). https://doi.org/10.1016/j.sna.2025.116307
5. Nguyen, T.D. Burgner-Kahrs, J.: A tendon-driven continuum robot with extensible sections. In: 2015 IEEE/RSJ International Conference on Intelligent Robots and Systems (IROS). pp. 2130–2135 (2015). https://doi.org/10.1109/IROS.2015.7353661
6. Deutschmann, B., Reinecke, J. Dietrich, A.: Open source tendon-driven continuum mechanism: a platform for research in soft robotics. In: 2022 IEEE 5th International Conference on Soft Robotics (RoboSoft), pp. 54–61 (2022). https://doi.org/10.1109/RoboSoft54090.2022.9762144
7. Lee, C. et al.: Soft robot review. Int. J. Control. Autom. Syst. **15**(1), 3–15 (2017). https://doi.org/10.1007/s12555-016-0462-3
8. Dou, W., et al.: Soft robotic manipulators: designs, actuation, stiffness tuning, and sensing. Adv. Mater. Technol. **6**(9), 2100018 (2021). https://doi.org/10.1002/admt.202100018
9. Jin, L., et al.: Finite element analysis, machine learning, and digital twins for soft robots: state-of-arts and perspectives. Smart Mater. Struct. **34**(3), 033002 (2025). https://doi.org/10.1088/1361-665X/adadcd
10. Armanini, C., et al.: Soft robots modeling: a structured overview. IEEE Trans. Robot. **39**(3), 1728–1748 (2023). https://doi.org/10.1109/TRO.2022.3231360
11. Connolly, F., et al.: Mechanical programming of soft actuators by varying fiber angle. Soft Robot. **2**(1), 26–32 (2015). https://doi.org/10.1089/soro.2015.0001
12. El-Agroudy, M.N., Awad, M.I., Maged, S.A.: Soft finger modelling and co-simulation control towards assistive exoskeleton hand glove. Micromachines **12**(2), 181 (2021)
13. Soft Robotics Toolkit. https://softroboticstoolkit.com/book/export/html/453056. Last visit June 2025
14. Suzumori, K., et al.: Fiberless flexible microactuator designed by finite-element method. IEEE/ASME Trans. Mechatron. **2**(4), 281–286 (1997). https://doi.org/10.1109/3516.653052

15. Elsayed, Y., et al.: Finite element analysis and design optimization of a pneumatically actuating silicone module for robotic surgery applications. Soft Robot. **1**(4), 255–262 (2014). https://doi.org/10.1089/soro.2014.0016
16. Polygerinos, P., et al.: Modeling of soft fiber-reinforced bending actuators. IEEE Trans. Robot. **31**(3), 778–789 (2015). https://doi.org/10.1109/TRO.2015.2428504
17. Lamping, F., Müller, D. de Payrebrune, K.M.: A systematically derived design for a modular pneumatic soft bending actuator. In: 2022 IEEE 5th International Conference on Soft Robotics (RoboSoft). IEEE, Edinburgh, United Kingdom, pp. 41–47 (2022). https://doi.org/10.1109/RoboSoft54090.2022.9762087
18. Pagoli, A., et al.: Review of soft fluidic actuators: Classification and materials modeling analysis. Smart Mater. Struct. **31**(1), 013001 (2021)
19. Faure, F., et al.: SOFA: a multi-model framework for interactive physical simulation. In: Payan Y. (ed.) Soft Tissue Biomechanical Modeling for Computer Assisted Surgery. Springer, Berlin, Heidelberg, pp. 283–321 (2012). https://doi.org/10.1007/8415_2012_125
20. Lamping, F., De Payrebrune, K.M.: A virtual work model for the design and parameter identification of cylindrical pressure-driven soft actuators. J. Mech. Robot. **14**(3), 031004 (2022). https://doi.org/10.1115/1.4052849
21. Tonyali, S. et al.: Simulation and quantitative evaluation of three surgical techniques of endoscopic enucleation of prostate on a realistic phantom model. World J. Urol. **43**(1), 39 (2024). https://doi.org/10.1007/s00345-024-05404-4

Switchable Strain-Limiting Structures in Soft Actuators: Principles, Mechanisms, and Applications

Jan Peters, Cora Maria Sourkounis, Johann Licher, and Annika Raatz

Abstract Switchable strain-limiting structures (SSLS) offer a powerful means of enhancing the functional versatility of soft robotic actuators by enabling programmable deformation and tunable stiffness. This chapter presents and compares three distinct SSLS implementations based on form-locking, force-locking, and material-locking principles. Each mechanism is introduced through its physical concept and technical realization, and illustrated by a representative use case: MRI-guided percutaneous tumor biopsy and ablation, minimal-invasive surgical manipulators, and soft actuators for deep-sea applications. A structured comparison highlights their respective trade-offs in mechanical performance, response dynamics, control effort, and environmental suitability. The chapter concludes with a discussion of application potentials beyond the presented examples and outlines future directions for hybrid and adaptive SSLS-enabled soft robotic systems.

1 Introduction

Soft robotic systems have emerged as a transformative technology in fields like medical interventions [1], human-machine interaction [2], and deep-sea exploration [3]. By leveraging compliant materials and fluidic actuation, these robots offer inherent safety, adaptability to unstructured environments, and high mechanical compliance—capabilities that rigid robotic systems often lack. However, these advantages come

J. Peters (✉) · C. M. Sourkounis · J. Licher · A. Raatz
Institute of Assembly Technology and Robotics (match), Leibniz University Hannover, Garbsen, Germany
e-mail: peters@match.uni-hannover.de

C. M. Sourkounis
e-mail: sourkounis@match.uni-hannover.de

J. Licher
e-mail: licher@match.uni-hannover.de

A. Raatz
e-mail: raatz@match.uni-hannover.de

A. Raatz et al. (eds.), *Soft Material Robotic Systems*,
https://doi.org/10.1007/978-3-032-22453-8_9

at a cost: precise control over motion and force transmission remain core challenges. These limitations are particularly critical when a robot must switch between compliant and stiff behavior during operation, or when accurate positioning is required under external load.

Strain-limiting structures (SLS) are a key design element in soft robotic systems, enabling the transformation of actuation inputs—such as fluid pressure—into directed, functional motion. By locally constraining material deformation (expansion, contraction or bending), these structures dictate how an actuator behaves, thereby influencing bending direction, elongation range, and force transmission. Traditionally, structures like fibers [4] or textiles [5] are permanently embedded in the material, resulting in a fixed kinematic behavior [6, 7]. For multiple degrees of freedom, typically multiple actuation chambers are needed which leads to high hardware requirements. For example, many soft robots follow a three degree of freedom design by using three or more pneumatic chambers that can independently pressurized to allow the actuator to bend continuously in all directions [8]. This leads to the need of one valve per pressure chamber [9]. While many soft robotic systems are designed to be small, lightweight, and cheap to manufacture, the needed hardware for actuation is often bulky and expensive [10]. Additionally, each actuation chamber adds to the diameter of the robot which contradicts miniaturization efforts [11].

While SLS being traditionally passive and permanently embedded, recent advancements have introduced mechanisms that allow SLS to be selectively activated or deactivated, forming the class of switchable strain-limiting structures (SSLS). SSLS offer a powerful alternative to the paradigm of one valve per pressure chamber: they enable reconfigurable kinematics by selectively activating or deactivating the strain-limiting effect of structures embedded in the soft actuator. This introduces a new design space for soft robotics, allowing one and the same actuator to exhibit a wide range of motions and stiffness profiles [12, 13]. SSLS can be categorized according to the physical principle by which the strain limitation is achieved. In this chapter, we present and compare the three different implementations:

- **Form-locking**: A connection where parts are held together by their interlocking shapes, preventing relative movement
- **Force-locking**: A connection maintained by frictional forces generated through clamping or pressing
- **Material-locking**: A connection formed by the fusion or bonding of materials, creating a continuous material joint.

Each switching principle enables distinct functional capabilities and is suited to different application scenarios. Drawing on three recent actuator designs developed by the authors, this chapter systematically explores the mechanical principles, control strategies, and performance characteristics of form-, force-, and material-locking SSLS. A detailed comparison highlights their trade-offs, and real-world applications in medical and marine domains illustrate their practical relevance.

2 Theoretical Background of Switchable Strain-Limiting Structures

At their core, SSLS serve a similar but more versatile role to bearings in classical rigid-body robotics by guiding the motion in one direction while restricting other movement directions. Through decoupling the control of motion direction and the actuation, previously single direction actuators can become programmable robots with reconfigurable kinematics. The SSLS can be realized through high-stiffness fibers, textiles, or embedded mechanical features that prevent strain in designated directions. Switchability can be achieved by different physical principles. In this chapter, we distinguish between three fundamental categories based on the nature of the force transmission mechanism involved:

Form-locking (closure by form): Form-locking SSLS rely on geometric engagement between structural features—e.g., teeth or latches—that physically block motion when engaged [14, 15]. When disengaged these features decouple, allowing free deformation. The approach is generally robust, fast to switch, and energy-efficient.

Force-locking (closure by friction force): Force-locking structures operate by applying a normal force to generate friction at a contact interface—e.g., clamping tendons or surfaces. When engaged, the interface resists relative motion, effectively shortening or fixing an otherwise compliant element such as tendons [12] or sheets [16]. The locking mechanism can be pneumatically triggered but mechanically realized, providing strong holding forces with minimal continuous energy input required. Another example are electroactive limiting structures that apply opposite electric charge on two parallel pads which are then adhered preventing relative motion [17, 18].

Material-locking (closure by adhesive force): Material-locking SSLS use a material whose mechanical properties change upon external stimulation. Typical implementations involve low melting point alloys (LMPAs) [19, 20], wax or shape memory polymers [21] which are solid under ambient conditions and soften when melted. In their solid form, these materials form rigid internal structures that limit deformation; while when heated, they are compliant.

Each of the three mechanisms provides a difference balance between complexity, response time, energy consumption, and mechanical performance. The next section introduces specific actuator designs exemplifying each principle, serving as the basis for the comparative analysis that follows.

3 Mechanism Description and SSLS Applications

Switchable strain-limiting structures (SSLS) enhance the functional diversity of soft robotic actuators by enabling programmable deformation modes and tunable stiffness. This section presents the design principles, mechanical implementation, and actuation strategies behind three representative SSLS mechanisms, each corresponding to a different physical switching principle: form-locking, force-locking, and material-locking. To illustrate their functional integration and practical relevance, this section briefly summarizes one exemplary application for each principle. While these examples are specific and grounded in experimental implementation, they represent only a subset of the possible domains in which such mechanisms may prove beneficial.

3.1 Form-Locking SSLS: Interlocking Comb Structures

The form-locking mechanism utilizes geometrically interlocking components to create a reversible mechanical constraint. The actuator consists of multiple pneumatic chambers for motion and separate chambers containing comb-like structures embedded in the silicone main body as shown in Fig. 1a). Each comb chamber contains pairs of interlocking teeth that engage when not pressurized and disengage when

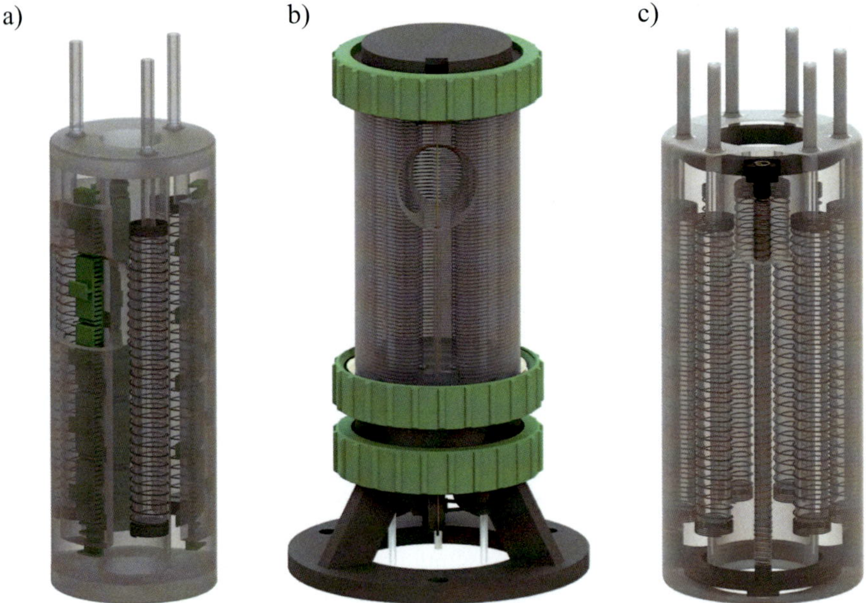

Fig. 1 **a** Form-locking SSLS actuator with three pneumatic chambers for actuation and three chambers containing interlocking comb-like structures [22]. **b** Force-locking SSLS actuator utilizing a clamping mechanism to guide the pneumatic actuation pressure by fixing the length of tendons routed through the actuator [23]. **c** Force-locking SSLS actuator with six actuation chambers and three chambers containing a LMPA for selective motion and stiffness control [24]. For detailed descriptions of the actuator please refer to the respective works

pressurized. By inflating the comb chamber (positive pressure), the combs are radially displaced, breaking the interlock and allowing free deformation. When deflated, the elastic recoil of the silicone brings the combs back into contact, re-establishing the strain-limiting constraint. Selectively locking one or more of the comb chambers allows the actuator to bend directionally over these locked chambers, while unlocking all combs permits axial elongation. In addition, the SSLS can be used to position lock the actuator. When all comb chambers are engaged, the actuation pressure can be released and the actuator stays in its current configuration without the need for further energy input. The system requires only positive pressure and no electrical components, making it well-suited for the sensitive environment of interventional magnetic resonance imaging (iMRI).

In this context, the form-locking actuator was developed for a soft robotic manipulator designed to assist with percutaneous tumor biopsy and ablation procedures inside the MRI bore [22]. Conventional rigid robots are unsuitable in such settings due to electromagnetic interference, spatial constraints, and safety concerns. The use of pneumatically activated interlocking combs allows the actuator to be configured and locked in multiple positions without compromising MRI compatibility and maintaining a low hardware footprint.

Key features of the presented form-locking SSLS include:

- Position locking (failsafe locked state)
- Uses only positive pressure for both actuation and SSLS activation
- Fast response (>1 s for locking and unlocking)
- MRI-compatible design without metal components and electronics.

3.2 *Force-Locking SSLS: Tendon Locking Mechanism*

Force-locking is realized through frictional clamping of pre-tensioned tendons. The actuator segment features three longitudinal pneumatic chambers for elongation and three tendons routed through embedded channels as illustrated in Fig. 1b). At the base of each segment, a tendon-locking module uses miniature pneumatic chambers to rotate rigid levers that clamp the tendon against a high-friction surface. When the locking chambers are activated (via positive pressure), the tendons are held fixed in length, and further elongation of the actuator causes it to bend in the direction opposite to the locked tendon(s). A tendon-retraction mechanism (a spring-loaded pulley) ensures slack tendon management during reconfiguration. This setup enables independent stiffness and configuration control in multi-segment soft robots, allowing selective stiffening of individual segments without influencing adjacent ones.

This SSLS principle was implemented in a multi-segment soft robotic manipulator prototype for minimally invasive surgery (MIS) [23]. Each segment integrates the tendon-based antagonistic actuation that allows tendons to be locked independently, enabling the robot to stiffen or hold its shape locally. This setup allows precise spatial configuration of each segment while preserving compliance where needed—for example, to accommodate anatomical constraints or respiration-induced motion. Experimental results demonstrated a more than 200% increase in stiffness, as well as independent control over base and distal segments. Due to the SSLS mechanism installed in series and otherwise only thin tendons routed through the actuators, there is particular potential for miniaturising the diameter of the robot. While this actuator is an example for a medical use case of SSLS actuators, the same principle is transferable to industrial inspection systems for confined spaces. The SSLS principle does not allow for position locking as the tendons routed through the actuator can only absorb tensile forces.

Key features of the presented force-locking SSLS include:

- Modular and continous tendon-locking positions in each segment
- High stiffness modulation range (up to 200%)
- MR-compatible design without metal components and electronics
- Especially suited for diameter miniaturization.

3.3 Material-Locking SSLS: Low-Melting-Point Alloy Chambers

Material-locking is implemented by embedding chambers filled with a low-melting-point alloy (LMPA) alongside pneumatic actuation chambers as described in Fig. 1c). In the solid state, the LMPA acts as a force-translating structure, constraining elongation and bending on the respective side. When heated via integrated nichrome wires, the alloy melts, rendering the structure flexible again. The actuator contains three such LMPA chambers. Selective melting enables controlled bending (one or two rigid sides), while melting all three allows axial elongation. Steel helices within the LMPA chambers ensure structural continuity and force transmission when the alloy is solid. Like the presented form-locking actuator, this SSLS principle allows position locking of the actuator. For that, the actuation chambers are actuated while one, two or all LMPA chambers are in their liquid state to reach a designated configuration. The heating of all chambers is then switched off. Once all LMPA chambers are solidified, the actuator maintains its configuration without power, offering energy-efficient position locking.

This material-locking SSLS actuator was developed as a part of a deep-sea suction sampling system [24]. Designed to operate at depths of up to 6000 m, the actuator would be driven hydraulically by seawater while using the LMPA as the SSLS material. With just one necessary pressure pump for actuation and only electronic components used for guiding the movement direction the hardware footprint is reduced. It was shown, that the actuator can be stiffened by more than 300% between the unstiffened and stiffened state. Additionally, the actuator can be locked in position without consuming energy, an essential advantage for battery-constrained subsea vehicles. This enables the robotic structure to maintain a prescribed bending configuration while resisting external loads from turbulent flows or seabed contact. The cold environment accelerates the cooling process of the LMPA - the major drawback of heating-based SSLS.

Key features of the presented material-locking SSLS include:

- High programmable stiffness modulation via thermal phase change
- Energy efficient passive stiffened state
- Beneficial in the deep-sea environment due to faster cooling of the LMPA.

Each of these mechanisms exemplifies a distinct approach to switchable strain limitation, offering unique trade-offs in terms of complexity, controllability, energy efficiency, and application scope. In the next section, these mechanisms are compared qualitatively.

4 Comparison and Discussion

The form-, force-, and material-locking strategies described in the previous section each offer distinct advantages depending on application requirements, actuator architecture, and system constraints. In this section, a structured comparison of the three switchable strain-limiting structures, focusing on six key criteria is presented.

4.1 Comparison of Switchable Strain-Limiting Structures

SSLS can be compared with each other based on various criteria, which are briefly explained below. The activation method is one of the most important criteria, as it significantly influences other properties such as response time and hardware footprint, and of course has a major impact on energy consumption. The response time is crucial for the intended application of the robot. Many applications that require rapid configuration changes are usually difficult to implement with thermally triggered SSLS. The hardware footprint, one of the main motivations for using SSLS, is also controversial. The SSLS mechanism itself can contribute significantly to the hardware footprint. In most cases, however, the pressure control valves used in fluidically actuated actuators are the most expensive component, which can be reduced by SSLS. Stiffening performance and the possibility of position locking are features that are particularly relevant for the application. The properties listed in Table 1 refer explicitly to the actuators presented in this chapter.

1. **Activation Method**: The type of energy or signal required to switch between states (e.g., pneumatic, thermal, mechanical, electrical).
2. **Response Time**: Time required to switch between flexible and constrained states.

Table 1 Comparison of the switchable strain-limiting structures presented in this work

Criterion	Form-locking [22]	Force-locking [23]	Material-locking [24]
Activation method	Pneumatic	Pneumatic/Mech.	Thermal
Response time	Fast	Fast	Slow
Hardware footprint	Medium	Medium	Low
Cross section	Big	Small	Medium
Stiffening ratio	Medium	Medium[a]	High
Position locking	Yes (axially)	No	Yes (uni-directional)

[a] Highly depending on the direction of force corresponding to the tendon direction

3. **Cross-Sectional Space Requirements**: Additional space that is used by the SSLS in the radial direction of the actuator.
4. **Hardware Footprint**: Number of valves, regulators, or electronic components needed.
5. **Stiffening Performance**: Qualitative increase in mechanical resistance or stiffness (i.e. ratio between stiffened and unstiffened state).
6. **Position Locking**: Ability to passively lock the actuator in its current configuration without the need for further actuation.

4.2 Discussion

The three analyzed switchable strain-limiting structures (SSLS) exhibit distinct mechanical behaviors, constraints, and application trade-offs, which become evident when comparing their performance in real-world prototypes. The presented form-locking SSLS, based on interlocking combs, is characterized by fast pneumatic switching and a medium hardware footprint. While position locking in the elongated configuration is reliable and passive, it is limited to discrete intervals defined by the comb teeth geometry. Moreover, the relatively small maximum bending angle (16° at 110 kPa) and inability to hold bent configurations limit the range of controllable shapes. The additional comb chambers increase radial bulk, making the actuator less suited for space-constrained environments.

The Force-locking system, using frictionally clamped tendons, excel in bending performance. The reported bending angle of up to 143.6° and stiffness increase of up to 3.0 (direction-dependent) make this approach suitable for surgical manipulators requiring segment-specific behavior. However, tendons do not support axial load retention, making position locking infeasible and the stiffening capabilities highly directional. While the tendon routing maintains a compact cross-section, the need for three solenoid valves per segment increases control complexity. Still, this method remains MR-compatible and highly modular.

Material-locking using low-melting-point alloys offers the highest degree of versatility. It enables both axial and bent position locking—continuously and passively—after cooling. The actuator can achieve 39.7° bending or nearly 30% elongation at only 45 kPa, and demonstrates a stiffness increase of up to 4.1 in axial loading. The system requires only one actuation line and compact LMPA chambers, making it spatially efficient. However, long transition times (180 s melt, even longer solidify) restrict its use to slowly changing configurations or thermally supportive environments like the deep sea.

A cross-mechanism comparison reveals that no solution is universally optimal. Form-locking is well-suited for discrete MRI tasks with limited deformation needs. Force-locking is ideal where fast actuation and high maneuvrability is critical and passive holding is not. Material-locking is the only option for passive locking for straight and bent configurations, but its time consuming switching mechanism must be considered.

5 Conclusion

Switchable strain-limiting structures represent a possible advancement in the development of soft robotic systems for real world applications. By enabling selective restriction or release of deformation, they unlock new possibilities for actuation efficiency, shape programmability, and adaptive stiffness modulation—all without abandoning the core benefits of soft robotics: compliance, safety, and versatility. While all the presented mechanisms were developed to decouple motion control from permanently embedded structures and to avoid the need for additional pressure regulators for each motion direction, the SSLS can also be used in combination with separately actuated chambers to increase the overall workspace.

This chapter has presented three fundamentally different approaches to achieving switchability—form-locking, force-locking, and material-locking—each grounded in a distinct physical principle and demonstrated through a dedicated actuator design. Their comparison shows that no single approach is universally superior; instead, each offers trade-offs suited to different environments, control architectures, and functional demands. The accompanying application cases—from MRI-guided interventions to deep-sea sampling—illustrate both the practical viability and the domain-specific optimization of each method. At the same time, these examples are not exclusive: the core mechanisms presented here are broadly transferable and hold promise for a wide range of future applications in medicine, industry, exploration, and beyond.

Acknowledgements This study was funded by Deutsche Forschungsgemeinschaft (DFG, German Research Foundation) under grant no. 498342743. We would like to thank all our colleagues and students who helped in conducting experiments, creating figures and writing this chapter.

References

1. Arezzo, A., et al.: Total mesorectal excision using a soft and flexible robotic arm: a feasibility study in cadaver models. Surg. Endosc. **31**, 264–273 (2016)
2. Peters, J., Anvari, B., Licher, J., Wiese, M., Raatz, A., Wurdemann, H.A.: Acceptance and usability of a soft robotic, haptic feedback seat for autonomy level transitions in highly automated vehicles. IEEE Trans. Haptics **18**(1), 58–72 (2025)
3. Li, G., Wong, T.W., Shih, B., et al.: Bioinspired soft robots for deep-sea exploration. Nat. Commun. **14**, 7097 (2013)
4. Peters, J., et al.: Actuation and stiffening in fluid-driven soft robots using low-melting-point material. In: 2019 IEEE/RSJ International Conference on Intelligent Robots and Systems (IROS), Macau, China, pp. 4692–4698 (2019)
5. Peters, J., Anvari, B., Chen, C., Lim, Z., Wurdemann, H.A., Hybrid fluidic actuation for a foam-based soft actuator. In: 2020 IEEE/RSJ International Conference on Intelligent Robots and Systems (IROS), Las Vegas, NV, USA, pp. 8701-8708 (2020)
6. Connolly, F., Walsh, C.J., Bertoldi, K.: Automatic design of fiber-reinforced soft actuators for trajectory matching. Proc. Natl. Acad. Sci. U.S.A. **114**, 51–56 (2017)
7. Kim, S.Y., Baines, R., Booth, J., et al.: Reconfigurable soft body trajectories using unidirectionally stretchable composite laminae. Nat. Commun. **10**, 3464 (2019). https://doi.org/10.1038/s41467-019-11294-7
8. Xavier, M.S., Tawk, C.D., Yong, Y.K., Fleming, A.J.: 3D-printed omnidirectional soft pneumatic actuators: design, modeling and characterization. Sens. Actuators A Phys. **332**(2), 113199 (2021). https://doi.org/10.1016/j.sna.2021.113199
9. Lee, K., Bayarsaikhan, K., Aguilar, G., Realmuto, J., Sheng, J.: Design and characterization of soft fabric omnidirectional bending actuators. Actuators **13**(3), 112 (2024). https://doi.org/10.3390/act13030112
10. Jung, Y., Kwon, K., Lee, J., et al.: Untethered soft actuators for soft standalone robotics. Nat. Commun. **15**, 3510 (2024). https://doi.org/10.1038/s41467-024-47639-0
11. Elsayed, Y., Vincensi, A., Lekakou, C., Geng, T., Saaj, C., Ranzani, T., Cianchetti, M., Menciassi, A.: Finite element analysis and design optimization of a pneumatically actuating silicone module for robotic surgery applications. Soft Robot. **1**, 255–262 (2014)
12. Yang, B. et al.: Reprogrammable soft actuation and shape-shifting via tensile jamming. Sci. Adv. **7**, eabh2073 (2021)
13. McDonald, K, Ranzani, T.: Hardware methods for onboard control of fluidically actuated soft robots. Front. Robot. AI **8**, 720702 (2021). https://doi.org/10.3389/frobt.2021.720702
14. Liu, J., Yin, L., Chandler, J., Chen, X., Valdastri, P., Zuo, S.: A dual-bending endoscope with shape-lockable hydraulic actuation and water-jet propulsion for gastrointestinal tract screening. Int. J. Med. Robot. Comput. Assisted Surg. **17** (2021)
15. Xiong, Q., Zhou, X., Yeow, C.-H.: A soft pneumatic actuator with multiple motion patterns based on length-tuning strain-limiting layers. In: 2023 IEEE International Conference on Soft Robotics (RoboSoft), pp. 1–6 (2023)
16. Narang, Y.S., Vlassak, J.J., Howe, R.D.: Adv. Funct. Mater. **28**, 1707136 (2018). https://doi.org/10.1002/adfm.201707136
17. Xiong, Q., Ang, B.W.K., Jin, T., Ambrose, J.W., Yeow, R.C.H.: Earthworm-inspired multimaterial, adaptive strain-limiting, hybrid actuators for soft robots. Adv. Intell. Syst. **5**, 2200346 (2023)
18. Campbell, G.M., Yin, J., Song, Y., Gandhi, U., Yim, M., Pikul, J.: Electroadhesive clutches for programmable shape morphing of soft actuators. In: 2022 IEEE/RSJ International Conference on Intelligent Robots and Systems (IROS), Kyoto, Japan, pp. 11594–11599 (2022). https://doi.org/10.1109/IROS47612.2022.9982131
19. Buckner, T.L., Yuen, M.C., Kim, S.Y., Kramer-Bottiglio, R.: Enhanced variable stiffness and variable stretchability enabled by phase-changing particulate additives. Adv. Funct. Mater. **29**, 1903368 (2019)

20. Gunawardane, P.D.S.H., Budiardjo, N., Alici, G., de Silva, C.W., Chiao, M.: Thermoelastic strain-limiting layers to actively-control soft actuator trajectories. In: 2022 IEEE 5th International Conference on Soft Robotics (RoboSoft), Edinburgh, United Kingdom, pp 48–53 (2022)
21. Firouzeh, A., Salerno, M., Paik, J.: Soft pneumatic actuator with adjustable stiffness layers for multi-DoF actuation. In: 2015 IEEE/RSJ International Conference on Intelligent Robots and Systems (IROS), Hamburg, Germany, pp. 1117–1124 (2015)
22. Peters, J., Licher, J., Hensen, B., Wacker, F., Raatz, A.: Soft robotic actuator leveraging switchable strain-limiting structures for tumor biopsy and ablation in MRI. In: 2024 IEEE International Conference on Soft Robotics (RoboSoft), pp. 983–989 (2024)
23. Licher, J., Peters, J., Raatz, A., Wurdemann, H.A.: Tendon locking for antagonistic configuration- and stiffness-control in soft robots. In: 2025 IEEE International Conference on Robotics and Automation (ICRA), Atlanta, GA, USA, pp. 15322–15328 (2025). https://doi.org/10.1109/ICRA55743.2025.11127937
24. Peters, J., Sourkounis, C.M., Wiese, M., Kwasnitschka, T., Raatz, A.: Single channel soft robotic actuator leveraging switchable strain-limiting structures for deep-sea suction sampling. In: IEEE/RSJ International Conference on Intelligent Robots and Systems (IROS), pp. 6484–6490 (2023)

Curved Rods as Building Blocks for Programmable Soft Robotics

Sophie Leanza and Ruike Renee Zhao

Abstract Curved rods offer a promising yet underutilized foundation for programmable soft robotics. When deformed, these rods store elastic energy that can be harnessed to drive dramatic, reversible shape changes. Through careful design of their curvature and arrangement, these structures can exhibit multistability and tunable energy landscapes, enabling precise control over transitions between distinct configurations. From rings to various 3D assemblies, curved rod-based architectures can be programmed to morph in shape, area, and volume in response to simple inputs. Embracing curved rods as a structural element can open the door to robotic systems with rich, adaptive capabilities.

1 Introduction—The Curved Rod as a Soft Robotic Building Block

Soft robotic designs that exploit structural instabilities, for instance by incorporating beams, curved plates/domes, or origami to their structures, are advantageous in that they've been shown to achieve rapid movements, dramatic shape changes, and amplified force outputs [1]. One type of underexplored structural element that is promising for use in soft robotic design is the curved rod, or a rod that is curved in its natural, stress-free state (Fig. 1a). Once deformed from its stress-free state, for instance to a straightened configuration (Fig. 1a), the rod carries elastic energy which can be stored and later harnessed for soft robotic movements. Through appropriate constraints, these rods can be used to assemble larger pre-stressed structures, such as rings or 3D assemblies (Fig. 1b), which exhibit multistability and achieve significant shape and volume changes via buckling. Here, we argue that curved rods can serve as a minimalist and transformative building block for soft robotics.

S. Leanza · R. R. Zhao (✉)
Stanford University, Stanford, CA 94305, USA
e-mail: rrzhao@stanford.edu

S. Leanza
e-mail: leanza@stanford.edu

A. Raatz et al. (eds.), *Soft Material Robotic Systems*,
https://doi.org/10.1007/978-3-032-22453-8_10

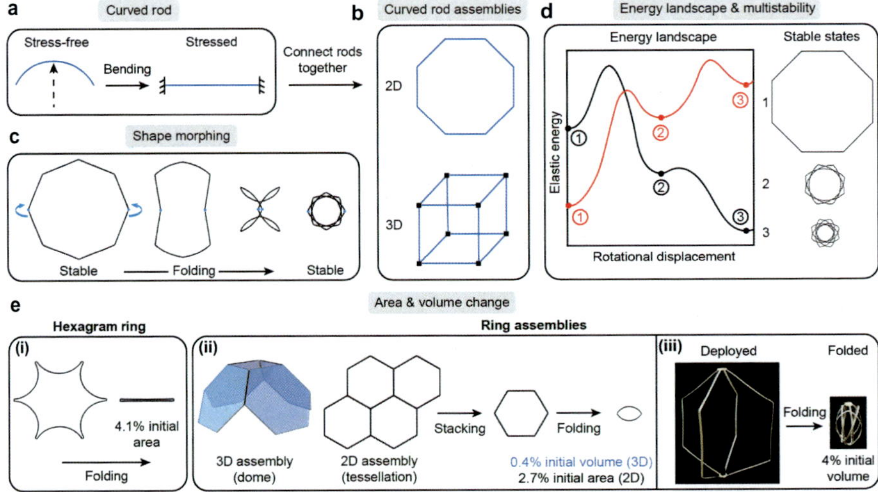

Fig. 1 **a** The curved rod, a potential building block for soft robotics. Deforming the rod from its stress-free state imparts elastic energy. **b** With appropriate constraints, multiple stressed rods can be assembled to generate 2D and 3D structures such as rings and cubes, respectively. **c** Shape morphing of a ring is achieved by applying appropriate loads, reconfiguring the ring to another stable state [2]. **d** The energy landscape of the curved rod assembly is determined by designing the initial curvature of the rods and their geometry. Through careful design, multiple stable states of the structure, corresponding to the energy minima, are achievable [2]. **e** Structures assembled from multiple rods are capable of significant area and volume change upon reconfiguration. Examples include (i) a hexagram ring which can fold into a straight-line shape [3], (ii) 3D and 2D assemblies of hexagonal rings that can be stacked and folded to greatly reduced sizes [4], and (iii) a ball-shaped assembly of hexagonal rings that can be folded to a ball of reduced size [4]

2 Design, Structural Simplicity, and Scalability

The design space for structures made of curved rods is quite extensive. They can be used to create rings, ranging from circular [5, 6] to regular polygonal rings [2, 7, 8] and those with curved geometries [3, 9, 10]. 3D structures can be made by assembling several rings together [4], or by directly assembling curved rods into various 3D geometries (Fig. 1b). Thus, the curved rod is a general building block that can be used to generate structures of different forms which can be tailored towards specific soft robotic applications. From a fabrication perspective, curved rods are appealing as their geometries are simple and slender, making them lightweight and modular, ideal for assembling more complex structures. The rods can be made from a variety of materials, including polymers and metals, and can potentially be scaled up or down, to millimeter or meter scales, making them useful for a wide range of applications.

3 Shape Morphing, Energy Landscape, and Multistability

Rings of different geometries have been studied extensively and have been shown to snap/fold reversibly between stable states when bending or twisting loads are applied [7]. An example of the folding process of an octagonal ring is shown in Fig. 1c [2], in which the ring is reconfigured between two stable states when rotational displacements are applied at its corners. These shape changes arise from the ring's multistability, which is governed by the curvature and geometry of its constituent rods. By tailoring these design parameters, the energy landscape of the ring can be programmed (Fig. 1d), with stable states corresponding to local energy minima. For example, an octagonal ring can be designed to possess three stable configurations (Fig. 1d) [2], each with tunable relative energies. The black and red curves in Fig. 1d illustrate how altering the initial curvature of the rods reshapes the energy landscape—predefining which stable states have the highest and lowest energies. Thus, by designing the curvature of the rods, both the stable configuration and the energy at that state are designed. This multistability of rod-based structures would allow a single robot to adopt distinct postures or functional states. In addition, the programmable energy landscape is useful to soft robotics, in which one configuration might store high energy for sudden jumping or flipping while reconfiguration to a lower-energy, differently shaped stable state may be more suited for locomotion such as crawling or rolling.

4 Reconfiguration: Area and Volume Change

Rings and other assemblies of curved rods are capable of dramatic changes in volume [4] and shape [5–7]. By coordinating the transitions of multiple rods, these structures can reversibly collapse into compact forms or deploy back into their more voluminous configuration. Rings can be specially designed, for instance by programming the curvature of the ring's edges, to achieve significant area reduction from the deployed to folded state such as the hexagram (star-shaped) ring in Fig. 1e (i). Assemblies of rings, for instance into dome structures or tessellations (Fig. 1e (ii)), can achieve significant volume and area changes upon stacking and folding the constituent rings. In addition, other assemblies can transition between different 3D forms, such as the ball-like structure in Fig. 1e (iii). These transformations enable a wide range of robotic behaviors, including navigating tight spaces, forming protective shells, or adapting stiffness and contact area during interaction.

5 Conclusion

Curved rods represent a powerful, underutilized design element in soft robotics for creating pre-stressed and programmable structures. Their ability to store and release energy, transition between multiple stable states, and drive large-scale shape change positions them as a uniquely versatile soft robotic element. By embracing these principles, we can create soft robots in which motion, shape, and function emerge from structure—enabling minimalist designs with rich, reconfigurable behaviors.

References

1. Chi, Y., Li, Y., Zhao, Y., Hong, Y., Tang, Y., Yin, J.: Bistable and multistable actuators for soft robots: structures, materials, and functionalities. Adv. Mater. **34**(19), 2110384 (2022)
2. Lu, L., Leanza, S., Dai, J., Hutchinson, J.W., Zhao, R.R.: Multistability of segmented rings by programming natural curvature. Proc. Natl. Acad. Sci. **121**(31), e2405744121 (2024)
3. Dai, J., Lu, L., Leanza, S., Hutchinson, J., Zhao, R.R.: Curved ring origami: bistable elastic folding for magic pattern reconfigurations. J. Appl. Mech. 1–27 (2023)
4. Leanza, S., Wu, S., Dai, J., Zhao, R.R.: Hexagonal ring origami assemblies: foldable functional structures with extreme packing. J. Appl. Mech. **89**(8) (2022)
5. Mouthuy, P.-O., Coulombier, M., Pardoen, T., Raskin, J.-P., Jonas, A.M.: Overcurvature describes the buckling and folding of rings from curved origami to foldable tents. Nat. Commun. **3**(1), 1–8 (2012)
6. Audoly, B., Seffen, K.A.: Buckling of naturally curved elastic strips: the ribbon model makes a difference. J. Elast. **119**(1), 293–320 (2015)
7. Wu, S., Yue, L., Jin, Y., Sun, X., Zemelka, C., Qi, H.J., Zhao, R.: Ring origami: snap-folding of rings with different geometries. Adv. Intell. Syst. **3**(9), 2100107 (2021)
8. Wu, S., Dai, J., Leanza, S., Zhao, R.R.: Hexagonal ring origami–Snap-folding with large packing ratio. Extreme Mech. Lett. **53**, 101713 (2022)
9. Lu, L., Dai, J., Leanza, S., Zhao, R.R., Hutchinson, J.W.: Multiple equilibrium states of a curved-sided hexagram: part I—stability of states. J. Mech. Phys. Solids 105406 (2023)
10. Lu, L., Dai, J., Leanza, S., Hutchinson, J.W., Zhao, R.R.: Multiple equilibrium states of a curved-sided hexagram: part II–transitions between states. J. Mech. Phys. Solids **180**, 105407 (2023)

Soft Tendon Driven Manipulators: From Continuum Joints to Dexterous Robotic Systems

Oliver Sebastian Neumann, Jens Reinecke, Lena Sophie Ewering, and Bastian Deutschmann

Abstract Soft tendon-driven robotic systems provide exceptional flexibility and adaptability, enabling safe interaction with dynamic environments and delicate object manipulation, surpassing the limitations of rigid robotic frameworks. This report investigates advancements in the design, control, and application of soft manipulators, tackling challenges in energy efficiency, structural durability, and precise motion. A significant contribution is the development of multi-segmented continuum manipulators with optimized tendon routing, reducing energy loss and unintended deformation while enhancing precision. The study introduces hybrid soft-rigid robotic hands, incorporating articulated palms and embedded stiffeners, markedly improving grasp stability and dexterity. Drawing from human biomechanics, a tendon-driven flexible spine and torso, guided by spinal engine theory, integrate passive compliance and tunable stiffness for superior mobility and resilience. Proprioceptive sensing, decentralized on-edge computing via a multiprocessor system-on-chip (MPSoC), and bio-inspired learning algorithms enable real-time adaptability and control efficiency. These innovations advance the development of robust, energy-efficient soft robotic systems, with potential applications in healthcare, industrial automation, and human-robot interaction.

1 Introduction

Soft manipulators and grippers are emerging as promising solutions for dexterous robotic tasks, leveraging morphological intelligence and compliance to adapt to unstructured environments [1, 2]. These systems exploit embodied intelligence to simplify control and improve performance in tasks such as grasping and manipulation [3, 4]. Soft robots can be designed using various materials and actuation methods, including fluidic elastomers and pneumatic systems [4, 5]. However, challenges remain in developing proprioception for soft robots [6] and optimizing their

O. S. Neumann (✉) · J. Reinecke · L. S. Ewering · B. Deutschmann
German Aerospace Center, Weßling, Germany
e-mail: oliver.neumann@dlr.de

A. Raatz et al. (eds.), *Soft Material Robotic Systems*,
https://doi.org/10.1007/978-3-032-22453-8_11

design to maximize beneficial morphological computation while minimizing undesirable interactions [7]. Ongoing research focuses on improving modeling, simulation, and optimization techniques for soft robotic systems [8]. As the field advances, soft manipulators show potential for applications in agriculture, healthcare, and human-robot interaction [1, 6]. Conventional control paradigms need to be reconsidered, so that the full potential of such a system can be utilised [9].

This chapter is organized as follows. In Sect. 2, the modular joints developed in this line of research are presented which are made of soft materials. Afterwards in Sect. 3, the application of such modular joints in present or future robotic systems is described. The chapter closes with a summary of the presented and a summary of topics that will be addressed in the future.

2 Modular Continuum Joints

Modular continuum joints serve as the building blocks for soft tendon-driven manipulators, enabling flexible and adaptive motion. This chapter explores their design, focusing on optimized tendon routing, resonance tuning, and workspace shaping to enhance performance. It then delves into control strategies, covering model-based approaches, proprioceptive sensing, and recent advancements in learning-based control for improved adaptability and precision.

2.1 Design

The design of the soft manipulators in this study focuses on tendon-driven continuum joints, which utilize the intrinsic compliance of soft materials to achieve flexible and adaptable motion. A continuum joint consists of a soft material structure placed between two rigid platforms, providing a deformable but stable motion frame. An important aspect of the design is the integration of tendon guide channels into the soft structure, as suggested in the computational approach to tendon routing [10]. The tendons are routed from the lower to the upper platform and attached to the upper platform. By pulling on these tendons, the upper platform moves when the soft material structure is loaded and deforms accordingly. In contrast to conventional methods of external tendon routing, which increase the risk of tendon damage and unwanted friction, this method calculates predefined internal channels that allow the tendons to move freely without affecting the surrounding material. These channels, designed using a finite element model, ensure that the tendons follow optimal paths that minimize unwanted deformation and energy loss. Tendons can be either external or internal, allowing for flexible design depending on the requirements of the application.

Another crucial consideration in the design is the resonance behavior of the continuum joint, as investigated experimentally [11]. The study demonstrated that tendon pretension significantly affects the joint's natural frequency and damping properties, which is vital for optimizing energy efficiency in cyclic motions. By adjusting tendon tension, the resonance frequency can be tuned, enabling energy-efficient periodic actuation, such as in humanoid torso rotation. The static workspace of the joint can be shaped by the fixation points of the tendons, as well as the geometric and material properties of the soft material structure, allowing for tailored motion profiles. Additionally, the dynamic workspace can be shaped using the nonlinear elasticity of the joint, providing further adaptability to complex tasks that require variable stiffness or compliance.

To validate these design principles and ensure reproducibility of experimental results in research, an open-source testbed was developed [12]. This modular platform allows for different tendon configurations, material choices, and actuation strategies to be tested systematically. The design incorporates 3D-printed molds for soft material casting, customizable tendon routing setups, and an experimental framework for evaluating bending, torsion, and actuation performance. Mechanical testing, including bending and torsion experiments, confirmed the robustness of the generic hard-soft interface between the rigid platforms and the soft material structure, ensuring that the continuum joint can withstand significant loads without failure. This interface is critical for maintaining structural integrity under the stresses induced by tendon actuation and external forces.

By combining computational modeling, experimental resonance analysis, and an open-source test platform, this study establishes a systematic approach for designing soft manipulators with high dexterity, tunable mechanical properties, and improved longevity. These findings contribute to the advancement of soft robotics, enabling safer and more efficient interactions between robots, humans, and dynamic environments. The ability to precisely shape both static and dynamic workspaces through tendon fixation, material selection, and nonlinear elasticity opens new possibilities for applications requiring adaptive, resilient, and energy-efficient robotic systems.

2.2 *Control of Continuum Joints*

Model-based control of continuum joints can be effectively implemented by integrating classical methods from nonlinear control theory with dynamic equations that describe the behavior of soft robots, which are derived from mechanical modeling. This approach has been successfully applied to both fully actuated dynamics, as demonstrated in [13], and underactuated dynamics, as shown in [14]. In recent developments, closed-loop control strategies for underactuated systems have been further enhanced by incorporating fractional-order controllers, which improve robustness and stability, as highlighted in [15].

One of the key challenges in implementing nonlinear model-based control strategies is the necessity of accurately estimating the generalized coordinates of the soft robot. This estimation process can be achieved through external vision systems, such as cameras positioned outside the robot's workspace, or through onboard sensor systems that provide real-time feedback on the robot's configuration and movement. Studies such as [16, 17] have demonstrated the effectiveness of these techniques in obtaining precise state estimations necessary for control using model-based and model-free approaches respectively.

A fundamental assumption in many model-based control approaches is that the tendons used to actuate the soft robot are inelastic, meaning they do not stretch significantly under tension. Under this assumption, achieving a high level of model accuracy is crucial for effective control. However, recent research has explored possible extensions to this assumption. One such extension considers the case where tendons exhibit elasticity. In this scenario, modifications to the control law allow for compensation of the elastic behavior, enabling stable and predictable motion control even for high dynamic motions [18].

Another significant advancement in the field involves the use of onboard cameras for visual servoing. This approach requires knowledge of the forward Jacobian, which relates changes in actuation inputs to the observed movement of the robot. By leveraging real-time visual feedback, onboard cameras can enhance the accuracy and adaptability of control strategies, as demonstrated in [19].

Furthermore, it has been shown that achieving precise dynamic modeling may not always be necessary when model-based control is combined with learning-based techniques. Instead of relying on a highly accurate dynamic model, data-driven approaches can compensate for modeling uncertainties, improving overall system performance. This hybrid methodology, which integrates learning with traditional model-based control, has been successfully explored in [20].

3 Robust and Dexterous Robotic Systems

Achieving dexterity in robotic systems requires a combination of structural adaptability, efficient actuation, and intelligent control strategies. This section presents key design approaches that enhance both robustness and manipulation capabilities in robotic hands and continuum mechanisms.

We explore multi-segmented continuum manipulators that balance flexibility and control, articulated palm structures that improve grasping adaptability, and hybrid-continuum mechanisms incorporating passive gravity compensation. Additionally, we discuss the integration of articulated spine structures for humanoid robots, enabling increased agility and human-like movement. These developments contribute to the design of robotic systems capable of precise and reliable interaction with dynamic environments.

3.1 Soft Articulated Neck

In our research, as detailed in [10, 21], we explored the development of a structurally flexible humanoid spine, with a specific emphasis on the cervical region, utilizing a tendon-driven elastic continuum mechanism, see Fig. 1. The primary research direction is to enhance the mechanical robustness and human-like performance of humanoid robots operating in unpredictable environments by incorporating passive compliance and a large range of motion. The approach utilizes a silicone-based elastic backbone actuated by tendons, contrasting with the human spine's muscle and ligament stabilization, to achieve durability and flexibility. Main findings from the study include the successful validation of a planar prototype [21] demonstrating a workspace of at least $\pm 30°$, dynamic motion capabilities up to 1.5 Hz, and significant robustness against impacts, with tendon forces reduced by nearly half compared to a rigid mechanism. These results informed the design of a modular three degree of freedom (DoF) neck prototype [21], providing a versatile testbed for exploring various tendon routings and continuum shapes, ultimately aiming to contribute to the development of a robust humanoid robot torso and neck with improved energy efficiency and gravity compensation.

The prototype shown in Fig. 1 is utilized as a neck system of the soft articulated humanoid David [21] and possess $\pm 30°$ of tilt motion and $\pm 50°$ bending motion.

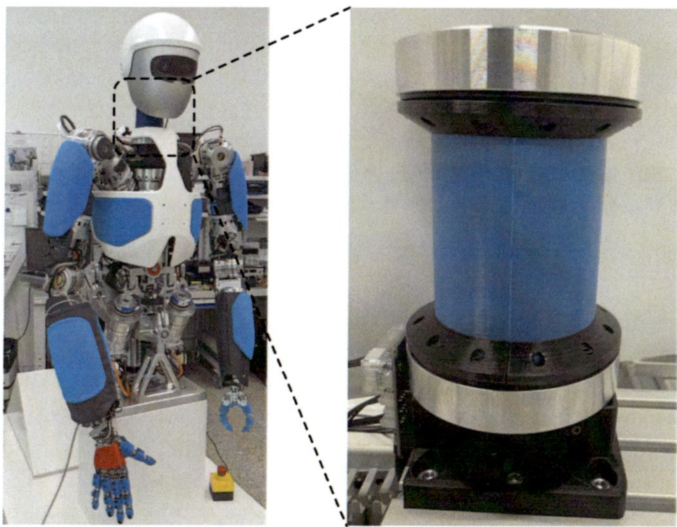

Fig. 1 (Left) Integrated module is mounted as neck for the humanoid DAVID. (Right) Integrated module as single component with power electronic

3.2 Multi-segmented Continuum Manipulators

For this research direction, we derived workspace requirements from anthropomorphic counterparts, specifically the finger and thumb [22]. We measured the tip force of the finger, a key parameter for robotic in-hand manipulation, and determined its resulting pose using AprilTags, which are easily attachable and open-source. To minimize distal inertia and enhance finger dynamics, motors were positioned away from the actuated joints, utilizing tendon-driven mechanisms to reduce moving mass at the extremities and enable more agile movements. To compare configurations, we used two metrics: the reachable workspace for a given tendon actuation force and the resulting fingertip force in an empirically chosen pose.

Given the requirements defined above, we conducted a range of motion experiment to evaluate different structural configurations of the prototypes. The tendon force applied was incrementally increased, first for individual joints and then in a coupled manner. The following structural parameters were systematically investigated:

- Direction-dependent stiffness for bending is achieved through various cross-sections of the continuum, with flexion designed to be less stiff than adduction in the medial and distal links, see Fig. 3.
- Embedded rigid stiffener, allowing for constraining motion along a certain axis but also serve as carrier for Bowden tubes or sensors, see Fig. 2 (left) and Fig. 3.
- Internal routing within the segmented continuum structure using Bowden tubes to reduce friction and guide tendons through bent silicone, see Fig. 2 (left).
- External routing was explored for ease of manufacturing, incorporating passive tendons to achieve a nonlinear transmission ratio over the bending range, see Fig. 2 (right).
- Trimming and cut-away sections to enable more compact bending radii of the continuum structure and improve grasp stability by enhancing its ability to wrap around objects, see Fig. 2 (right).

A comparison of the reachable workspace for two selected variants is shown in Fig. 2.

The external tendon routing offers ease of manufacturing but introduces trade-offs in performance and control. Since the tendons for the distal links are not guided close to the neutral axis, this configuration results in segment coupling, leading to buckling. While the external routing allows for higher lever arms and consequently higher joint torque, it lacks shielding and protection, making it more prone to interference with objects during manipulation tasks. Additionally, the angle of attack for the applied force changes over the bending angle of the segments, complicating control compared to the internal routing, which provides a more guided force introduction.

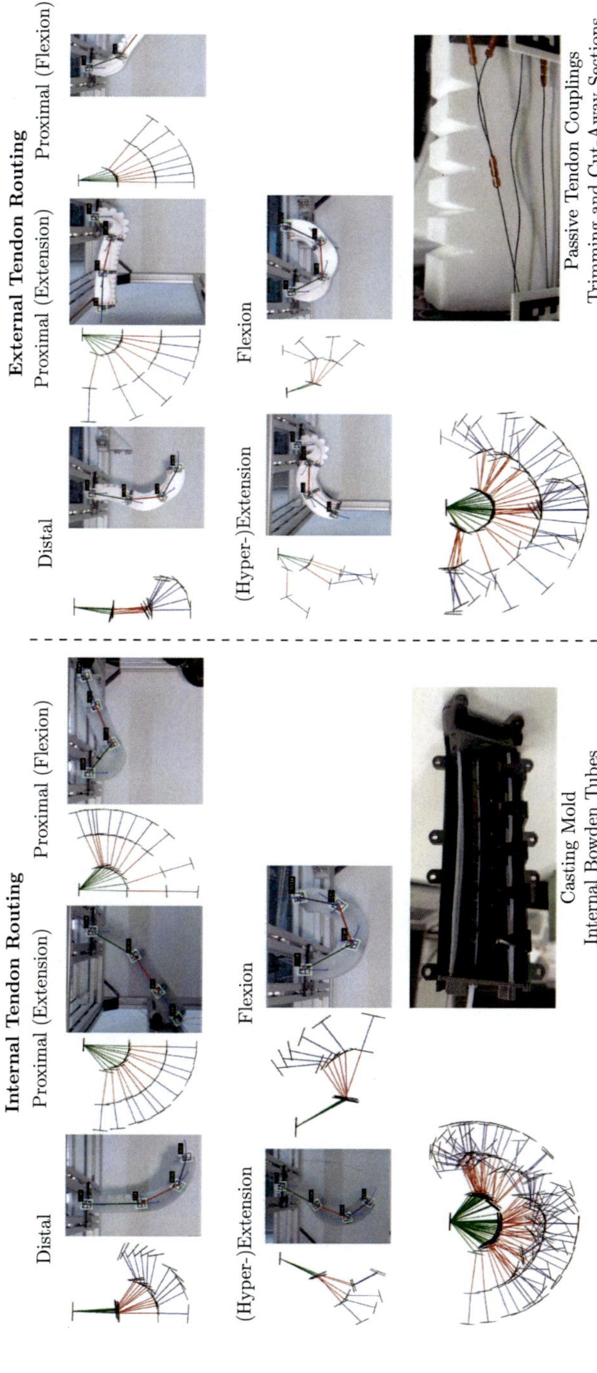

Fig. 2 Range of motion of two selected prototypes under various actuation patterns, including distal, proximal, and coupled actuation scenarios. The left tree shows the range of motion for internal tendons and a integrated rigid stiffener. The right tree with external routing and passive tendon couplings

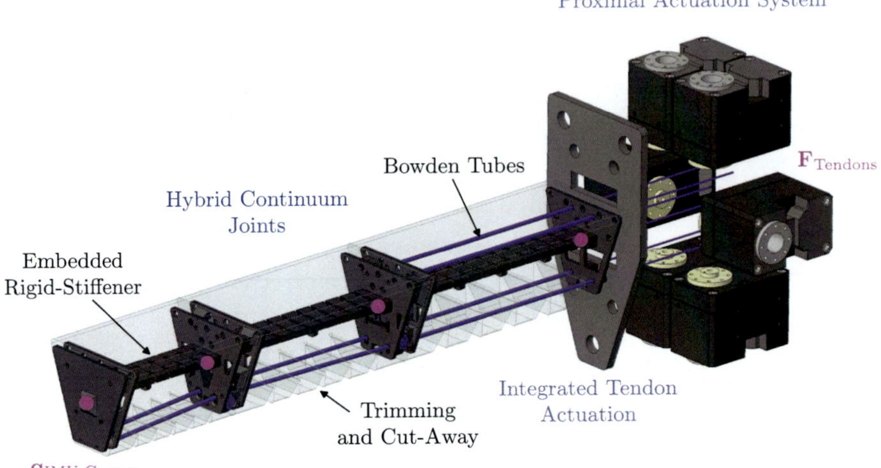

Fig. 3 Prototype for multi-segmented continuum finger, featuring embedded rigid bone-like elements and a internal actuation to analyze stiffener mechanisms and dynamic behavior

3.3 Flexible Articulated Palm for Robotic Hands

Recent studies emphasize the role of flexible, actuated palms in robotic hands to enhance grasping and manipulation capabilities. Research demonstrates that soft, concave palms improve finger workspace, manipulability, and adaptability [23, 24]. Innovative designs feature dual-layered structures for surface conformity and adjustable stiffness [25], alongside pneumatic actuation for metacarpal movement and thumb opposition [26, 27]. These developments seek to emulate the multifaceted functionality of the human hand, which evolved for manipulation and protective support during striking actions [28–30]. The human hand's distinct morphology, encompassing soft tissues and neurological integration, enables precision and power grips critical for throwing and clubbing [29]. This evolutionary framework guides the design of more adaptable, human-like robotic hands, integrating soft materials, active palms, and distributed tactile sensing to optimize dexterity and object interaction.

In the human hand, palm flexion results from the palmar displacement of the little finger's metacarpal bone (MC) distal end [31], as depicted in Fig. 4a. This motion partially transfers to the ring finger's MC, yielding a smooth curvature rather than abrupt, hinge-like bending [23]. Quantitative physiological data on the little finger's MC displacement are absent; thus, we estimated it as equivalent to the MC diameter, based on personal observation and tracking by [32]. The palm must also exhibit passive compliance to mitigate collision severity independent of control system response time. If the flexible material doubles as enclosing soft tissue with embedded bones, it acts as a skin-like interface, requiring resilience against contact with both large, smooth objects and small, sharp ones at typical grip forces of approximately 30 N.

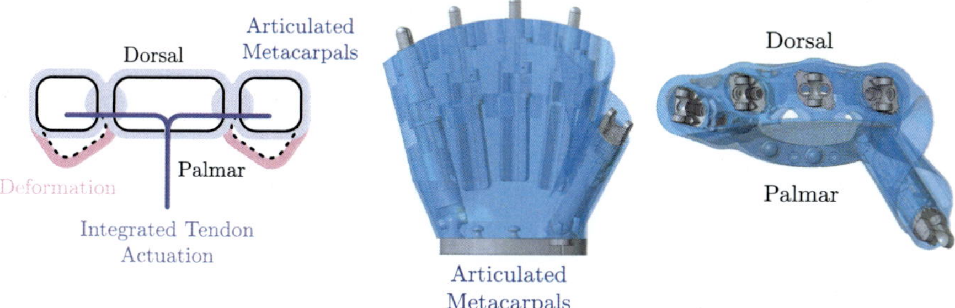

(a) Sketch of an articulated and segmented flexible robotic palm **(b)** Prototype with segmented, actuated palm structures

Fig. 4 Design and development of a flexible robotic palm. The left sketch illustrates the initial concept of an articulated and segmented palm structure. The right depicts a prototype, integrating flexible materials and actuated elements to enhance robotic grasping and manipulation

Tendon-driven actuation, with a motor in the forearm, facilitates palm flexion and enables mechanical couplings, such as linking little finger MC motion to thumb abduction or finger splaying.

Two-dimensional finite element method (FEM) analysis characterized the behavior of a soft, silicone-like material ($E = 1.5\,\text{MPa}$, $\sigma_m = 5\,\text{MPa}$) under simple contact with rigid objects and unidirectional force, then in an idealized palm configuration. Grasped objects were modeled as circular contact bodies of varying radii. Large bodies ($r = 10\,\text{mm}$) induced negligible stresses at typical grip forces, sustaining up to fivefold increases before exceeding modeling assumptions, with stresses well below material limits. Small objects ($r = 0.5 - 1\text{mm}$) generated significantly higher stresses with a steeper gradient, yet remained subcritical; however, the smallest ($r = 0.5\,\text{mm}$) failed to reach 30 N within model constraints. Stress escalated sharply with decreasing radius, suggesting grip force adjustments for object size and caution with small, sharp objects. When sharp features protruded from larger smooth surfaces, forces exceeding twice the typical grip force were achieved without critical stress. Stress also rose steeply with reduced skin thickness, with a critical threshold near 2 mm, independent of the thickness-to-radius ratio.

For the idealized palm, a hand cross-section was simplified to five planar metacarpal bones embedded in soft material, halved at the middle finger for computational efficiency. Tendon actuation was modeled by applying force at the little finger MC palmar surface toward the hand's center, targeting a minimum grip force of 30 N, as shown in Fig. 5a. Geometric parameters were varied to adjust force application and stiffness, influencing deformation and grip force. Four parameters, outlined in Fig. 5b, were analyzed:

1. Force direction: pertinent to tendon routing design.
2. Inter-bone soft material thickness: balancing stability and compliance.
3. Material enclosure versus interosseal filling: impacting stress distribution.
4. Tendon insertion site: leveraging biological protrusions for improved mechanical advantage.

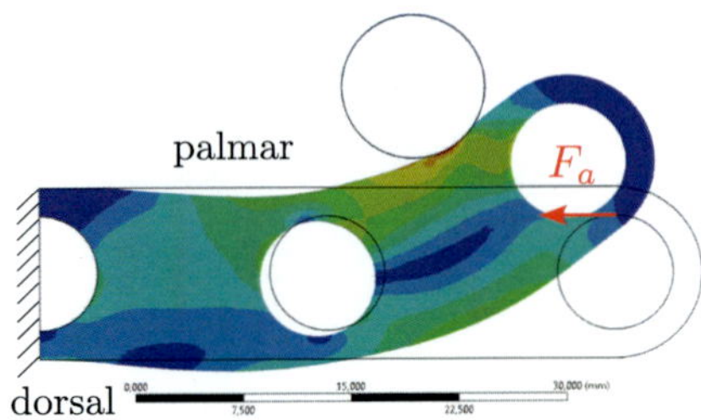

(a) FEM model, the palm is fixed at the left edge and produces a reaction force in the fixed contact body. Black lines show the undeformed body, the contour plot shows the Equivalent von Mises stress, max. 0.46 MPa, for the body deforming under $F_a = 100\,\text{N}$

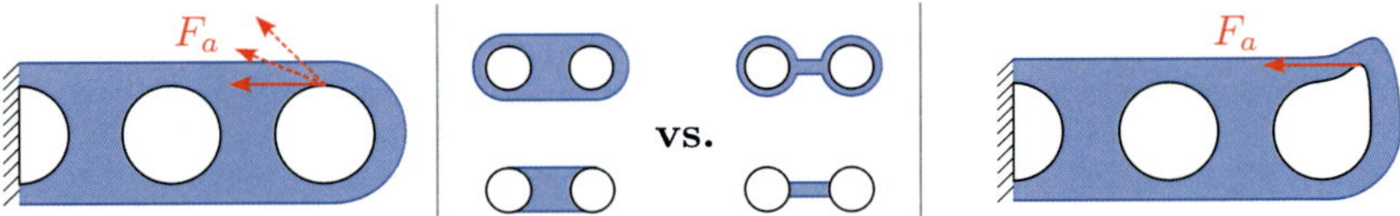

(b) Schematic illustrations of parameters to be varied in the FEM model of the idealised palm, proportions not to scale

Fig. 5 Model of an idealised flexible palm, cut in half due to symmetry

4 Summary and Future Research

This chapter consolidates the key findings of the research on soft tendon-driven manipulators, highlighting their advantages in dexterity, adaptability, and energy-efficient motion. The study has demonstrated how modular continuum joints with optimized tendon routing improve flexibility while minimizing energy loss and unwanted deformations. Computationally designed internal tendon guide channels have been shown to reduce friction and enhance the longevity of the system. Additionally, resonance tuning has been identified as a crucial factor in improving energy efficiency for cyclic motions, such as those required in humanoid applications. Experimental validation using an open-source testbed has confirmed the structural integrity of these designs under various loading conditions.

In terms of control, nonlinear model-based approaches have been successfully applied, with sensor-based state estimation—using external vision systems or onboard sensors—playing a vital role in improving accuracy. Recent advances, including the modeling of elastic tendons, visual servoing, and learning-based control strategies, have further enhanced the adaptability of these manipulators. The development of multi-segmented continuum manipulators and flexible articulated palms has expanded the functionality of soft robotic hands, enabling more precise and stable grasping through embedded rigid elements and compliant structures. These findings contribute to the creation of robotic systems capable of human-like dexterity, improved interaction with dynamic environments, and safe, energy-efficient operation.

Despite these advancements, several challenges remain, particularly in control precision, proprioceptive sensing, and integration of intelligent learning methods. The following sections outline future research directions aimed at addressing these limitations. Section 4.2 explores decentralized, on-edge control, drawing inspiration from biological motor systems to enable real-time sensorimotor integration. By distributing computation across embedded processors, this approach seeks to improve responsiveness and robustness in manipulation tasks. Section 4.3 presents a vision for hybrid robotic hands, integrating soft and rigid materials to balance flexibility and structural stability. This approach aims to enhance dexterity, impact resistance, and functional versatility, ultimately bridging the gap between robotic and human hand performance.

4.1 Design of a Continuum Robotic Torso

The spinal engine theory, proposed by Gracovetsky and Iacono [33], posits that the spine plays a central role in locomotion, acting as an active engine rather than a passive structure. It suggests that coupled motions of the spine—such as lateral bending, axial rotation, and flexion/extension—generate and transfer energy efficiently through the body during walking or running, with the pelvis and shoulders rotating in opposition to amplify movement. This theory emphasizes the spine's ability to store and release elastic energy, absorbing impacts and driving limb motion via its dynamic, impulse-loaded behavior. Future research on continuum robotic torsos should focus on integrating adjustable parallel stiffness elements to enhance stability and load-bearing capacity while preserving compliance. Bio-inspired actuation strategies, such as tendon-driven mechanisms and variable impedance actuation, could improve dynamic control and energy efficiency by leveraging resonant behavior, drawing from the spinal engine theory's coupled motion principles. Real-time adaptive control using proprioceptive sensor fusion and machine learning can optimize movement coordination, ensuring robust responses to external disturbances in line with the spinal engine's dynamic adaptability. Additionally, evaluating continuum torsos in dynamic walking robots could enhance agility, energy efficiency, and impact absorption, reflecting the spinal engine's role in locomotion. The design, see Fig. 6, of such torsos, inspired by anthropomorphic principles, features a lightweight, modular structure with a central continuum joint—a flexible, cylindrical material—supported by a tripod joint for rotational torque transmission and cable-driven actuation for bending motions. This setup allows for underactuated yet robust behavior, enabling experimentation with varying stiffness and geometries to mimic human-like spinal dynamics while maintaining structural integrity.

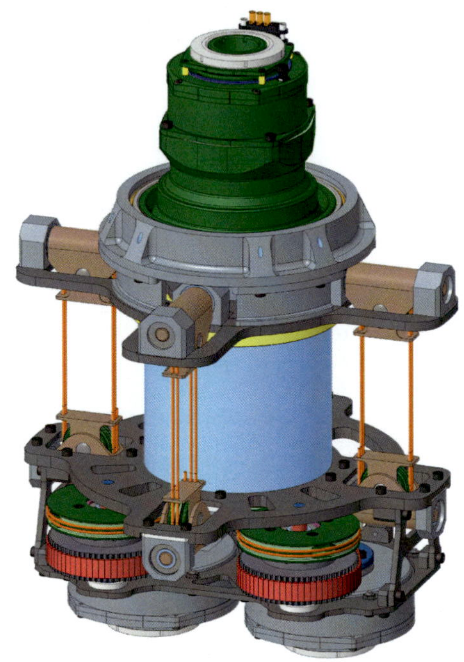

Fig. 6 Prototype of a continuum robotic torso inspired by anthropomorphic spinal dynamics. This lightweight, modular design features a central flexible continuum joint supported by a tripod joint for rotational torque transmission and cable-driven actuation for bending motions. The setup enables underactuated yet robust behavior, facilitating experimentation with variable stiffness and geometries to emulate human-like movement patterns, as explored in future research directions for dynamic walking robots

4.2 Decentralized On-Edge Control

Recent investigations have concentrated on developing neurologically inspired control architectures for dexterous grasping in robotic systems. These methodologies integrate perception, planning, and motion generation [34], frequently employing adaptive learning techniques [35, 36]. Certain models leverage deep neural networks to forecast human-like grasping actions from visual inputs [37], while others utilize biologically inspired visual exploration for grasp synthesis [38]. Tactile sensing is pivotal in grasp regulation, emulating human tactile pathways [39]. Sensory-motor integration strategies, merging vision, electromyography (EMG), and force optimization, have been devised for manipulation and grasping tasks [40]. Advanced prosthetic limbs have demonstrated simultaneous neural control of reaching and grasping via intracranial electroencephalography (IEEG) signals [41]. These bio-inspired strategies seek to narrow the disparity between human dexterity and robotic grasping proficiency, enhancing flexibility and adaptability in object manipulation.

In our study, a multi-processor system on chip (MPSoC), depicted in Fig. 7, serves as a versatile and high-performance embedded platform.

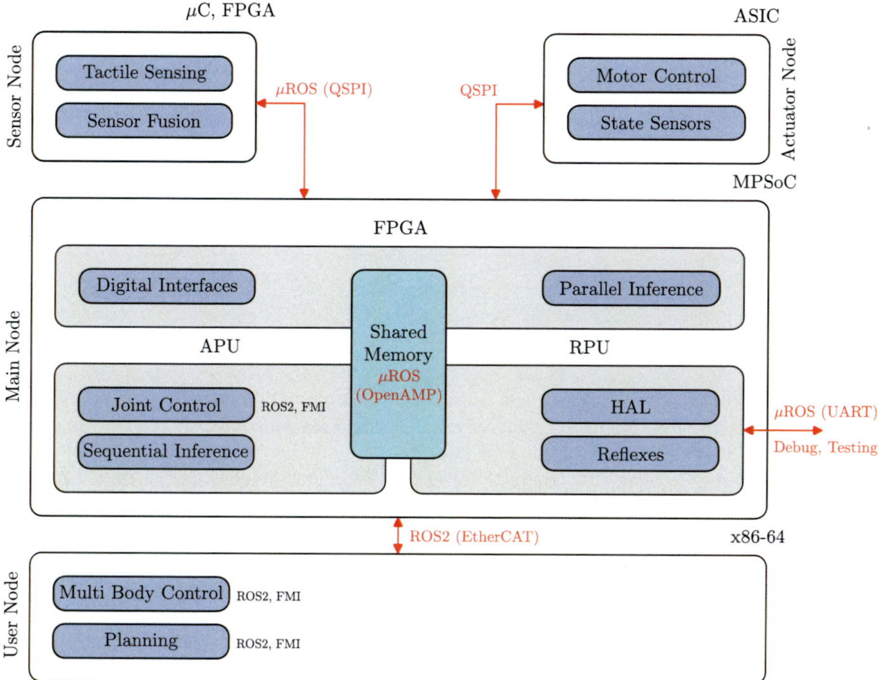

Fig. 7 MPSoC-based control architecture facilitating decentralized sensing and actuation. The system employs μROS as a communication layer for reliable and efficient data exchange over various physical interfaces, according to bandwidth and latency requirements. The integration of real-time processors, application processors, and an FPGA enables deterministic sensor-actuator communication, parallel computation, and flexible data transmission

4.3 Hybrid Design Vision for Dexterous Robotic Hands

Designing robotic hands that are dexterous, robust, and intuitive is fundamental to advancing human-robot collaboration [42, 43]. Recent research highlights the importance of hybrid designs combining soft materials with rigid backbones to mimic the stability and adaptability of human hands [1, 2, 44, 45]. The integration of sensory systems, such as distributed tactile sensors, has further advanced perception and control in robotic hands [46].

However, significant challenges remain. Current designs often lack cohesive structural elements that integrate sensors while maintaining robustness and integrity [32, 44, 47]. Proprioception—the ability to sense self-movement, force, and position—is unexplored in soft robotics, limiting precise pose estimation and dynamic manipulation [48]. Despite decades of advancements, human-level dexterity in robotic hands remains elusive, with ongoing efforts centered on sensing, actuation, and control [49,

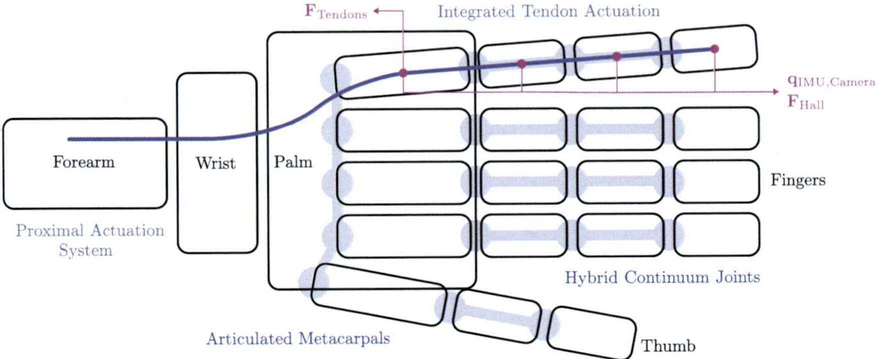

Fig. 8 Vision of a robotic hand system consisting of hybrid continuum structures, including a flexible palm with embedded rigid metacarpals for mounting soft fingers. The fingers combine embedded rigid bone-like elements to constrain motion with a soft surrounding skin to increase contact area during grasping. Softness is utilized as a feature to enhance robustness, providing built-in overload protection and adaptability to dynamic interactions

50]. Future directions may involve integrating advanced learning algorithms and bio-inspired designs to overcome current limitations in dexterous manipulation and focus on robust and maintainable systems [46].

To address these gaps, we want to exploit multi-material joints that integrate rigid and soft elements, as shown in Fig. 8. This approach aims to design a dexterous robotic hand that mimics and exceeds human-like properties [51, 52]. Our research focuses on developing robotic hands capable of dexterous manipulation beyond power grasps, leveraging the inherent advantages of softness—such as built-in overload protection, controllable compliance, and dynamic robustness—to create hybrid systems. These systems combine the precision of traditional robotics with the adaptability of soft robotics, enabling advanced applications in precision tasks and collaborative environments.

Acknowledgements This research has been funded by the German Research Foundation (DFG), grant number 405032572, as part of the priority program 2100 Soft Material Robotic Systems.

Competing Interests The authors have no conflicts of interest to declare that are relevant to the content of this chapter.

References

1. Hughes, J., Culha, U., Giardina, F., Guenther, F., Rosendo, A., Iida, F.: Soft manipulators and grippers: a review. Front. Robot. AI **3** (2016), ISSN: 2296-9144. https://doi.org/10.3389/frobt.2016.00069
2. Laschi, C., Mazzolai, B.: Lessons from animals and plants: the symbiosis of morphological computation and soft robotics. IEEE Robot. Autom. Mag. **23**(3), 107–114 (2016), ISSN: 1070-9932. https://doi.org/10.1109/mra.2016.2582726
3. Manti, M., et al.: Exploiting morphology of a soft manipulator for assistive tasks. In: Lecture Notes in Computer Science. Springer International Publishing, pp. 291–301, ISBN: 978-3-319-63536-1 (2017). https://doi.org/10.1007/978-3-319-63537-8_25
4. Marchese, A.D., Rus, D.: Design, kinematics, and control of a soft spatial fluidic elastomer manipulator. Int. J. Robot. Res. **35**(7), 840–869, ISSN: 0278-3649 (2015). https://doi.org/10.1177/0278364915587925
5. Milana, E., Raemdonck, B.V., Peerlinck, S., Reynaerts, D., Gorissen, B.: Advances in soft robot actuation and their morphological control (2020)
6. Scharff, R.: Soft robotic manipulators with proprioception. Ph.D. dissertation, Delft University of Technology (2021). https://doi.org/10.4233/UUID:2A2A3B0D-7DEE-4518-B96D-42DD58492FFD
7. Ghazi-Zahedi, K., Deimel, R., Montufar, G., Wall, V., Brock, O.: Morphological computation: the good, the bad, and the ugly. In: 2017 IEEE/RSJ International Conference on Intelligent Robots and Systems (IROS), pp. 464–469. IEEE (2017). https://doi.org/10.1109/iros.2017.8202194
8. Nurzaman, S.G., Wang, L., Iida, F., Lipton, J., Floreano, D., Rus, D.: Design optimization of soft robots. IEEE Robot. Autom. Mag. **27**(4), 10–11 (2020), ISSN: 1070-9932. https://doi.org/10.1109/mra.2020.3028658
9. Santina, C.D., Duriez, C., Rus, D.: Model based control of soft robots: a survey of the state of the art and open challenges. IEEE Control Syst. **43**(3), 30–65, ISSN: 1066-033X, 1941-000X (2023). https://doi.org/10.1109/MCS.2023.3253419. arXiv: 2110.01358 [eess]
10. Reinecke, J., Deutschmann, B., Dietrich, A., Eugster, S.R., Hutter, M.: A aomputational approach for internal tendon routing channels in a tendon-driven continuum joint. Soft Robotics, soro.2023.0029, Dec 2024, ISSN: 2169-5172, 2169-5180. https://doi.org/10.1089/soro.2023.0029. (visited on 03/07/2025)
11. Neumann, O., Deutschmann, B., Reinecke, J.: Utilization of the resonance behavior of a tendon-driven continuum joint for periodic natural motions in soft robotics. Appl. Sci. **14**(20), 9532, Oct 2024, ISSN: 2076-3417. https://doi.org/10.3390/app14209532. (visited on 03/07/2025)
12. Deutschmann, B., Reinecke, J., Dietrich, A.: Open source tendon-driven continuum mechanism: a platform for research in soft robotics. In: 2022 IEEE 5th International Conference on Soft Robotics (RoboSoft), Edinburgh, United Kingdom: IEEE, Apr pp. 54–61, ISBN: 978-1-6654-0828-8 (2022). https://doi.org/10.1109/RoboSoft54090.2022.9762144. (visited on 03/07/2025)
13. Della Santina, C., Katzschmann, R.K., Bicchi, A., Rus, D.: Model-based dynamic feedback control of a planar soft robot: trajectory tracking and interaction with the environment. Int. J. Robot. Res. **39**(4), 490–513, ISSN: 0278-3649, 1741-3176 (2020). https://doi.org/10.1177/0278364919897292. (visited on 03/11/2025)
14. Deutschmann, B.: Modeling and control for a class of tendon-driven continuum mechanisms, Ph.D. dissertation, Leibniz Universität Hannover (2020)
15. Monje, C.A., Deutschmann, B., Muñoz, J., Ott, C., Balaguer, C.: Fractional order control of continuum soft robots: combining decoupled/reduced-dynamics models and robust fractional order controllers for complex soft robot motions. IEEE Control Syst. **43**(3), 66–99, ISSN: 1066-033X, 1941-000X (2023). https://doi.org/10.1109/MCS.2023.3253420. (visited on 03/07/2025)

16. Deutschmann, B., Chalon, M., Reinecke, J., Maier, M., Ott, C.: Six-DoF pose estimation for a tendon-driven continuum mechanism without a deformation model. In: IEEE Robot. Autom. Lett. **4**(4), 3425–3432, ISSN: 2377-3766, 2377-3774 (2019). https://doi.org/10.1109/LRA.2019.2927943. (visited on 03/11/2025)
17. Raffin, A., Deutschmann, B., Stulp, F.: Fault-tolerant Six-DoF pose estimation for tendon-driven continuum mechanisms. Front. Robot. AI **8**, 619 238, ISSN: 2296-9144 (2021). https://doi.org/10.3389/frobt.2021.619238. (visited on 03/07/2025)
18. Ribeiro, L.N., Borja, P., Santina, C.D., Deutschmann, B.: Singular-perturbation control of a tendon-driven soft robot: theory and experiments. IEEE Trans. Control. Syst. Technol. 1–8, ISSN: 1063-6536, 1558-0865, 2374-0159 (2025). https://doi.org/10.1109/TCST.2025.3546564. (visited on 03/14/2025)
19. Deutschmann, B., Akim, M., Della-Santina, C.: Visual servoing control of a tendon-driven soft robotic neck. In: Submitted to IFAC Joint Conferrence on MECHATRONICS and ROBOTICS 2025 IEEE Transactions on Control Systems Technology (2025)
20. Pierallini, M., et al.: A provably stable iterative learning controller for continuum soft robots. In: IEEE Robotics and Automation Letters, **8**(10), 6427–6434, ISSN: 2377-3766, 2377-3774 (2023). https://doi.org/10.1109/LRA.2023.3307007. (visited on 03/07/2025)
21. Reinecke, J., Deutschmann, B., Fehrenbach, D.: A structurally flexible humanoid spine based on a tendon-driven elastic continuum. In: 2016 IEEE International Conference on Robotics and Automation (ICRA), Stockholm, Sweden: IEEE, pp. 4714–4721, ISBN: 978-1-4673-8026-3 (2016). https://doi.org/10.1109/ICRA.2016.7487672. (visited on 03/07/2025)
22. Schünke, M., Schulte, E., Schumacher, U., Voll, M., Wesker, K. (eds.): Prometheus LernAtlas der Anatomie - Allgemeine Anatomie und Bewegungssystem (Thieme eRef), 5, vollständig, überarbeitete Georg Thieme Verlag, Stuttgart New York (2018)978-3-13-139524-5 978-3-13-242085-4 978-3-13-242084-7. 10.1055/b-006-149643
23. Capsi Morales, P., Grioli, G., Piazza, C., Bicchi, A., Catalano, M.G.: Exploring the role of palm concavity and adaptability in soft synergistic robotic hands. IEEE Robot. Autom. Lett. 1–1, ISSN: 2377-3766, 2377-3774 (2020). https://doi.org/10.1109/lra.2020.3003257. (visited on 01/14/2025)
24. Pozzi, M., Malvezzi, M., Prattichizzo, D., Salvietti, G.: Actuated palms for soft robotic hands: review and perspectives. IEEE/ASME Trans. Mechatron. **29**(2), 902–912, ISSN: 1083-4435, 1941-014X (2024). https://doi.org/10.1109/tmech.2023.3328944. (visited on 01/14/2025)
25. Lee, J., Kim, J., Park, S., Hwang, D., Yang, S.: Soft robotic palm with tunable stiffness using dual-layered particle jamming mechanism. IEEE/ASME Trans. Mechatron. **26**(4), 1820–1827, ISSN: 1083-4435, 1941-014X (2021). https://doi.org/10.1109/TMECH.2021.3077941. (visited on 01/29/2025)
26. Liu, Y., Xiao, H., Hao, T., Pang, D., Wang, F., Liu, S.: Dexterous all-soft hand (DASH) with active palm: multi-functional soft hand beyond grasping. Smart Mater. Struct. **32**(12), 125012, ISSN: 0964-1726, 1361-665X (2023). https://doi.org/10.1088/1361-665X/ad07a3. (visited on 01/29/2025)
27. Shorthose, O., Albini, A., He, L., Maiolino, P.: Design of a 3D-printed soft robotic hand with integrated distributed tactile sensing. IEEE Robot. Autom. Lett. **7**(2), 3945–3952, ISSN: 2377-3766, 2377-3774 (2022). https://doi.org/10.1109/LRA.2022.3149037. (visited on 01/29/2025)
28. Kivell, T.L., Baraki, N., Lockwood, V., Williams-Hatala, E.M., Wood, B.A.: Form, function and evolution of the human hand. Am. J. Biol. Anthropol. **181**(S76), 6–57, ISSN: 2692-7691, 2692-7691 (2023). https://doi.org/10.1002/ajpa.24667. (visited on 01/29/2025)
29. Young, R.W.: Evolution of the human hand: The role of throwing and clubbing. J.Anat. **202**(1), 165–174, ISSN: 0021-8782, 1469-7580 (2003). https://doi.org/10.1046/j.1469-7580.2003.00144.x. (visited on 01/14/2025)
30. Morgan, M.H., Carrier, D.R.: Protective buttressing of the human fist and the evolution of hominin hands. J. Exp. Biol. **216**(2), 236–244, ISSN: 1477-9145, 0022-0949 (2013). https://doi.org/10.1242/jeb.075713. (visited on 01/29/2025)
31. Marzke, M.W., Wullstein, K.L., Viegas, S.F.: Evolution of the power ("squeeze") grip and its morphological correlates in hominids. Am. J. Phys. Anthropol. **89**(3), 283–298, ISSN: 0002-9483, 1096-8644 (1992). https://doi.org/10.1002/ajpa.1330890303. (visited on 02/12/2025)

32. Grebenstein, M.: Approaching human performance: the functionality-driven awiwi robot hand. In: Springer Tracts in Advanced Robotics. Springer International Publishing, Cham, vol. 98, ISBN: 978-3-319-03592-5 978-3-319-03593-2 (2014). https://doi.org/10.1007/978-3-319-03593-2. (visited on 01/16/2025)
33. Gracovetsky, S., Iacono, S.: Energy transfers in the spinal engine. J. Biomed. Eng.**9**(2), 99–114, ISSN: 01415425 (1987). https://doi.org/10.1016/0141-5425(87)90020-3. (visited on 03/11/2025)
34. Knips, G., Zibner, S.K.U., Reimann, H., Schöner, G.: A neural dynamic architecture for reaching and grasping integrates perception and movement generation and enables on-line updating. Front. Neurorobotics **11**, ISSN: 1662-5218 (2017). https://doi.org/10.3389/fnbot.2017.00009. (visited on 02/03/2025)
35. Zollo, L., et al.: An anthropomorphic robotic platform for progressive and adaptive sensorimotor learning. Adv. Robot. **22**(1), 91–118, ISSN: 0169-1864, 1568-5535 (2008). https://doi.org/10.1163/156855308X291854. (visited on 02/03/2025)
36. Ciancio, A.L., Zollo, L., Baldassarre, G., Caligiore, D., Guglielmelli, E.: The role of learning and kinematic features in dexterous manipulation: a comparative study with two robotic hands. Int. J. Adv. Robot. Syst. **10**(10), 340, ISSN: 1729-8806, 1729-8814 (2013). https://doi.org/10.5772/56479. (visited on 02/03/2025)
37. Santina, C.D., et al.: Learning from humans how to grasp: a data-driven architecture for autonomous grasping with anthropomorphic soft hands. IEEE Robot. Autom. Lett. **4**(2), 1533–1540, ISSN: 2377-3766, 2377-3774 (2019). https://doi.org/10.1109/LRA.2019.2896485. (visited on 02/03/2025)
38. Recatalá, G., Chinellato, E., del Pobil, Á, P., Mezouar, Y., Martinet, Y.: Biologically-inspired 3D grasp synthesis based on visual exploration. Auton. Robot. **25**(1–2), 59–70, ISSN: 0929-5593, 1573-7527 (2008). https://doi.org/10.1007/s10514-008-9086-7. (visited on 02/03/2025)
39. Romano, J.M., Hsiao, K., Niemeyer, G., Chitta, S., Kuchenbecker, K.J.: Human-inspired robotic grasp control with tactile sensing. IEEE Trans. Robot. **27**(6), 1067–1079, ISSN: 1552-3098 (2011). https://doi.org/10.1109/TRO.2011.2162271. (visited on 02/03/2025)
40. Hu, Y., Li, Z., Li, G., Yuan, P., Yang, C., Song, R.: Development of sensory-motor fusion-based manipulation and grasping control for a robotic hand-eye system. IEEE Trans. Syst., Man, Cybern.: Syst. 1–12 ISSN: 2168-2216, 2168-2232 (2016). https://doi.org/10.1109/TSMC.2016.2560530. (visited on 02/03/2025)
41. Fifer, M.S., et al.: Simultaneous neural control of simple reaching and grasping with the modular prosthetic limb using intracranial EEG. IEEE Trans. Neural Syst. Rehabil. Eng. **22**(3), 695–705, ISSN: 1534-4320, 1558-0210 (2014). https://doi.org/10.1109/TNSRE.2013.2286955. (visited on 02/03/2025)
42. Kim, S., Laschi, C., Trimmer, B.: Soft robotics: a bioinspired evolution in robotics. Trends Biotechnol. **31**(5), 287–294, ISSN: 0167-7799 (2013). https://doi.org/10.1016/j.tibtech.2013.03.002. (visited on 01/14/2025)
43. Rus, D., Tolley, M.T.: Design, fabrication and control of soft robots. Nature **521**(7553), 467–475, ISSN: 0028-0836, 1476-4687 (2015). https://doi.org/10.1038/nature14543. (visited on 01/14/2025)
44. Buchner, T.J.K., et al.: Vision-controlled jetting for composite systems and robots. Nature **623**(7987), 522–530, ISSN: 0028-0836, 1476-4687 (2023). https://doi.org/10.1038/s41586-023-06684-3. (visited on 01/16/2025)
45. Billard, A.G.: In good hands: A case for improving robotic dexterity. Science **386**(6727), ISSN: 0036-8075, 1095-9203 (2024). https://doi.org/10.1126/science.adu2950. (visited on 01/14/2025)
46. Li, Y., Wang, P., Li, R., Tao, M., Liu, Z., Qiao, H.: A survey of multifingered robotic manipulation: biological results, structural evolvements, and learning methods. Front. Neurorobotics **16**, 843267, ISSN: 1662-5218 (2022). https://doi.org/10.3389/fnbot.2022.843267. (visited on 01/16/2025)
47. Puhlmann, S., Harris, J., Brock, O.: *RBO Hand 3*: a platform for soft dexterous manipulation. IEEE Trans. Robot. **38**(6), 3434–3449, ISSN: 1552-3098, 1941-0468 (2022). https://doi.org/10.1109/tro.2022.3156806. (visited on 01/14/2025)

48. Gilday, K., Hughes, J., Iida, F.: Sensing, actuating, and interacting through passive body dynamics: a framework for soft robotic hand design. Soft Robotics **10**(1), 159–173, ISSN: 2169-5172, 2169-5180 (2023). https://doi.org/10.1089/soro.2021.0077. (visited on 01/14/2025)
49. Tomović, R.: Advances in the design of autonomous dextrous hands. Robot. Comput.-Integr. Manuf. **7**(3–4), 381–385, ISSN: 07365845 (1990). https://doi.org/10.1016/0736-5845(90)90025-4. (visited on 01/16/2025)
50. Ozawa, R., Tahara, K.: Grasp and dexterous manipulation of multi-fingered robotic hands: a review from a control view point. Adv. Robot. **31**(19–20), 1030–1050, ISSN: 0169-1864, 1568-5535 (2017). https://doi.org/10.1080/01691864.2017.1365011. (visited on 01/16/2025)
51. Cabibihan, J.J., Joshi, D., Srinivasa, Y.M., Chan, M.A., Muruganantham, A.: Illusory sense of human touch from a warm and soft artificial hand. IEEE Trans. Neural Syst. Rehabil. Eng. **23**(3), 517–527, ISSN: 1534-4320, 1558-0210 (2015). https://doi.org/10.1109/TNSRE.2014.2360533. (visited on 01/15/2025)
52. Ueno, A., Hlavac, V., Mizuuchi, I., Hoffmann, M.: Touching a human or a robot? investigating human-likeness of a soft warm artificial hand. In: 2020 29th IEEE International Conference on Robot and Human Interactive Communication (RO-MAN), Naples, Italy: IEEE, pp. 14–20, ISBN: 978-1-7281-6075-7 (2020). https://doi.org/10.1109/RO-MAN47096.2020.9223523. (visited on 01/15/2025)

Integrating Unconventional Approaches for Intelligent and Adaptive Soft Robotic Systems

Hwayeong Jeong and Jamie Paik

Natural organisms interact with their environments in highly sophisticated and adaptive ways [1–3]. As robots are increasingly expected to integrate seamlessly into the real world, a critical requirement for their success is the ability to interact intelligently with complex natural environments that include human presence [4–7]. These dynamic and often unpredictable settings demand heightened adaptability, responsiveness to environmental stimuli, and the capacity for effective coexistence and collaboration with humans.

What will it take for robotics to fundamentally transform how robots interact with humans and the natural world? Robots must be seamlessly integrated into their environments, achieving a form of omnipresence. They should exhibit morphological intelligence, allowing them to adapt their physical form and behavior to shifting conditions, and possess deep contextual awareness to enable informed and responsive interactions. Soft robotics presents a promising pathway toward this vision [8–10]. By leveraging embodied intelligence through compliant, deformable structures, soft robots have the potential to interact with their surroundings in a safe, intuitive manner [11–13]. However, despite notable progress, soft robotics still faces significant barriers to widespread adoption. Realizing its full potential demands an integrated, system-level architecture in which all functional components are cohesively agglomerated within compliant bodies. Paradoxically, the very property that gives soft robots their strengths to their inherent softness to also presents fundamental challenges to achieving such agglomerate integration. In contrast, natural organisms possess highly complex body structures, often composed in part of soft materials, yet they exhibit exceptional adaptability through precise and effective control of their

H. Jeong · J. Paik (✉)
Reconfigurable Robotics Lab, ÉcolePolytechnique Fédérale de Lausanne (EPFL),Lausanne, Switzerland
e-mail: jamie.paik@epfl.ch

H. Jeong
e-mail: hwayeong.jeong@epfl.ch

A. Raatz et al. (eds.), *Soft Material Robotic Systems*,
https://doi.org/10.1007/978-3-032-22453-8_12

bodies. Drawing inspiration from such systems, researchers are developing novel frameworks that incorporate AI-driven learning to address similar challenges in engineered systems, enabling robots to navigate and respond intelligently to dynamic, unpredictable environments.

The rapid progress of modern AI has been largely fueled by its ability to collect and learn from vast and diverse datasets sourced from across the digital landscape [14, 15]. Today, AI systems can transfer knowledge by leveraging previously acquired data—for instance, interpreting new images by referencing patterns learned from earlier datasets [16–18]. This capability extends beyond visual data to include textual information as well, exemplified by the success of large language models [19, 20]. However, this raises an important question: **What new domains must be explored to drive the next wave of AI innovation?** One critical limitation in current AI systems is the absence of physical interaction data [21, 22]. Unlike visual or textual data, AI lacks access to the rich, multi-modal information that arises from direct interaction with the physical world. This includes sensory modalities such as contact force, texture, friction, temperature, and even smell, all of which are critical dimensions for natural and human interactions that remain largely missing from current datasets. Most AI systems remain confined to visual or textual data. Addressing the absence of physical interaction data presents new opportunities in robotics, haptics, and embodied learning, paving the way for AI systems capable of sensing, interpreting, and responding to their environments with greater contextual awareness. These capabilities are essential for the next generation of AI, particularly for systems designed to function effectively within and alongside the physical world.

Soft robots are increasingly recognized as ideal platforms for acquiring rich, context-sensitive data through interactive scenarios [23–26]. Their ability to maneuver through confined or dynamic spaces and operate safely in hazardous environments makes them particularly well-suited for collecting large-scale datasets that are otherwise difficult to access. These capabilities are also advantageous in human-robot interaction, where soft robots can safely engage in close physical contact, capturing fine-grained motion, tactile, and behavioral data without risk. Moreover, they can be designed to mimic human movement, providing a scalable, reproducible source of human-like motion data. This is especially valuable in fields such as rehabilitation, haptics, and human-robot collaboration, where obtaining consistent, high-fidelity datasets from human subjects can be challenging [27, 28].

How, then, are researchers advancing core technologies to unlock soft robots' unique interaction capabilities that support data collection? Researchers have explored a range of innovative and unconventional strategies to enhance the performance and autonomy of soft robots. Smart materials capable of muscle-like actuation enable decentralized, adaptive behaviors, while soft actuators embedded with localized control logic promote scalable and compact designs through distributed decision-making [29–32]. This decentralization extends to sensing, where morphological computation leverages the robot's physical structure to process information inherently, reducing reliance on centralized control systems. Concurrently, advancements in power sources, including flexible batteries, energy harvesting technologies, and integrated energy systems, are facilitating the development of fully soft, untethered robotic platforms capable of long-term, autonomous operation in complex environments.

Pushing the boundaries of conventional design, biohybrid robots offer distinct benefits for data collection by integrating living cells, tissues, and neural components into robotic platforms [33–35]. These systems exhibit lifelike motion and high energy efficiency through muscle-based actuators, while biologically derived sensors can detect subtle chemical or mechanical cues beyond the reach of conventional devices. By capturing the complexity of biological processes, biohybrid robots generate rich, high-dimensional datasets. Their potential is particularly evident in sensitive domains such as internal medicine and ecological monitoring, where soft, efficient, and minimally invasive solutions are essential. Ranging from single-cell constructs to organism-integrated systems, biohybrid robotics represents a compelling frontier for biologically integrated automation and environmental data collection.

While recent progress in soft and biohybrid robotics has been substantial, several persistent and foundational challenges continue to impede widespread implementation. Power distribution continues to be a major hurdle, as traditional centralized power systems are incompatible with flexible architectures. Embedding energy sources within the robot's body presents a promising alternative, yet current technologies face limitations in energy density and durability. Sensor integration presents another layer of complexity—soft sensors must operate reliably under constant deformation, while signal noise from compliant materials or biological variability further complicates performance. Tackling these challenges will require advances in sensor fusion, error correction, and adaptive learning. Control also poses significant difficulties. Soft robots exhibit high degrees of freedom, nonlinear behavior and operate under uncertain environmental conditions—all of which demand sophisticated control algorithms. While AI offers powerful tools, real-time learning and adaptation remain limited, especially when faced with variability and unpredictability. A further challenge lies in the lack of integrated, system-level design approaches. Many soft robotic systems remain component-based, hindering scalability and seamless functionality. To move forward, there is a pressing need for standardized, modular building blocks, combining actuators, sensors, and controllers into cohesive, scalable architectures.

The future of soft robotics does not lie in the mere integration of discrete components but in the development of fully unified, system-level architectures that fundamentally rethink how robots are conceived, built, and interact. By embracing unconventional design strategies, the field is progressing toward a new paradigm of soft intelligence—one that is embodied, decentralized, and inherently responsive to the complexities of dynamic environments, including those shaped by nature and human interaction. In this framework, intelligent behavior arises not merely from passive adaptability, but through the active integration of distributed sensing, localized control, and morphological computation. Soft robots leverage their physical form and material properties to interpret and respond to their surroundings, enabling contextually meaningful interactions. In this evolution, soft robots offer not only novel ways of engaging with the world but also powerful platforms for generating rich, high-resolution datasets essential for training the next generation of intelligent systems. Moving forward, progress will hinge on a sustained commitment to pushing the boundaries of design, materials science, and computational models. This will ultimately shaping a new class of robotics that is soft in structure and equally soft in intelligence, adaptability, and interaction.

References

1. Polzin, M., Guan, Q., Hughes, J.: Robotic locomotion through active and passive morphological adaptation in extreme outdoor environments. Sci. Robot. **10**(99), 6419 (2025)
2. Shah, D.S., Powers, J.P., Tilton, L.G., Kriegman, S., Bongard, J., Kramer-Bottiglio, R.: A soft robot that adapts to environments through shape change. Nat. Mach. Intell. **3**(1), 51–59 (2021)
3. Richter, J.N., Hochner, B., Kuba, M.J.: Octopus arm movements under constrained conditions: adaptation, modification and plasticity of motor primitives. J. Exp. Biol. **218**(7), 1069–1076 (2015)
4. Cully, A., Clune, J., Tarapore, D., Mouret, J.-B.: Robots that can adapt like animals. Nature **521**(7553), 503–507 (2015)
5. Mon-Williams, R., Li, G., Long, R., Du, W., Lucas, C.G.: Embodied large language models enable robots to complete complex tasks in unpredictable environments. Nat. Mach. Intell. 1–10 (2025)
6. Yano, Y., Mizutani, A., Fukuda, Y., Kanaoka, D., Ono, T., Tamukoh, H.: Unified understanding of environment, task, and human for human-robot interaction in real-world environments. In: 2024 33rd IEEE International Conference on Robot and Human Interactive Communication (ROMAN), pp. 224–230. IEEE (2024)
7. Nygaard, T.F., Martin, C.P., Howard, D., Torresen, J., Glette, K.: Environmental adaptation of robot morphology and control through real-world evolution. Evol. Comput. **29**(4), 441–461 (2021)
8. Sun, J., Lerner, E., Tighe, B., Middlemist, C., Zhao, J.: Embedded shape morphing for morphologically adaptive robots. Nat. Commun. **14**(1), 6023 (2023)
9. van Laake, L.C., Overvelde, J.T.B.: Bio-inspired autonomy in soft robots. Commun. Mater. **5**(1), 198 (2024)
10. Kortman, V.G., Mazzolai, B., Sakes, A., Jovanova, J.: Perspectives on intelligence in soft robotics. Adv. Intell. Syst. **7**(1), 2400294 (2025)
11. Sadati, S.H., El Diwiny, M., Nurzaman, S., Iida, F., Nanayakkara, T.: Embodied intelligence & morphological computation in soft robotics community: collaborations, coordination, and perspective. In: IOP Conference Series: Materials Science and Engineering, vol. 1261, p. 012005. IOP Publishing (2022)
12. Mengaldo, G., Renda, F., Brunton, S.L., Bächer, M., Calisti, M., Duriez, C., Chirikjian, G.S., Laschi, C.: A concise guide to modelling the physics of embodied intelligence in soft robotics. Nat. Rev. Phys. **4**(9), 595–610 (2022)
13. Cianchetti, M., Follador, M., Mazzolai, B., Dario, P., Laschi, C.: Design and development of a soft robotic octopus arm exploiting embodied intelligence. In: 2012 IEEE International Conference on Robotics and Automation, pp. 5271–5276. IEEE (2012)
14. Rawas, S.: Ai: the future of humanity. Discov. Arti. Intell. **4**(1), 25 (2024)
15. Roh, Y., Heo, G., Whang, S.E.: A survey on data collection for machine learning: a big data-AI integration perspective. IEEE Trans. Knowl. Data Eng. **33**(4), 1328–1347 (2019)
16. Ahmed, M., Zhang, X., Shen, Y., Ali, N., Flah, A., Kanan, M., Alsharef, M., Ghoneim, S.S.: A deep transfer learning based convolution neural network framework for air temperature classification using human clothing images. Sci. Rep. **14**(1), 31658 (2024)

17. Kadeethum, T., O'Malley, D., Choi, Y., Viswanathan, H.S., Yoon, H.: Progressive transfer learning for advancing machine learning-based reduced-order modeling. Sci. Rep. **14**(1), 15731 (2024)
18. Shaha, M., Pawar, M.: Transfer learning for image classification. In: 2018 Second International Conference on Electronics, Communication and Aerospace Technology (ICECA), pp. 656–660. IEEE (2018)
19. Naveed, H., Khan, A.U., Qiu, S., Saqib, M., Anwar, S., Usman, M., Akhtar, N., Barnes, N., Mian, A.: A comprehensive overview of large language models. ACM Trans. Intell. Syst. Technol. (2023)
20. Huang, H., Zheng, O., Wang, D., Yin, J., Wang, Z., Ding, S., Yin, H., Xu, C., Yang, R., Zheng, Q., et al.: Chatgpt for shaping the future of dentistry: the potential of multi-modal large language model. Int. J. Oral Sci. **15**(1), 29 (2023)
21. Zhang, P., Zhang, H., Xu, H., Xu, R., Wang, Z., Wang, C., Garg, A., Li, Z., Ajoudani, A., Liu, X.: Scaling laws in scientific discovery with ai and robot scientists. *arXiv preprint* arXiv:2503.22444 (2025)
22. Ahmad, S.F., Han, H., Alam, M.M., Rehmat, M., Irshad, M., Arraño-Muñoz, M., Ariza-Montes, A., et al.: Impact of artificial intelligence on human loss in decision making, laziness and safety in education. Human. Social Sci. Commun. **10**(1), 1–14 (2023)
23. Xu, Y., Zhang, S., Li, S., Wu, Z., Li, Y., Li, Z., Chen, X., Shi, C., Chen, P., Zhang, P., et al.: A soft magnetoelectric finger for robots' multidirectional tactile perception in non-visual recognition environments. NPJ Flexible Electron. **8**(1), 2 (2024)
24. Jiang, Y., Yin, S., Dong, J., Kaynak, O.: A review on soft sensors for monitoring, control, and optimization of industrial processes. IEEE Sens. J. **21**(11), 12868–12881 (2020)
25. Yoon, S.H., Huo, K., Zhang, Y., Chen, G., Paredes, L., Chidambaram, S., Ramani, K.: isoft: a customizable soft sensor with real-time continuous contact and stretching sensing. In: Proceedings of the 30th Annual ACM Symposium on User Interface Software and Technology, pp. 665–678 (2017)
26. Thuruthel, T.G., Shih, B., Laschi, C., Tolley, M.T.: Soft robot perception using embedded soft sensors and recurrent neural networks. Sci. Robot. **4**(26), 1488 (2019)
27. Garcia-Sosa, A., Quintana-Hernandez, J.J., Ballester, M.A.F., Carmona-Duarte, C.: Exploring the potential of robot-collected data for training gesture classification systems. *arXiv preprint*arXiv:2405.04241 (2024)
28. Bui, V., Alaei, A.: Virtual reality in training artificial intelligence-based systems: a case study of fall detection. Multimedia Tools Appl. **81**(22), 32625–32642 (2022)
29. Seong, M., Sun, K., Kim, S., Kwon, H., Lee, S.-W., Veerla, S.C., Kang, D.K., Kim, J., Kondaveeti, S., Tawfik, S.M., et al.: Multifunctional magnetic muscles for soft robotics. Nat. Commun. **15**(1), 7929 (2024)
30. Wang, J., Gao, D., Lee, P.S.: Recent progress in artificial muscles for interactive soft robotics. Adv. Mater. **33**(19), 2003088 (2021)
31. Xu, K., Pérez-Arancibia, N.O.: Electronics-free logic circuits for localized feedback control of multi-actuator soft robots. IEEE Robot. Autom. Lett. **5**(3), 3990–3997 (2020)
32. Van Raemdonck, B., Milana, E., De Volder, M., Reynaerts, D., Gorissen, B.: Nonlinear inflatable actuators for distributed control in soft robots. Adv. Mater. **35**(35), 2301487 (2023)
33. Guix, M., Mestre, R., Patiño, T., De Corato, M., Fuentes, J., Zarpellon, G., ánchez, S.S.: Biohybrid soft robots with self-stimulating skeletons. Sci. Robot. **6**(53), 7577 (2021)
34. Filippi, M., Yasa, O., Kamm, R.D., Raman, R., Katzschmann, R.K.: Will microfluidics enable functionally integrated biohybrid robots? Proc. Natl. Acad. Sci. **119**(35), e2200741119 (2022)
35. Appiah, C., Arndt, C., Siemsen, K., Heitmann, A., Staubitz, A., Selhuber-Unkel, C.: Living materials herald a new era in soft robotics. Adv. Mater. **31**(36), 1807747 (2019)

cc
BY NC ND

Soft Tensegrity Structures with Variable Stiffness and Shape Changing Ability

David Herrmann, Lukas Merker, Leon Schaeffer, Lukas Lehmann, Christoph Hemeling, Hannes Jahn, Lena Zentner, and Valter Böhm

Abstract Soft tensegrity structures with variable stiffness and shape changing ability hold significant potential for soft robotic applications. With few exceptions, this class of structures primarily consists of tensioned and compressed members (at least some of which are compliant) forming a prestressed, stable equilibrium state. Consequently, changes in prestress of the structures enable independent or combined adjustments of shape and stiffness, which can be achieved actively or passively. This article explores the fundamental realization principles of soft tensegrity structures focusing on application examples, including manipulators, force and contact sensors, multistable soft grippers, and locomotion systems. Additionally, it highlights the potential of smart materials for passive stiffness modulation. The findings highlight that soft tensegrity structures, characterized by variable stiffness and shape changing capabilities, play a crucial role in advancing the performance and adaptability of soft robotic systems.

Keywords Soft robotics · Tensegrity structure · Mechanical compliance · Manipulator · Sensor · Locomotion

1 Introduction

To enhance the performance of soft robotic systems, it is essential to explore new realization possibilities for these systems. In numerous applications, the utilization of mechanically prestressed compliant structures in soft robotic systems proves advantageous. The targeted stiffness of these structures can be reversibly adjusted without altering their shape, a process that can be generated only by a limited number of actuators. Soft tensegrity structures (TSs), which are based on highly elastic materials,

D. Herrmann (✉) · L. Schaeffer · L. Lehmann · V. Böhm
Regensburg University of Applied Sciences, Regensburg, Germany
e-mail: david.herrmann@st.oth-regensburg.de

L. Merker (✉) · C. Hemeling · H. Jahn · L. Zentner
Technische Universität Ilmenau, Ilmenau, Germany
e-mail: lukas.merker@tu-ilmenau.de

A. Raatz et al. (eds.), *Soft Material Robotic Systems*,
https://doi.org/10.1007/978-3-032-22453-8_13

correspond to a special class of mechanically prestressed structures. The structural configuration of these systems is characterized by the presence of a set of disconnected compressed members, which are interconnected by a continuous net of compliant tensioned members (Fig. 1). The resulting shape of these structures is defined by the prestress. Soft robots based on these structures offer several advantageous properties, such as foldability/deployability, low mass, high strength-to-weight ratio and shock-absorbing capabilities [1]. It is notable that these structures possess a significant capacity for both shape and stiffness modification. Intrinsically compliant TSs exhibit substantial distinctions from other compliant structures, a consequence of their prestress and distinctive morphology. In consideration of the two primary characteristics, these structures hold considerable promise for the advancement of sophisticated soft material robotic systems, characterized by their highly adaptable mechanical behavior. The integration of smart materials within these structures, equipped with simultaneous sensing and actuating capabilities, facilitates a substantial degree of functional integration within the systems. The potential applications of these structures within soft robotic systems are manifold, including but not limited to variable stiffness structures (which can be actively and passively modified in terms of mechanical compliance), actuator systems (for manipulation and locomotion purposes), and fully or partially compliant passive structures with predefined (adaptable) directional stiffnesses. In recent years, there has been a notable increase in the research focus on robotic systems based on these structures.

To realize compliant TSs, an innovative yet complex one-step manufacturing process was introduced in [2]. In the context of mobile robots, intensive investigations were carried out on cable-actuated systems for terrestrial locomotion based on the tensegrity icosahedron [3, 4]. In [5], a modular TS is considered as an actuation unit for a robotic fish. In [6–11], manipulation systems are presented, with a focus on determining movement behavior and control aspects. The actuation of these systems is facilitated by cables connecting the members or by the application of McKibben artificial muscles, positioned between the endpoints of compressed members. In [12], McKibben-type pneumatic actuators were employed to facilitate the locomotion of a mobile robot equipped with robotic skins. In the context of investigating alternative actuation methods for tensegrity-based robotic systems, the utilization of smart materials, such as dielectric elastomer actuators (DEA) [13, 14], liquid crystal elastomers

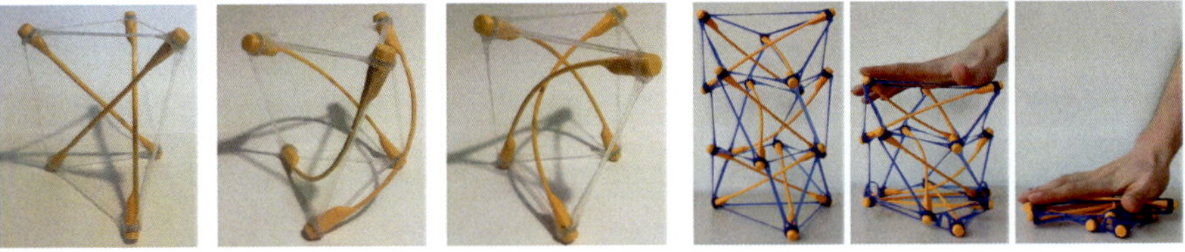

Fig. 1 Compliant soft TS based on flexible compressed members and highly elastic tensioned members

(LCE) [15] and magneto-sensitive elastomers [2], has been examined through preliminary investigations. Due to their mechanical compliance and stiffness adaptability, we have recently been explored compliant TS for use in biomedical applications beyond robotics (e.g. [16, 17]). From the literature review of the considered field, it can be concluded that soft material robotic systems based on soft, fully compliant TSs were not investigated in detail and systematic studies on these systems are not available. In this paper, we discuss realization possibilities in Sect. 2 and outline various application examples for these systems in Sect. 3 including manipulators, sensors, multistable soft grippers, and locomotion systems. Finally, we provide an evaluation and conclusion in Sect. 4.

2 Realization

Due to their specific morphology, intrinsically compliant TSs can serve as a bridge between fully and partially compliant soft robots. By altering only a few member parameters, a transition between the two sets of characteristics can be observed. In both cases, the identical design methodology and calculation methods can be employed, including partially and fully compliant soft robots. Furthermore, with only slight topology modifications (addition of members to a given structure), the overall characteristic of the structures can be influenced essentially. As a basis for soft robots, conventional and non-conventional TSs, based on partly or entirely highly elastic members, made of elastomer materials can be used (Fig. 2).

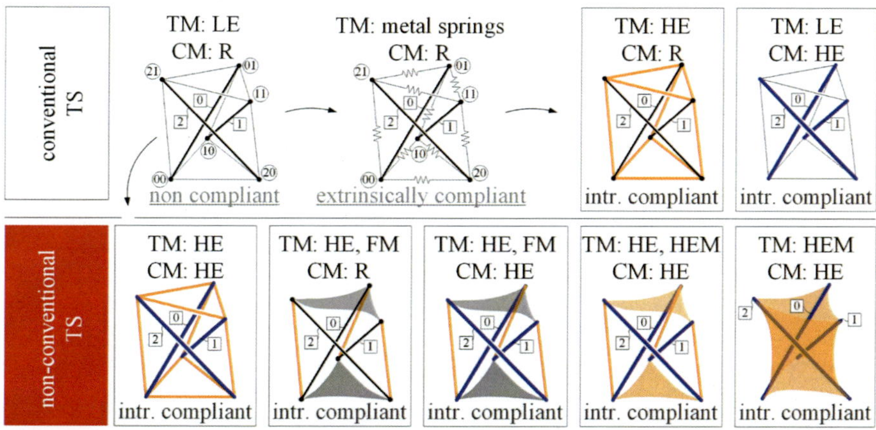

Fig. 2 Conventional and non-conventional TS with different member properties (orange/blue: intrinsically compliant tensioned/compressed members, TM: tensioned member, CM: compressed member)

Conventional intrinsically compliant TS consist of straight compressed members with negligible elasticity and highly elastic tensioned members [18]. Non-conventional intrinsically compliant TS, with increased shape and stiffness variability in comparison with conventional TS, can be realized by using

- highly elastic tensioned members and highly elastic compressed members,
- tensioned members with negligible elasticity and highly elastic compressed members,
- highly elastic thin membranes as tensioned members and rigid compressed members,
- thin flexible membranes with small elasticity as tensioned members and compliant compressed members.

In conventional TSs each member is connected to another member at those end points (two connecting points per each member). In non-conventional TSs, however, the potential use of members with more than two connecting points may also be considered. The integration of such components in conjunction with internal actuators, which facilitate shape modification, enables the realization of TSs with variable topologies, thereby enhancing their shape and stiffness modulation capabilities (Fig. 3).

The present focus of tensegrity research is on structures with minimal configurations, in which a minimal number of tensioned members are applied to realize self-tensioning. However, in soft robotic applications, TS with non-minimal configurations, involving more than the minimum necessary number of tensioned members, can also be targeted. The incorporation of additional tensioned members within these structures facilitates greater flexibility in modulating their stiffness and geometry, a phenomenon that is accompanied by an expected augmentation in magnitude. These supplementary tensioned members are positioned either between compressed members or between a compressed member and an additional node. In the latter scenario, the additional nodes are connected exclusively to tensioned members, a configuration

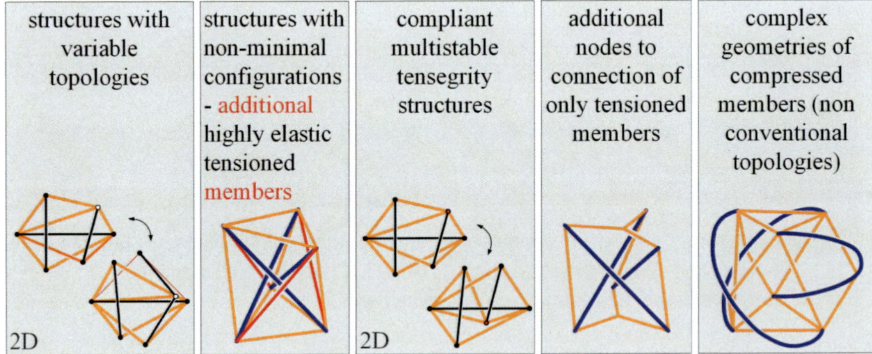

Fig. 3 TS with specific morphological properties and behavior (orange/blue: intrinsically compliant tensioned/compressed members)

analogous to their implementation in star-shaped TS [19]. In addition to elementary tensegrity units, consideration can be given to clustered TSs. These are constructed by means of the repetitive connection of several elementary tensegrity units. Due to their modular design, soft material robotic systems can be realized, which can be highly adapted to specific individual application tasks [20, 21]. The utilization of compliant multistable elastomer TS in soft robotic applications holds considerable promise. These structures possess multiple stable equilibrium configurations, defined as states of minimal overall local potential energy. Through the employment of varying prestress states and, consequently, disparate overall stiffnesses within these structures' different equilibrium configurations, the development of variable stiffness structures and actuators with discrete, switchable characteristics becomes feasible. In this instance, the alteration of stiffness and shape is contingent on the reconfiguration of constituent members. It is important to note that the energy expended is solely for the purpose of transitioning between these equilibrium configurations, rather than maintaining a constant prestress state within the structure. The main characteristics of the TS under consideration are:

- reduced morphological complexity—structures correspond to connections of members with simple geometries,
- global mechanical compliance—also in the case of using some stiff members,
- global complex shape change ability (whole body deformation) due to locally induced loads—possibility to create complex movement tasks with a small number of actuators,
- possibility to create variable stiffness structures—stiffness change of these structures can be realized without change of their shape, and furthermore, possibility of bidirectional stiffness change (increasing or decreasing stiffness),
- possibility to simultaneously change stiffness and shape—soft actuators with the feasibility of stiffness variability in their deformed states,
- reduced modelling complexity—due to uniaxial loading of straight members,
- foldability—small ratio between the sum of the member-volumes and the whole hull volume of the structure,
- redundancy—failure of actuators/sensors can be compensated by other actuators and sensors, placed on other members.

Furthermore, resulting from the use of soft intelligent materials with integrated sensing and actuation capabilities (e.g. polyborosiloxanes, thermo-magneto-sensitive elastomers) in TS a high degree of functional integration—simultaneous use of the members as sensors (detection of shape and/or stiffness change) and actuators as well as the possibility of passive adaptability to environmental/operational conditions (energy free adaptation capability) can result.

3 Application Examples

This section outlines application examples of applying shape and stiffness variable TSs as manipulators, sensors, multistable systems as well as locomotion systems.

3.1 Tensegrity-Manipulators

The development of planar and 3D tensegrity robotic arms (manipulators) is a current subject of research [22]. The conception of tensegrity manipulators is based on two design principles: either modular design, based on modules with identical topology, or the use of specific design for each individual functional system part [7, 23, 24]. The configuration of the majority of recognized tensegrity manipulators is predicated upon the utilization of stacked structures, a methodology that entails the arrangement of analogous or commensurate modules in a uniform configuration, with the modules themselves being either elementary 3D TSs or rigid or deformable frames. In [9, 14, 25–28], the focus is on tensegrity manipulators based on tensegrity-antiprisms and the tensegrity hexahedron, while [29–38] present systems comprising triangular or tetrahedral frames, or spatial star-shaped modules. The modules of stacked tensegrity manipulators for spatial movements are three-dimensional and interconnected directly or indirectly only by tensioned members. However, due to the complexity of their spatial geometry and the extensive use of tensioned members in their interconnection, the shape-changeability of these systems is constrained. The utilization of planar modules in 3D tensegrity manipulators has the potential to enhance the range of motion while maintaining a straightforward design and actuation process. These advantageous properties position a tensegrity robotic arm based on planar X-modules as a compelling solution. Our robotic arm, inspired by [39], consists of four identical system modules, each comprising a compressed X-member with two integrated actuator units [40]. Two aligned system modules are intrinsically compliantly connected by four intrinsically compliant passive tensioned members and four active inelastic tensioned members (red ropes in Fig. 4). The initial length of these ropes is adjusted by winding or unwinding them on a rope winch (P.1) powered by a DC motor (P.2). Starting from the motor, which is positioned at the center of the X-member, the rope extends outward through a guide in one of the two side covers (P.3) to reach the end of a compressed member on the opposite side. It is then fastened using a screw. The utilization of a solitary motor to concurrently wind and unwind a pair of tensioned members serves to streamline the actuation process. Consequently, the requirement for each compressed X-member is limited to a maximum of two motors with winches, thereby ensuring the attainment of the desired level of actuation. The experimental results demonstrate the system's ability to undergo manifold shape changes, thereby revealing a substantial workspace.

In the second example, the feasibility of employing flexible, controllable compressed members for actuation within a tensegrity manipulator is demonstrated, as

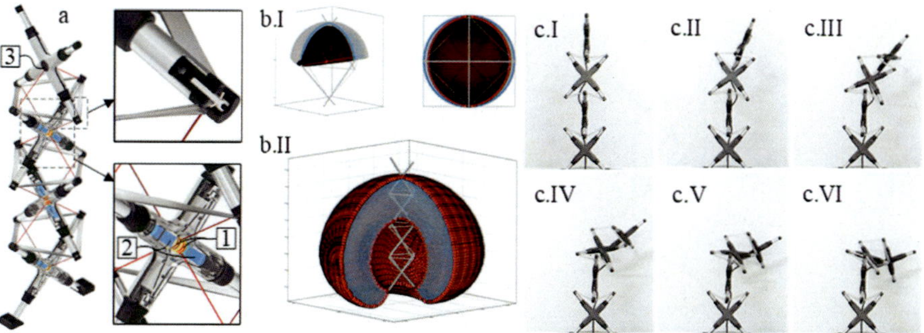

Fig. 4 Tensegrity manipulator (photos reprinted from [40] with permission, © 2024 IEEE): **a** CAD model of the system, **b** I: workspace of the base unit, II: workspace of the robotic arm with edge regions (red) and interspace (blue), **c** example image sequence

illustrated in Fig. 5. The chosen topology is based on three tensegrity anti-prisms stacked vertically with alternating torsional directions, consisting of 12 deformable compressed members (marked in black). These compressed members are indirectly connected by tensioned members, which are categorized into three distinct groups (see Fig. 5: red (1, 4) horizontal tensioned members at the top and bottom, blue (2) vertical tensioned members between layers, and orange (3) horizontal tensioned members at intermediate layers). Thanks to the symmetrical design of the deformable compressed members, they can be bent in two directions with the help of actuators, minimizing the risk of collisions with other members in the structure. Furthermore, the specific design of the compressed members prevents deformation due to the pre-stress induced by the tensioned members, ensuring that deformation only can occur under active control.

Each actuated compressed member consists of five distinct components: the symmetrical main body (P.1) with a cover (P.2), and the actuation unit comprising a DC motor (P.3), a winch (P.4), and a cable (P.5). Additionally, each side of the compressed member is equipped with a screw (P.6) and a washer (P.7) for mounting the tensioned members, as depicted in Fig. 5. The movement of the arm is induced by bending of selected compressed members in defined directions. Movements in the axial direction (stretching, compression) are also possible by simultaneous and uniform actuation of the compressed members of the structure.

3.2 *Sensor Application*

An inherent property of TSs is to transmit and convert external loads into uniaxial loading of the tensioned members. In turn, the resulting strains of the tensioned members can be detected and used to draw conclusions about external loads. In literature, this idea has been previously realized by equipping the tensioned members

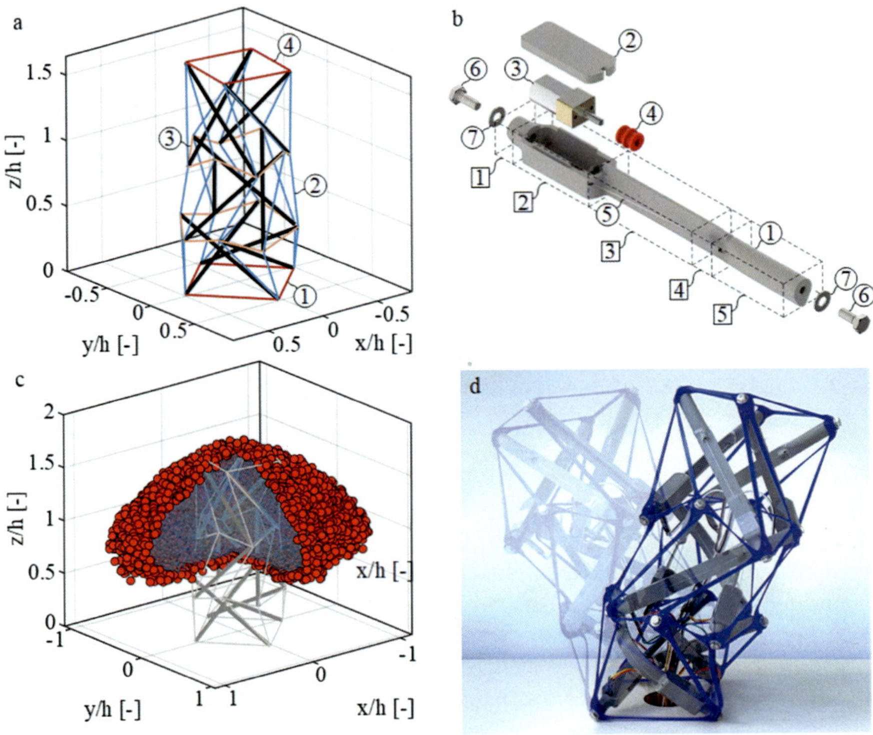

Fig. 5 Fully-compliant tensegrity robotic arm—actuation by bending of selected compressed members (photos reprinted from [41] with permission, © 2025 Springer): **a** topology of the considered tensegrity structures: compressed members (black); tensioned members: bottom (1)/top (4) circumferential (red), vertical (2, blue) and circumferential in the intermediate layer (3, orange), **b** CAD model of the compressed member with relevant system parts P.1 to P.7 and the different functional areas of the main body of the compressed member (framed in a rectangle), **c** workspace of the needle tower, **d** example of motion by activating a compressed member in each of the two lower layers

with load cells or strain gauges. A contrary approach is to not only extend the existing tensioned members with sensing capabilities, but to completely replace them using fiber optic sensors (FOS). In this way, a FOS multi-functionally serves both as a measuring device and load-bearing structural element, a concept embodied in the patent application [42].

Here, we employ Fiber Segment Interferometry (FSI), which utilizes range-resolved interferometry and sinusoidal laser diode modulation to demodulate signals based on optical path differences [43, 44]. This approach allows for range-resolved detection of strains within multiple segments of a single FOS with only one light input and output. The segments are defined by a pair of weak reflectors inscribed into the core of the fiber. Alternatively, the Fresnel reflections occurring at fiber end faces may be utilized. Both ideas are briefly demonstrated with two striking applications:

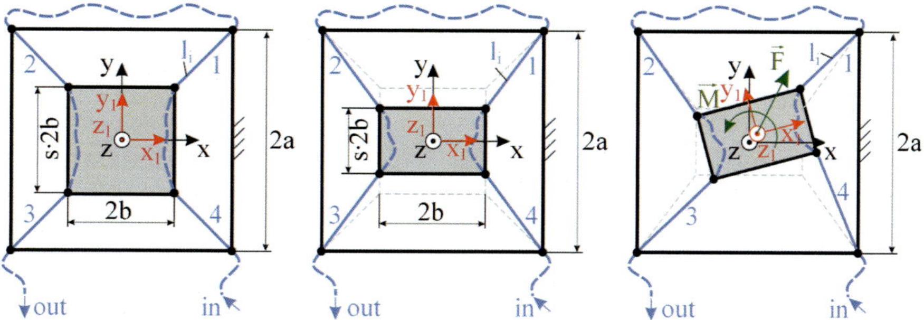

Fig. 6 Basic sensor design (photos reprinted from [45] with permission, © 2025 Springer): **a** relaxed configuration ($s = 1$), **b** pre-stressed reference configuration ($s < 1$), **c** loaded configuration

a theoretical proof of concept of a planar structure for force-torque measurement using a single multi-segment FOS [45], and an experimental one of a spatial structure using three individual FOS for force measurement and localization [44]. In both cases, pre-stress is a key aspect to change the stiffness behavior, measuring range, and sensitivity.

3.2.1 Example 1: Planar Structure for Force-Torque Measurement

The structure comprises two rigid bodies (black frames): an outer reference frame with edge length $2a$ and a floating body with edge lengths $2b$ and $s \cdot 2b$ (Fig. 6). These are connected at their four vertices by strain-sensitive segments (red lines) of a single FOS running through the structure. In the relaxed state (Fig. 6a), the pre-stress parameter is $s = 1$. Any adjustment mechanism setting $0 < s < 1$ establishes the pre-stressed reference configuration (Fig. 6b), where all tensioned fiber segments experience identical strain $0 < \varepsilon_i|_{i=1,...,4} = \varepsilon_p < \varepsilon_{\max}$. In the loaded configuration (Fig. 6c), a force-torque pair causes displacement (rotation and translation) of the floating body, with measurable tensioned member lengths $l_i > 0$.

To infer the applied force-torque-pair, these lengths are first used to calculate the position vector $\vec{r} = (x, y)^T$ and rotation angle φ of the floating body, as well as the orientations of the tensioned members:

$$\begin{aligned} \varphi &= \arcsin\left(\frac{C_0}{8\lambda a^2(s-1)}\right) \\ x &= \frac{C_1\left(1-s\lambda\cos(\varphi)\right)+C_2\lambda\sin(\varphi)}{-8a\left(s\lambda^2-(1+s)\lambda\cos(\varphi)+1\right)} \\ y &= \frac{C_2\left(1-\lambda\cos(\varphi)\right)-C_1 s\lambda\sin(\varphi)}{-8a\left(s\lambda^2-(1+s)\lambda\cos(\varphi)+1\right)} \end{aligned} \quad \text{with} \quad \begin{cases} C_0 = l_1^2 - l_2^2 + l_3^2 - l_4^2 \\ C_1 = l_1^2 - l_2^2 - l_3^2 + l_4^2 \\ C_2 = l_1^2 + l_2^2 - l_3^2 - l_4^2 \end{cases} \tag{1}$$

$$\vec{l}_i = \vec{r} + \mathbf{T}(\varphi)\,\vec{b}_i - \vec{a}_i, \quad l_i = |\vec{l}_i| = l_0(1+\varepsilon_i), \quad \vec{e}_i = \frac{\vec{l}_i}{l_i} \tag{2}$$

In (2), $\mathbf{T}$ is a standard rotation matrix, and $\vec{a}_i$ and $\vec{b}_i$ are vectors defined by a, b, and s. Using (2) and the material law of the FOS, the forces $\vec{F}_i$ in the tensioned members are used to infer the applied force-torque pair by the equilibrium conditions:

$$\boxed{\vec{F} = -\sum_{i=1}^{4} \vec{F}_i, \qquad \vec{M} = -\sum_{i=1}^{4} \mathbf{T}\,\vec{b}_i \times \vec{F}_i \quad \text{with} \quad \vec{F}_i = -EA\varepsilon_i\,\vec{e}_i} \tag{3}$$

The measuring range is limited by the strain restriction ($0 < \varepsilon_i < \varepsilon_{\max}$, $\forall i = 1, ..., 4$) and depends on the Young's modulus E, cross-sectional area A, and the ratio $\lambda = b/a > 0$. For $\lambda = 0.5$, the range for dimensionless forces $f = F/(EA)$ and torques $m = M/(EAa)$ is shown in Fig. 7.

3.2.2 Example 2: Spatial Structure for Force Measurement and Localization

In Fig. 8a, a schematic of the structure with the sensing setup is shown, along with a photograph of the actual setup, which consists of two rigid circular aluminum platforms with U-shaped bar assemblies connected by a central tension spring. Three peripheral FOS link the platforms at 120° intervals, counteracting the spring. The spring length is adjustable via a cable mechanism to control the structure's prestress and mechanical properties. The mechanical model and stiffness behavior are described in [46].

The experiments involve phase signal recording from the three FOS while the structure is loaded by a voice coil actuator in steps of 0.196 N, up to a maximum of 0.98 N, at positions marked in Fig. 8b. For loading position A3, the results are shown in Fig. 9a. The sum of the three filtered phase signals in Fig. 9c shows a linear correlation with the applied load at all positions. As the phase signal ratios $\Phi_i|_{i=1,2,3}$ vary with the loading position, they can be used to infer the loading point:

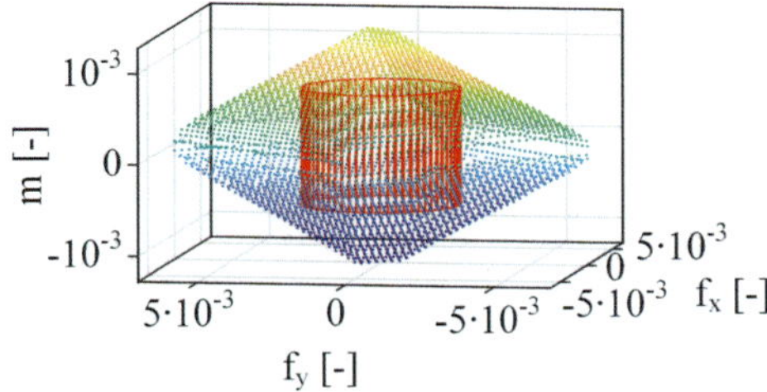

Fig. 7 Dimensionless combined force/torque measuring range for $\lambda = 0.5$ (photos reprinted from [45] with permission, © 2025 Springer)

$$x_\Phi = \frac{2\Phi_2 - \Phi_1 - \Phi_3}{2(\Phi_1 + \Phi_2 + \Phi_3)}, \qquad y_\Phi = \frac{\sqrt{3}(\Phi_1 - \Phi_3)}{2(\Phi_1 + \Phi_2 + \Phi_3))} \tag{4}$$

Plotting the reconstructed coordinates x_Φ and y_Φ against their respective reference values (x, y) in Fig. 9d reveals a linear correlation as well. The reconstructed points of force application for all load steps in Fig. 9b are generally in good agreement with the reference positions indicated by red markers.

3.3 Multistable Structures

Compliant multistable TS offer a great potential for applications in soft robotics. The main property of these structures is that their prestress states and therefore their overall stiffness differ in their equilibrium configurations [47]. In [47–52], we have shown that TSs, consisting of only few members, can have more than one stable equilibrium configuration and the overall stiffness of these structures differs significantly in these configurations. By proper design, a reversible change between the equilibrium configurations by internal reconfiguration of the members and therefore a discrete change of the overall stiffness can be induced. Based on these structures, e.g. soft gripper end effectors with discrete variable stiffness, to grip objects with preliminary defined different maximal gripping forces, can be realized (Fig. 10a). In [53], we have shown that TS with the simplest topologies, consisting of only two compressed member groups as equilateral triangles can show multistable characteristics (Fig. 10b).

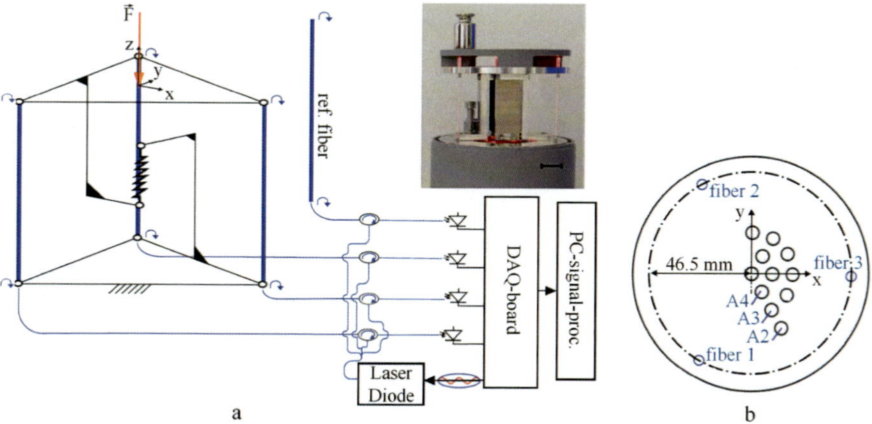

Fig. 8 Experimental setup (photos reprinted from [44] with permission, © 2025 SPIE): **a** schematic representation with photograph, **b** top platform with different loading positions

3.4 Locomotion

Previous research has shown that the application of curved members in TSs indicates their potential ability for use in rolling mobile robots. Here, the considered TS comprises two curved bending members, which are connected by eight tensioned members as shown in Fig. 11 [54, 55]. The theory of large deflections of rod-like structures [56] is used to analyze the deformation of the bending members [55]. Exploiting the symmetry of the structure, the system is reduced to a model of a one-sided-clamped beam of length L loaded at the free end (Fig. 11b). The mathematical treatment yields the following nonlinear boundary value problem, where s denotes the natural coordinate arc length parameter, θ and κ_0 the slope angle and pre-curvature of the beam axis, M_z the bending moment, EI_z the bending stiffness of the beam, and $\{F_x, F_y\}$ the force components:

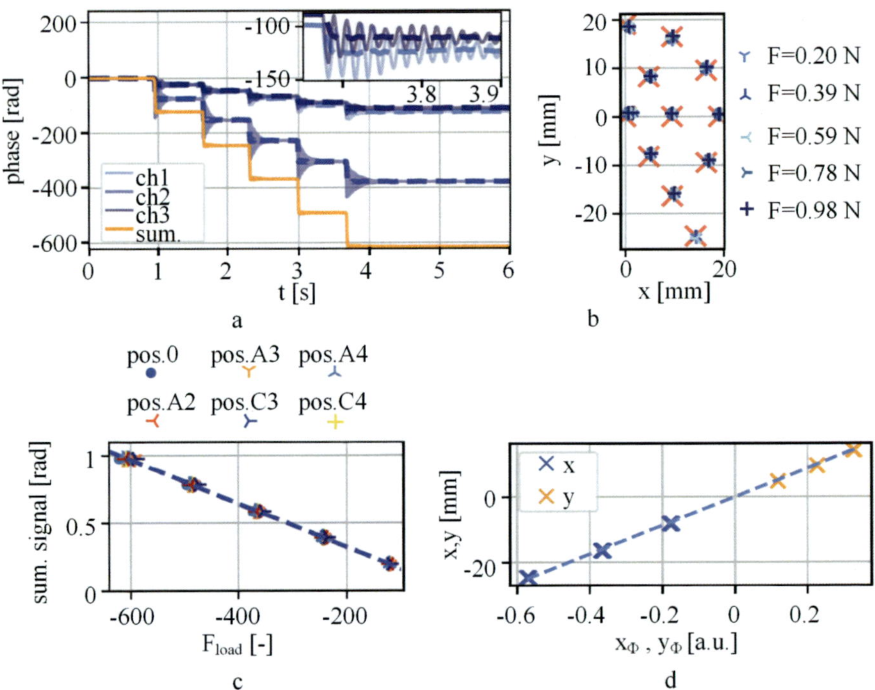

Fig. 9 Results (photos reprinted from [44] with permission, © 2025 SPIE): **a** separate phase signals with sum of phase signals for eccentric loading at A3 in steps of 0.196 N, **b** linear correlation between applied force and sum signal, **c** linear fit used to obtain real coordinates for loading point from x_Φ, y_Φ according to (4), **d** determination of loading points from measured phase signals, reference positions marked in red

$$\frac{dM_z}{ds} + F_y \cos\theta - F_x \sin\theta = 0 \qquad M_z(L) = 0 \tag{5}$$

$$\frac{d\theta}{ds} - \frac{M_z}{EI_z} - \kappa_0 = 0 \qquad \theta(0) = 0 \tag{6}$$

$$\frac{dx}{ds} - \cos\theta = 0 \qquad x(0) = 0 \tag{7}$$

$$\frac{dy}{ds} - \sin\theta = 0 \qquad y(0) = 0 \tag{8}$$

The deformation of one bending member was numerically calculated and validated in a parameter study using FEM with displacements diverting approximately 0.1 %. The analytical model is used with equilibrium conditions to find the relation between the forces of the tensioned members and the resulting stiffness and shape of the structure, which is sufficiently defined by the distance parameter d (Fig. 11). Affecting the curvature of the bending members, the distance parameter has significant impact on the system's locomotion behavior. This is exemplarily shown in Fig. 12, comparing the resulting trajectories of the structure with undeformed (rigid) bending members

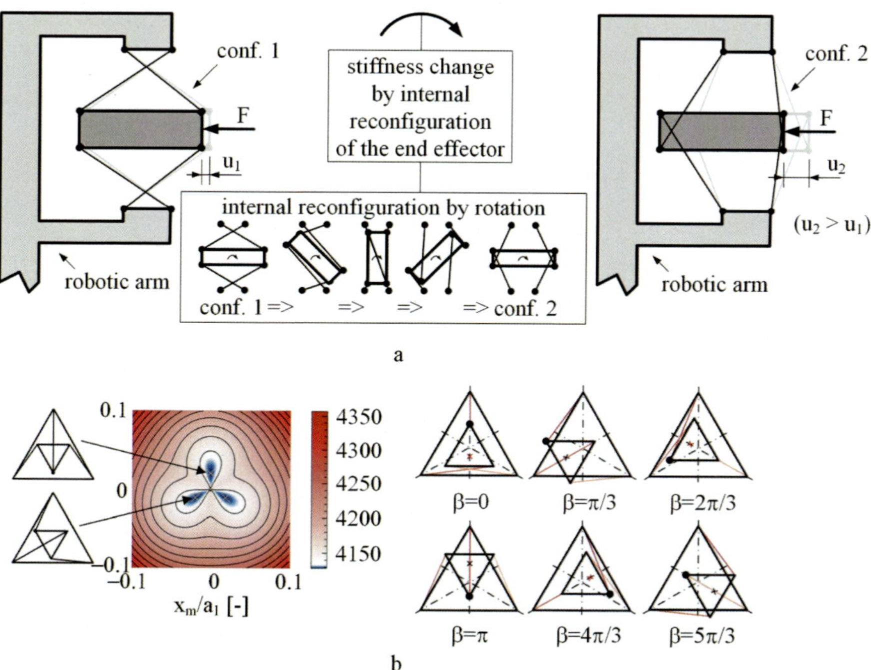

Fig. 10 Multistable structures: **a** use of stiffness change by reconfiguration of a multistable TS in an end effector application (photos reprinted from [47] with permission, © 2020 Springer), **b** multistable TS with three stable equilibrium configurations ($\beta=\pi/3$, $\beta=\pi$ and $\beta=5\pi/3$) with potential energy plot on the left based on two compressed member groups as equilateral triangles [based on [53]]

(initial structure) and deformed (flexible) bending members (deformed structure) for different ratios $\frac{d}{R}$, where R denotes the radius of the undeformed bending members.

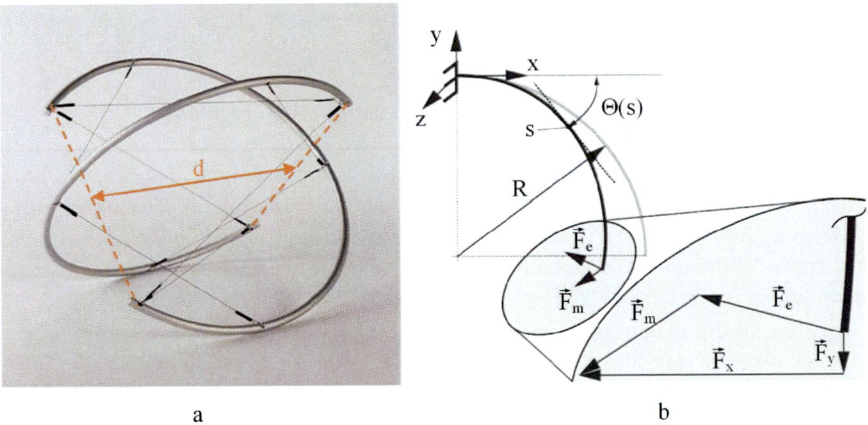

Fig. 11 Tensegrity structure with two curved members, connected by eight tensioned members (photos reprinted from [55] with permission, © 2024 Springer): **a** tensegrity structure with characteristic distance d, **b** one half of a curved element with the model parameters and the acting forces

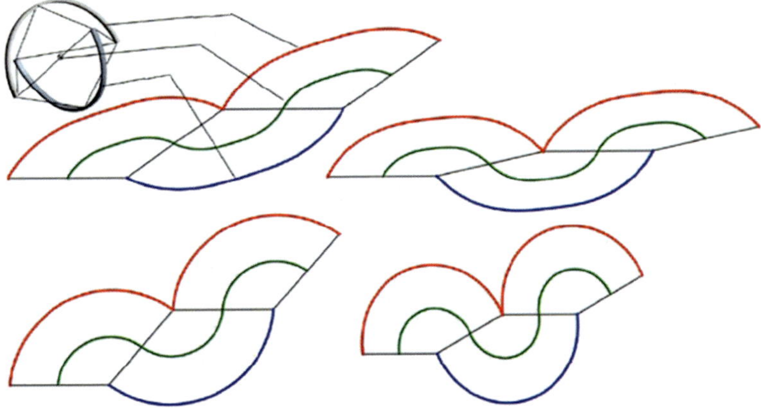

Fig. 12 Locomotion of initial and deformed structure (photos reprinted from [55] with permission, © 2024 Springer): Top view on trajectories of contact points of curved members (red and blue) and on trajectory of center of gravity (green). Upper row—initial and deformed structure with $\frac{d}{R} = 0.9$, lower row—initial and deformed structure with $\frac{d}{R} = 0.1$

3.5 *Towards Passive Stiffness Change Using Polyborosiloxane*

Beyond the active stiffness control of TSs, smart materials enable passive stiffness changes without external energy. These materials adapt to external stimuli, with polyborosiloxane (PBS) being a notable example. The synthesis of polydimethylsiloxane (PDMS) and boric acids exhibits viscoelastic behavior under shear stress as a function of load duration: it flows under slow-acting loads but remains solid and elastic under fast-acting loads, making it ideal for soft, impact-resistant structures. To explore tunable PBS properties, we prepared PBS samples using two PDMS types with distinct viscosities ($45 - 85 \cdot 10^{-6}\ \mathrm{m^2 s^{-1}}$ and $750 \cdot 10^{-6}\ \mathrm{m^2 s^{-1}}$). First, we synthesized PBS from each PDMS and mixed the final products in specific ratios. Then, we achieved the same ratios by blending PDMS precursors before PBS synthesis (pre-mixed samples) [57]. Additionally, we incorporated varying mass ratios of carbonyl iron powder as magnetically responsive particles [58].

Fourier Transform Infrared Spectroscopy, Raman Spectroscopy, and Scanning Electron Microscopy were used to assess the quality of synthesized PBS samples. Rheological analysis revealed their material properties and the ability to continuously tune the storage modulus, influenced by the intensity of Si-O-B bonds. Pre-mixed samples provided better predictability of rheological behavior across mixing ratios. Frequency sweep tests revealed the gel points, marking the transition from a viscous liquid to a solid rubbery state – critical for soft robotics applications. Additionally, shape-preserving tests clarified PBS flow behavior under gravity, which mechanical vibrations and external magnetic fields influenced shape retention duration [57, 58]. The Material characterization suggests potential for passive adaptability in TSs.

4 Conclusion

Soft tensegrity structures offer a versatile and promising approach for advancing soft robotic applications due to their intrinsic compliance, variable stiffness, and shape changing abilities. Their unique combination of tensioned and compressed members in a prestressed equilibrium state enables precise control over mechanical properties through active or passive adjustments of prestress. This ability allows for independent or coupled modifications of shape and stiffness, making them highly adaptable to dynamic and unstructured environments.

Through the exploration of various application examples, including manipulators, force and contact sensors, multistable grippers, and locomotion systems, this study underscores the broad applicability of soft tensegrity structures in robotics. These systems demonstrate enhanced adaptability, improved interaction capabilities, and functional versatility compared to traditional soft robotic designs. Furthermore, the integration of smart materials presents additional opportunities for passive stiffness modulation, reducing the need for complex actuation systems and expanding the range of potential applications. By bridging the gap between fully and partially compliant soft robots, tensegrity structures provide an innovative framework for the development of next-generation robotic systems. Their ability to combine flexibility with structural stability enables novel designs that balance compliance, load-bearing capacity, and responsive actuation. Future research should focus on refining fabrication techniques, optimizing control strategies, and exploring new material integrations to further enhance their performance and reliability.

In summary, soft tensegrity structures represent a significant step forward in soft robotics, offering an efficient and scalable means to achieve adaptable, multifunctional, and resilient robotic systems. Their potential to revolutionize soft robotic design highlights the need for continued interdisciplinary research at the intersection of materials science, mechanical engineering, and robotics.

Acknowledgements The authors gratefully acknowledge the support of the Deutsche Forschungsgemeinschaft (DFG, German Research Foundation) through Priority Program SPP 2100 "Soft Material Robotic Systems" (Projects BO4114/3, ZE714/14, Project number 405033228).

References

1. Paul, C., Roberts, J.W., Lipson, H., Valero Cuevas, F.J.: Gait production in a tensegrity based robot. In: ICAR '05. Proceedings., 12th International Conference on Advanced Robotics, 2005 (2005)
2. Lee, H., Jang, Y., Choe, J.K., Lee, S., Song, H., Lee, J.P., Lone, N., Kim, J.: 3d-printed programmable tensegrity for soft robotics. Sci. Robot. **5**(45), 9024 (2020)
3. Kim, K., Agogino, A.K., Agogino, A.M.: Rolling locomotion of cable-driven soft spherical tensegrity robots. Soft Rob. **7**(3), 346–361 (2020)
4. Zheng, Y., Li, Y., Lu, Y., Wang, M., Xu, X., Zhou, C., Luo, Y.: Robustness evaluation for rolling gaits of a six-strut tensegrity robot. Int. J. Adv. Rob. Syst. **18**(1), 1729881421993638 (2021)
5. Chen, B., Jiang, H.: Body stiffness variation of a tensegrity robotic fish using antagonistic stiffness in a kinematically singular configuration. IEEE Trans. Rob. **37**(5), 1712–1727 (2021)
6. Wei, D., Gao, T., Mo, X., Xi, R., Zhou, C.: Flexible bio-tensegrity manipulator with multi-degree of freedom and variable structure. Chinese J. Mech. Eng. **33**(3) (2020)
7. Li, W.-Y., Nabae, H., Endo, G., Suzumori, K.: New soft robot hand configuration with combined biotensegrity and thin artificial muscle. IEEE Robot. Automat. Lett. **5**(3), 4345–4351 (2020)
8. Zhou, H., Plummer, A.R., Cleaver, D.: Distributed actuation and control of a tensegrity-based morphing wing. IEEE/ASME Trans. Mechatron. **27**(1), 34–45 (2022)
9. Zappetti, D., Arandes, R., Ajanic, E., Floreano, D.: Variable-stiffness tensegrity spine. Smart Mater. Struct. **29**(7), 075013 (2020)
10. Li, W.-Y., Takata, A., Nabae, H., Endo, G., Suzumori, K.: Shape recognition of a tensegrity with soft sensor threads and artificial muscles using a recurrent neural network. IEEE Robot. Automat. Lett. **6**(4), 6228–6234 (2021)
11. Li, L., Kim, S., Park, J., Choi, Y., Lu, Q., Peng, D.: Robotic tensegrity structure with a mechanism mimicking human shoulder motion. J. Mech. Robot. **14**(2), 025001 (2021)
12. Booth, J.W., Cyr-Choinière, O., Case, J.C., Shah, D., Yuen, M.C., Kramer-Bottiglio, R.: Surface actuation and sensing of a tensegrity structure using robotic skins. Soft Rob. **8**(5), 531–541 (2021)

13. Boehler, Q., Abdelaziz, S., Vedrines, M., Poignet, P., Renaud, P.: From modeling to control of a variable stiffness device based on a cable-driven tensegrity mechanism. Mech. Mach. Theory **107**, 1–12 (2017)
14. Zappetti, D., Jeong, S.H., Shintake, J., Floreano, D.: Phase changing materials-based variable-stiffness tensegrity structures. Soft Rob. **7**(3), 362–369 (2020)
15. Wang, Z., Li, K., He, Q., Cai, S.: A light-powered ultralight tensegrity robot with high deformability and load capacity. Adv. Mater. **31**(7), 1806849 (2019)
16. Schaeffer, L., Herrmann, D., Schratzenstaller, T., Dendorfer, S., Böhm, V.: Preliminary theoretical considerations on the stiffness characteristics of a tensegrity joint for the use in dynamic orthoses. J. Med. Robot. Res. **08**(03n04), 2340008 (2023)
17. Schaeffer, L., Herrmann, D., Böhm, V.: Theoretical investigations on a dynamic hand orthosis based on a prestressed compliant structure with respect to stiffness and wrist-forces. In: 2024 International Symposium on Medical Robotics (ISMR) (2024)
18. Vega, J.C., Schorr, P., Kaufhold, T., Zentner, L., Zimmermann, K., Böhm, V.: Influence of elastomeric tensioned members on the characteristics of compliant tensegrity structures in soft robotic applications. Proc. Manuf. **52**, 289–294 (2020)
19. Jing Yao Zhang, M.O.: Tensegrity Structures—Form, Stability, and Symmetry. Springer, Japan (2015)
20. Herrmann, D., Schaeffer, L., Lehmann, L., Böhm, V., Rieffel, J.: Basic investigations on a compliant 2d tensegrity grid for the use in soft robotic applications. In: 2024 6th International Conference on Reconfigurable Mechanisms and Robots (ReMAR) (2024)
21. Rieffel, J., Herrmann, D., Lehmann, L., Schaeffer, L., Böhm, V.: Illuminating the morphological diversity of 2d tensegrity grids. In: Proceedings of Mechanism and Machine Theory Symposium, Book in Abstracts (2024)
22. Fasquelle, B., Furet, M., Khanna, P., Chablat, D., Chevallereau, C., Wenger, P.: A bio-inspired 3-dof light-weight manipulator with tensegrity x-joints. In: 2020 IEEE International Conference on Robotics and Automation (ICRA) (2020)
23. Lessard, S., Castro, D., Asper, W., Chopra, S.D., Baltaxe-Admony, L.B., Teodorescu, M., SunSpiral, V., Agogino, A.: A bio-inspired tensegrity manipulator with multi-dof, structurally compliant joints. In: 2016 IEEE/RSJ International Conference on Intelligent Robots and Systems (IROS) (2016)
24. Jung, E., Ly, V., Cessna, N., Ngo, M.L., Castro, D., SunSpiral, V., Teodorescu, M.: Bio-inspired tensegrity flexural joints. In: 2018 IEEE International Conference on Robotics and Automation (ICRA) (2018)
25. Fadeyev, D., Zhakatayev, A., Kuzdeuov, A., Varol, H.A.: Generalized dynamics of stacked tensegrity manipulators. IEEE Access **7**, 63472–63484 (2019)
26. Yoshimitsu, Y., Tsukamoto, K., Ikemoto, S.: Development of pneumatically driven tensegrity manipulator without mechanical springs. In: 2022 IEEE/RSJ International Conference on Intelligent Robots and Systems (IROS) (2022)
27. Kobayashi, R., Nabae, H., Suzumori, K.: Large torsion thin artificial muscles tensegrity structure for twist manipulation. IEEE Robotics and Automation Letters **8**(3), 1207–1214 (2023)
28. Yeshmukhametov, A., Koganezawa, K.: Design of a fully pulley-guided wire-driven prismatic tensegrity robot: Friction impact to robot payload capacity. IEEE Robotics and Automation Letters **8**(10), 6507–6514 (2023)
29. Tietz, B.R., Carnahan, R.W., Bachmann, R.J., Quinn, R.D., SunSpiral, V.: Tetraspine: Robust terrain handling on a tensegrity robot using central pattern generators. In: 2013 IEEE/ASME International Conference on Advanced Intelligent Mechatronics (2013)
30. Friesen, J., Pogue, A., Bewley, T., Oliveira, M., Skelton, R., Sunspiral, V.: Ductt: A tensegrity robot for exploring duct systems. In: 2014 IEEE International Conference on Robotics and Automation (ICRA) (2014)
31. Sabelhaus, A.P., Ji, H., Hylton, P., Madaan, Y., Yang, C., Agogino, A.M., Friesen, J., SunSpiral, V.: Mechanism Design and Simulation of the ULTRA Spine: A Tensegrity Robot. International Design Engineering Technical Conferences and Computers and Information in Engineering Conference, vol. Volume 5A: 39th Mechanisms and Robotics Conference, pp. 05–08059 (2015)

32. Naribole, S., Anil, R.K., Chakraborty, G.: Design, mathematical modelling and analysis of externally actuated somersaulting tensegrity spine. In: 2020 6th International Conference on Mechatronics and Robotics Engineering (ICMRE) (2020)
33. Shintake, J., Zappetti, D., Peter, T., Ikemoto, Y., Floreano, D.: Bio-inspired tensegrity fish robot. In: 2020 IEEE International Conference on Robotics and Automation (ICRA) (2020)
34. Sabelhaus, A.P., Zhao, H., Zhu, E.L., Agogino, A.K., Agogino, A.M.: Model-predictive control with inverse statics optimization for tensegrity spine robots. IEEE Trans. Control Syst. Technol. **29**(1), 263–277 (2021)
35. Gao, R., Liu, Y., Bi, Q., Yang, B., Li, Y.: Design of a novel quadruped robot based on tensegrity structures. In: 2021 IEEE International Conference on Mechatronics and Automation (ICMA) (2021)
36. Zhao, W., Pashkevich, A., Chablat, D.: Non-linear stiffness modeling of multi-link compliant serial manipulator composed of multiple tensegrity segments. 2021 IEEE 17th International Conference on Automation Science and Engineering (CASE), pp. 1636–1641 (2021)
37. Luo, J., Wu, Z., Xu, X., Chen, Y., Liu, Z., Ming, L.: Forward statics of tensegrity robots with rigid bodies using homotopy continuation. IEEE Robot. Automat. Lett. **7**(2), 5183–5190 (2022)
38. Ramadoss, V., Sagar, K., Ikbal, M.S., Calles, J.H.L., Siddaraboina, R., Zoppi, M.: Hedra: A bio-inspired modular tensegrity robot with polyhedral parallel modules. In: 2022 IEEE 5th International Conference on Soft Robotics (RoboSoft) (2022)
39. Snelson, K.: Art and Ideas. http://kennethsnelson.net/
40. Herrmann, D., Schaeffer, L., Schmitt, L., Körber, W., Merker, L., Zentner, L., Böhm, V.: Compliant robotic arm based on a tensegrity structure with x-shaped members. In: 2024 IEEE 7th International Conference on Soft Robotics (RoboSoft) (2024)
41. Herrmann, D., Lehmann, L., Schaeffer, L., Böhm, V.: Tensegrity manipulation using deformable, compressed members. In: 2025 CCToMM Symposium on Mechanisms, Machines, and Mechatronics (2025). submitted
42. Merker, L., Kissinger, T., Herrmann, D., Böhm, V., Zentner, L.: Self-supporting device (tensens platform) with prestressed optical fibers for determining an external force with associated point of application, patent application, document reference number (drn): 2023102717405500de (2023)
43. Kissinger, T., Charrett, T.O., Tatam, R.P.: Range-resolved interferometric signal processing using sinusoidal optical frequency modulation. Opt. Express **23**(7), 9415–9431 (2015)
44. Hemeling, C., Merker, L., Zentner, L., Fröhlich, T., Kissinger, T.: Optical fibers used as structural sensing elements in a tensegrity structure for force measurement and localization. In: International Conference on Optical Fibre Sensors 2025 (2025)
45. Merker, L., Kissinger, T., Böhm, V., Herrmann, D., Zentner, L.: Planar sensor design for force/torque measurement based on fiber optic sensing. In: Nguyen, D.-N., Tran, N.D.K., Huynh, V.T., Ono, T., Nguyen, V.H., Pandey, A.K. (eds.) Microactuators, Microsensors and Micromechanisms (2025)
46. Merker, L., Böhm, V., Zentner, L.: Modellbildung und steifigkeitsanalyse eines nachgiebigen tensegrity-ähnlichen koppelelements. In: 15. Kolloquium Getriebetechnik, September 13. – 15., 2023, Aachen, Germany (2023)
47. Böhm, V., Schorr, P., Feldmeier, T., Chavez-Vega, J.-H., Henning, S., Zimmermann, K., Zentner, L.: An approach to robotic end effectors based on multistable tensegrity structures. In: Pisla, D., Corves, B., Vaida, C. (eds.) New Trends in Mechanism and Machine Science (2020)
48. Herrmann, D., Schaeffer, L., Merker, L., Zentner, L., Böhm, V.: An approach to the realization of multistable tensegrity structures with deformable compressed members. In: Nguyen, D.-N., Tran, N.D.K., Huynh, V.T., Ono, T., Nguyen, V.H., Pandey, A.K. (eds.) Microactuators, Microsensors and Micromechanisms (2025)
49. Schorr, P., Schale, F., Otterbach, J.M., Zentner, L., Zimmermann, K., Böhm, V.: Investigation of a multistable tensegrity robot applied as tilting locomotion system. In: 2020 IEEE International Conference on Robotics and Automation (ICRA) (2020)
50. Schorr, P., Chavez, J., Zentner, L., Böhm, V.: Reconfigurable planar quadrilateral linkages based on the tensegrity principle. In: Zentner, L., Strehle, S. (eds.) Microactuators, Microsensors and Micromechanisms (2021)

51. Schorr, P., Chavez, J., Zentner, L., Böhm, V.: Reconfiguration of planar quadrilateral linkages utilizing the tensegrity principle. Mech. Mach. Theory **156**, 104172 (2021b)
52. Schorr, P., Zentner, L., Zimmermann, K., Böhm, V.: Jumping locomotion system based on a multistable tensegrity structure. Mech. Syst. Signal Process. **152**, 107384 (2021)
53. Herrmann, D., Schaeffer, L., Lehmann, L., Busch, T., Böhm, V.: Preliminary theoretical considerations on 2d multistable tensegrity structures based on equilateral triangles. In: Rosati, G., Gasparetto, A., Ceccarelli, M. (eds.) New Trends in Mechanism and Machine Science (2024)
54. Böhm, V., Jentzsch, A., Kaufhold, T., Schneider, F., Zimmermann, K.: An approach to compliant locomotion systems based on tensegrity structures. In: 56th International Scientific Colloquium, Ilmenau University of Technology (2011)
55. Jahn, H., Böhm, V., Zentner, L.: Analysis of deformation in tensegrity structures with curved compressed members. Meccanica **59**(9), 1369–1380 (2024)
56. Zentner, L.: Nachgiebige Systeme: Klassifikation, Modellbildung und Design Von Mechanismen und Aktuatoren. Walter de Gruyter GmbH & Co KG, Oldenbourg (2025)
57. Dante Aaron Ramirez Mestanza, M.A.V.B.: Rheological Properties of Polyborosiloxane (PBS) and Its Application in Tensegrity Structure. Project Seminar, Technische Universität Ilmenau, Germany (2023)
58. Bendezu, M.A.V.: Investigation of the Rheological Behavior of Polyborosiloxane with Magnetically Responsive Particles. Master Thesis, Technische Universität Ilmenau, Germany (2023)

Modelling and Control of Soft Robots

Dielectric Elastomer Soft Robots: Design, Modeling, Control, and Self-Sensing

Gianluca Rizzello, Stefan Seelecke, Matthias Baltes, Julian Kunze, Johannes Prechtl, and Giovanni Soleti

Abstract Dielectric elastomer (DE) transducers consist of soft capacitors made of a highly deformable polymeric membrane coated with compliant electrodes on both sides. DEs can work either as soft actuators or as sensors, and are also capable to perform the two tasks at the same time (self-sensing), thus they appear highly suitable for applications in soft robotics. This chapter explores the potential of silicone-based DE transducers in intelligent soft robotic systems. A DE-driven soft tentacle arm is chosen as target system, as it represents a challenging platform to demonstrate new design, fabrication, modeling, control, and self-sensing concepts. Leveraging a bi-stable design supported by numerical simulations, we achieve a remarkable bending performance magnification in comparison to mono-stable layouts. The combination of robust control algorithms with self-sensing proprioception schemes further allows the robot's position to be controlled autonomously and without external sensors.

G. Rizzello (✉) · S. Seelecke · M. Baltes · J. Kunze · J. Prechtl · G. Soleti
Department of Systems Engineering, Saarland University, Saarbrücken, Germany
e-mail: gianluca.rizzello@uni-saarland.de

S. Seelecke
e-mail: stefan.seelecke@uni-saarland.de

M. Baltes
e-mail: matthias.baltes@uni-saarland.de

J. Kunze
e-mail: julian.kunze@uni-saarland.de

J. Prechtl
e-mail: johannes.prechtl@uni-saarland.de

G. Soleti
e-mail: giovanni.soleti@uni-saarland.de

A. Raatz et al. (eds.), *Soft Material Robotic Systems*,
https://doi.org/10.1007/978-3-032-22453-8_14

1 Introduction

Common actuator solutions for soft robots, based on either pneumatic drives [1] or tendon-driven mechanisms [2], rely on rigid and cumbersome components (compressors, electric motors) which unavoidably affect the system weight, size, and energy efficiency. A potential means to overcome those limitations is offered by dielectric elastomer (DE) transducers, i.e., highly deformable capacitors made of a thin polymeric membranes sandwiched between compliant electrodes [3]. A DE can be operated as an actuator [4], by converting applied stimuli into motion, as well as a sensor [5], since its electrical capacitance depends on its current geometry. Additionally, a DE can also work as an actuator and as a sensor at the same time, a feature often referred to as self-sensing [6]. All these characteristics make DE transducers highly suitable for soft robotic applications. In this context, DEs allow to replace bulky elements with softer, simpler, lighter, more efficient, and self-sensing artificial muscles directly integrated within the robot structure [7, 8]. Example of DE-based soft robots which have been presented in the scientific literature include grippers [9], artificial muscles [10], tunable lenses [11], as well as robots capable to crawl [12], jump [13], swim [14], and fly [15], to mention a few. Even though the existing prototypes have succeeded in showcasing the high potential of DE technology for the realization of soft robots, in most of the cases the authors did not perform systematic studies aimed at investigating system scalability, modeling, and design optimization. Moreover, the development of control and self-sensing algorithms which allow complex DE soft robots to operate autonomously and an intelligent way has not been systematically investigated so far.

In this chapter, we present the development of an intelligent soft robotic tentacle arm with multiple degrees-of-freedom (DoF) based on DE technology. The main idea is to develop individual soft robotic systems capable of bending in one or more directions when actuated by DEs, and use them as building blocks for a modular soft tentacle design. First, we investigated the possibility of realizing such bending module based on established DE actuator (DEA) layouts, namely double-cone DEAs with segmented electrodes previously presented in [16]. As this concept turned out to lack scalability for our envisioned application, we then developed a new type of actuator concept more suitable for bio-inspired soft robot designs, namely coreless rolled DEAs. After discussing the rolled DEAs operating principle and manufacturing process, their electro-mechanical performance is characterized, and a physics-based model is derived. The developed actuators are then employed to develop a soft robotic planar bending module, consisting of two plates connected by a soft fexible backbone compressed by an antagonistic pair of rolled DEAs. The application of high voltage to one rolled DEA causes the structure to bend towards the opposite side, resulting in a bi-directional bending motion. The main advantage of the proposed structure, other than its modularity, is the fact that it exploits the buckling instability of the pre-compressed beam as a means to magnify the bending displacement, thus extending established stiffness compensation design principles used in single-DoF DEA systems [17, 18] to multi-DoF ones. Closed-loop motion control algorithms are then developed to stabilize the robot configuration in the open-loop unstable region in spite of the challenges posed by nonlinearities and underactuation, thus eliminating the main drawback of the bi-stable design (i.e., loss of proportional regulation) while keeping its benefit (i.e., large bending displacement). Self-sensing proprioception algorithms are then proposed, which enable real-time estimation of the robot configuration solely based on measurement of DEAs voltage and current signals performed during high-voltage actuation. By combining closed-loop control and self-sensing, sensorless stabilization of the robot is achieved. To conclude the chapter, the first steps towards the extension of the developed concept to three-dimensional soft tentacle arms are discussed. All the material presented hereafter is based on the authors' results published in papers [19–41].

The remainder of this chapter is organized as follows. Section 2 provides an overview on the operating principle of DE transducers. A preliminary concept for the soft tentacle arm, based on double-cone DEAs, is described in Sect. 3. The new coreless rolled DEA concept is then introduced in Sect. 4, and used to develop the planar soft robotic module described in Sect. 5. First steps towards the three-dimensional DE soft tentacle arm are then described in Sect. 6. Finally, Sect. 7 concludes the chapter and provides potential outlook for future works.

2 Dielectric Elastomer (DE) Transducers

A DE consists of a highly stretchable elastomer film (e.g., silicone, VHB acrylic, natural rubber) coated on both surfaces by compliant electrodes (e.g., carbon grease, carbon nanotubes, corrugated thin metal films), forming a flexible capacitor [3]. When a voltage difference is applied between the electrodes, electrostatic forces cause them to attract each other, resulting in a thinning of the membrane followed by an in-plane expansion (due to incompressibility of the elastomer). This effect, illustrated in Fig. 1(a), represents the main principle which allows DEs to be operated as actuators. This phenomenon can be quantified via the Maxwell stress equation [3], which states that the mechanical stress σ_e compressing the electrodes can be computed as follows

$$\sigma_e = -\epsilon_0 \epsilon_r E^2, \tag{1}$$

where E is the electric field caused by the externally applied voltage, while ϵ_0 and ϵ_r are the vacuum and DE relative permittivities, respectively. For typical membrane thickness on the order of $20 - 100\ \mu$m, a voltage on the order of 1–10 kV is required to generate a meaningful actuation.

Other than working as actuators, DEs can also be used as sensors. By referring to the upper part of Fig. 1b, the undeformed DE capacitance C_0 can be computed as

$$C_0 = \epsilon_0 \epsilon_r \frac{A_0}{z_0}, \tag{2}$$

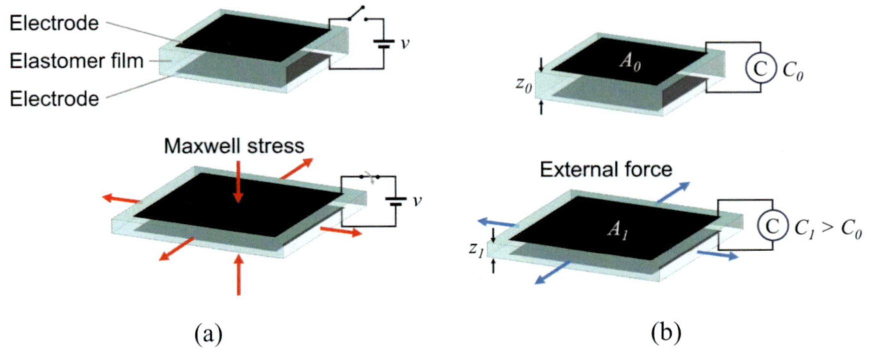

Fig. 1 Operating principle of dielectric elastomer transducers: **a** dielectric elastomer actuator; **b** dielectric elastomer sensor

where A_0 and z_0 represent the electrodes surface area and distance, respectively. When the membrane is deformed by an external force, as depicted in the lower part of Fig. 1b, the area increases to A_1 while the thickness decreases to z_1. Both effects result in the capacitance C_1 being larger than C_0. Therefore, changes in capacitance can be related to changes in the membrane geometric state. Eventually, actuation and sensing can be performed simultaneously, achieving the so-called self-sensing operation mode [6]. Self-sensing is a highly attractive feature of DE technology, since it permits to implement feedback control architectures without the need for additional electromechanical transducers, thus allowing to reduce cost, size, and complexity of the overall system [42].

DE transducers are characterized by large deformations ($> 100\%$), high compliance (Young's modulus within the range 0.1–10 MPa), high energy density (0.4 J/g), low energy consumption (on the order of milliwatts), high energy efficiency ($> 80\%$), fast response (bandwidth of several kilohertz), high flexibility and scalability, silent operations, and low cost [3]. All those features make them highly attractive for various applications in mechatronics as well as soft robotics [7, 8].

3 Preliminary DEA-Based Soft Bending System Concept

The first concept for the development of a DE-driven soft bending module is based on a double-cone DEA system [19]. The main actuating element consists of the annular DEA membrane shown in Fig. 2a, also referred to as cone DEA. In this type of actuator, the electrodes are segmented into four parts that can be activated independently, causing a local softening of the membrane. In our design, two cone DEAs are rigidly connected along their outer diameter, and pre-compressed out-of-plane against each other via a rigid cylindrical spacer. The obtained system, sketched in Fig. 2a, can undergo different actuation modes depending on which combinations of

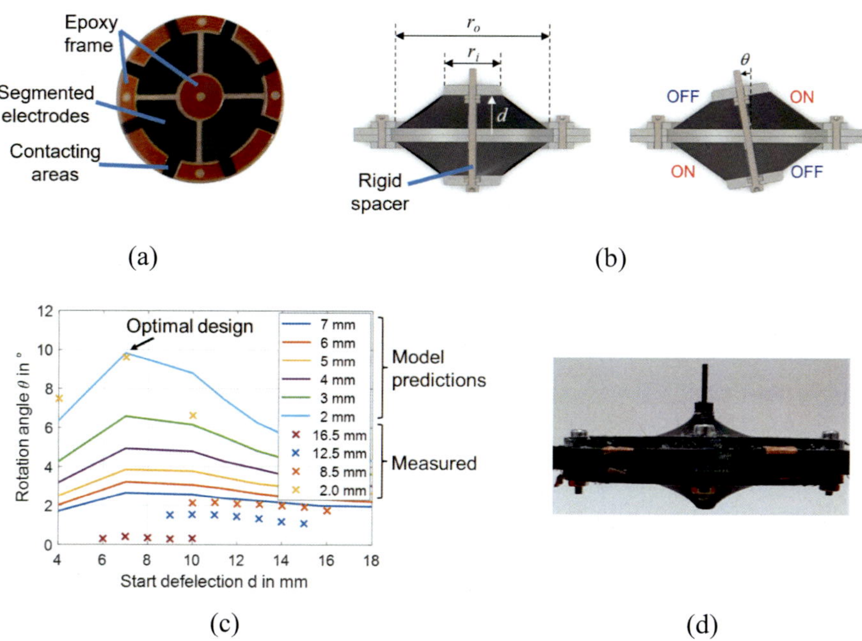

Fig. 2 Double-cone DE soft robot: **a** segmented membrane concept; **b** double-cone module operating principle; **c** model-based optimization; **d** optimized experimental prototype. Pictures **a**–**d** are based on publication [19], licensed under CC-BY 4.0

electrodes are activated simultaneously. For instance, actuating opposing electrodes on the upper and lower DEAs results in a bending motion, as shown Fig. 2b.

A physics-based model was derived for the double-cone DEA module, and used to perform a simulation study with the aim of finding the best geometric parameters that allow maximizing the bending displacement for a given actuation voltage [19]. The results of this simulation study are shown in Fig. 2c. Here, the abscissa axis denotes different rigid spacer lengths d, while the different plots represent the inner radius r_i (cf. Fig. 2b). The results of the parameter study reveal that a maximal bending angle of about 9.8° is obtained for $d = 7$ mm and $r_i = 2$ mm. A prototype is then built according to the predicted geometry, and shown in Fig. 2d. Experimental characterization revealed a bending angle of about 9.6°, sufficiently close to the theoretical prediction. Due to the resulting large aspect ratio, however, it is expected that this optimal design will scale poorly in terms of force and stacking when employed as building block for a soft tentacle arm.

4 Rolled DEA

This section introduces the novel rolled DEA concept.

4.1 Actuator Concept and Manufacturing

To overcome the limitations of the double-cone DEA module, we developed a new rolled DEA layout which is similar to a soft muscle fiber that expands when activated [20, 21]. The actuator is manufactured starting from two 50 μm-thick silicone membranes (Wacker ELASTOSIL® 2030 250/50) with carbon-based electrodes screen-printed onto one side, as shown in Fig. 3a. The two films are then put in contact (Fig. 3b), and tightly rolled in a spiral-like structure (Fig. 3c).

Ferrules are then crimped to the ends of the rolled DEAs, which allow to implement the electrical contacts as well as to connect the actuator to an external load or structure. To simplify the mounting, a nylon rod with a M2 thread is inserted at the free end of each ferrule. The resulting actuator resembles a thin muscle fiber, as shown in Fig. 4a. By inspecting the inner structure of the roll via transmitted light microscopy, cf. Fig. 4b, it can be seen that it has a smooth outer surface and an interior structure without wrinkles or hollow regions. Upon application of high voltage, the roll contracts in thickness and expands in length, as shown in Fig. 4c. Compared to other existing rolled DEAs [43, 44], the proposed design is characterized by higher slenderness, energy density, lightweight, and ease of manufacturing.

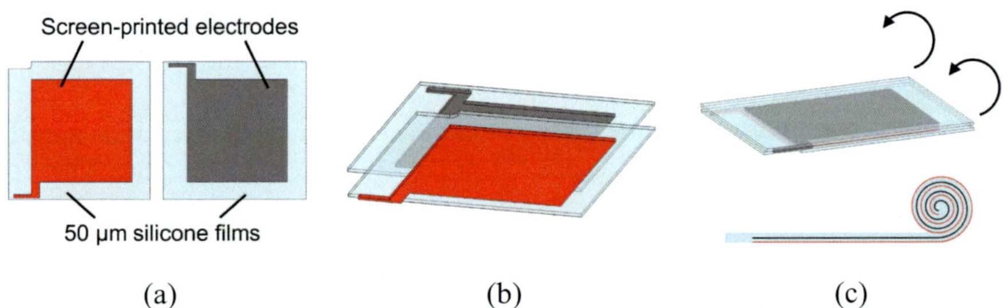

Fig. 3 Rolled DEA manufacturing: **a** an electrode pattern is screen printed onto two silicone membranes; **b** the two membranes are joined together; **c** the resulting structure is tightly rolled. Pictures **a–c** are based on publication [21], licensed under CC-BY 4.0

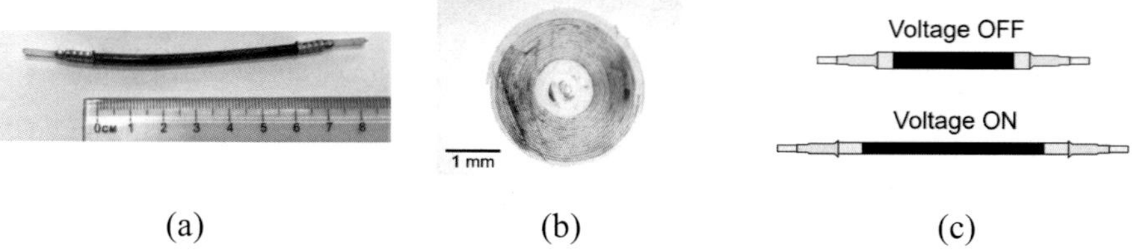

(a) (b) (c)

Fig. 4 Manufactured rolled DEA: **a** picture of the rolled DEA; **b** picture of the actuator cross-section; **c** actuation principle. Pictures **a–c** are based on publication [21]

4.2 Actuator Modeling and Validation

A physics-based, lumped-parameter model of the rolled DEA is developed for simulation and control applications [22, 23]. A sketch of the rolled DEA in undeformed and deformed configurations is provided in Fig. 5a. The model predicts both the axial force F and the electrical capacitance C based on inputs given by the actuator axial length l_1 and applied voltage v, while also accounting for material and geometry parameters, cf. the block diagram in Fig. 5b. Leveraging on a thermodynamic framework [23], the model of the rolled DEA can be formulated as follows:

$$\begin{cases} \dot{\xi}_{vj} = \dfrac{k_{vj}}{\eta_{vj}}\left(\dfrac{l_1}{L_1} - \xi_{vj}\right), \quad j = 1, \dots, M \\ F = \dfrac{2L_1L_2L_3}{l_1}\left(\displaystyle\sum_{i=1}^{3} 2C_{i0}i\left(\dfrac{l_1^2}{L_1^2} - \dfrac{L_1}{l_1}\right)\left(\dfrac{l_1^2}{L_1^2} + \dfrac{2L_1}{l_1} - 3\right)^{i-1}\right. \\ \qquad \left. - 0.5\alpha_e\epsilon_0\epsilon_r\dfrac{l_1}{L_1L_3^2}v_{DE}^2 + \dfrac{\eta_{v0}}{L_1}\dot{l}_1 + \displaystyle\sum_{j=1}^{M} k_{vj}\left(\dfrac{l_1}{L_1} - \xi_{vj}\right)\right) \\ C = 2\alpha_e\epsilon_0\epsilon_r\dfrac{L_2}{L_3}l_1 \end{cases} . \tag{3}$$

Quantities ξ_{vj} for $j = 1, \dots, M$ represent internal states describing the material viscoelasic dynamics, L_1, L_2, L_3 are the principal lengths of the unrolled DE membrane in undeformed configuration, C_{i0} for $i = 1, 2, 3$ define Yeoh parameters describing the hyperelastic material response, α_e is the electrode-to-membrane surface area ratio (cf. Fig. 3a), ϵ_0 and ϵ_r are the vacuum permittivity and DE relative permittivity, respectively, while η_{v0}, η_{vj}, and k_{vj} for $j = 1, \dots, M$ denote viscoelastic material parameters. The factor 2 appearing in both force F and capacitance C stems from the fact that the roll consists of a stack of 2 elastomer layers (cf. Fig. 3b), while L_1, L_2, and L_3 refer to a single layer.

The developed model is able to successfully reproduce the experimental response of three rolled DEA specimens with different geometries, i.e., with membrane surface area $L_1 \times L_2$ equals to 34 mm × 58 mm, 34 mm × 99 mm, and 91 mm × 58 mm,

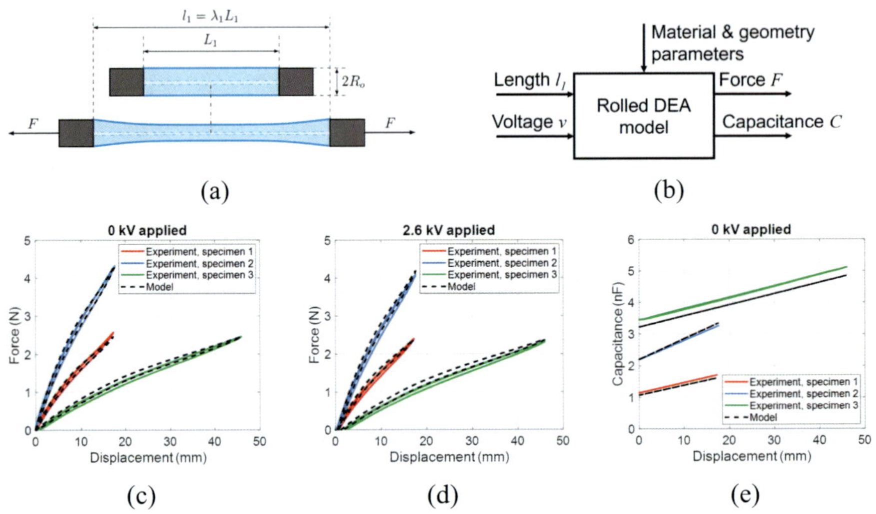

Fig. 5 RDEA modeling: **a** actuator model schematic; **b** actuator model block diagram; **c** low-voltage force prediction; **d** high-voltage force prediction; **e** electrical capacitance prediction. Picture **a** is based on publication [23]

respectively (for all cases, we have $L_3 = 48\ \mu$m). The model accurately predicts the force-displacement curves of the rolled DEA, including their nonlinearity and rate-dependent hysteretic effects, when the applied voltage is either low (Fig. 5c) or high (Fig. 5d). At the same time, the model also well reproduces the linear capacitance-displacement response of the transducer (Fig. 5e). Further extensions of the model allow to incorporate both rate-dependent and rate-independent hysteretic effects of the material within the same thermodynamic framework, enabling a more accurate description of the actuator dynamic behavior in a broader frequency range. Details are omitted for conciseness, the reader may refer to [24, 25] for details.

5 Rolled DEA-Based Planar Soft Tentacle Arm

Aspects related to design, modeling, control, and self-sensing proprioception of the rolled DEA-driven planar bending module are described in this section.

5.1 Rolled DEA-Driven Bi-Dimensional Soft Bending Module

The rolled DEAs are used as main actuator components to develop a second generation of soft planar bending modules [26, 27]. The system concept is illustrated in Fig. 6. The structure consists of two rigid plates connected by a flexible backbone. Two rolled DEAs are mounted in the structure in a pre-tensioned state, connecting the two rigid plates as shown in Fig. 6a. When a rolled DEA is activated via high voltage, its force decreases and the structure bends towards the unactuated DEA, as shown in Fig. 6b. This way, a bi-directional planar bending motion is achieved. A picture of the real-life system prototype is shown in Fig. 6c.

5.2 System Modeling and Validation

The presented soft bending module is characterized by several free design parameters. To guide the design, a physics-based simulation model is developed and used to analyze the impact of key geometric parameters on the resulting actuation displacement [28, 29]. For modeling purpose we define two reference frames, i.e., inertial frame $O - xy$ depicted in blue in Fig. 7a and attached to the base of the robot, and frame $O' - x'y'$ depicted in red in Fig. 7a and attached to the rigid top plate with origin located in the center of mass of the plate. We describe the robot configuration as $q = \left[\, q_x \; q_y \; \alpha \right]^{\mathsf{T}}$, where $\left(q_x,\; q_y\right)$ is the position of the origin of the frame $O' - x'y'$ expressed in the inertial frame $O - xy$, while α denotes the orientation of frame $O' - x'y'$ with respect to frame $O - xy$. The control input is given by vector $u = \left[u_{DE1} \; u_{DE2}\right]^{\mathsf{T}}$, where u_{DEi} is a control signal related to the voltage applied to the ith rolled v_{DEi} according to $u_{DEi} = L_2 L_3^{-1} \alpha_e \epsilon_0 \epsilon_r v_{DEi}^2$ (the meaning of each parameter is explained in Sect. 4.2). The model, based on Lagrangian mechanics, describes how the robot configuration variables q are affected by the control inputs

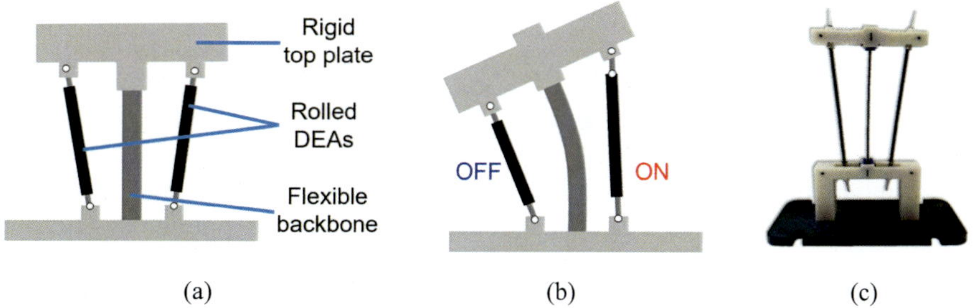

Fig. 6 Planar bending soft-robotic module driven by rolled DEAs: **a** system structure; **b** actuation of the rolled DEA on the right-hand side causes the structure to bend to the left-hand side; **c** picture of the experimental system prototype. Picture **c** is based on publication [27]

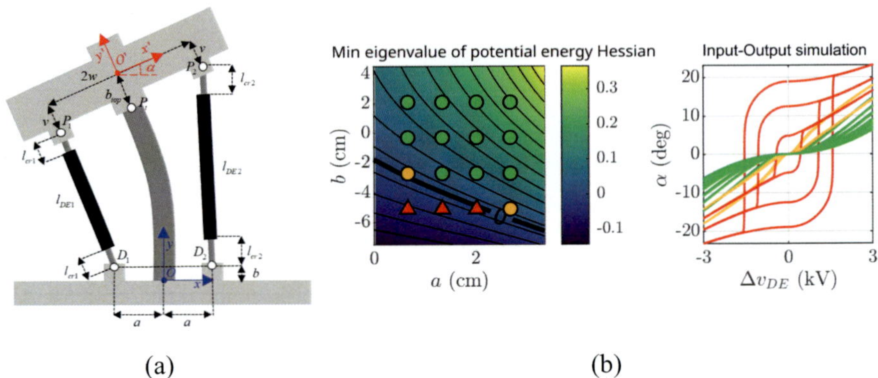

Fig. 7 Planar bending module simulation and optimization: **a** model layout; **b** simulation-based parameter study showing transitions from mono-stable to bi-stable operations. Picture **a** is based on publication [32], licensed under CC-BY 4.0

u, and is expressed in compact form as follows:

$$M\ddot{q} + D(q)\dot{q} + G(q) = J^{\mathsf{T}}(q)u. \tag{4}$$

Quantity $M \in \mathbb{R}^{3\times3}$ is the positive-definite and symmetric inertia tensor which is constant due to the adopted choice of q, $D(q) : \mathbb{R}^3 \to \mathbb{R}^{3\times3}$ is the positive-definite and symmetric damping matrix accounting for both structural dissipations and DE material viscoelastic losses, $G(q) : \mathbb{R}^3 \to \mathbb{R}^3$ is the vector of potential forces given as the gradient of the total potential energy $\mathcal{V}(q) : \mathbb{R}^3 \to \mathbb{R}$ (accounting for elastic energy of both beam and DEs as well as gravitational potential energy), i.e., $G(q) = \frac{\partial \mathcal{V}(q)}{\partial q}$, and $J(q) : \mathbb{R}^3 \to \mathbb{R}^{2\times3}$ is the actuation Jacobian given by $J^{\mathsf{T}}(q) = \frac{\partial l_{DE}^{\mathsf{T}}(q)}{\partial q}$, where $l_{DE}(q) : \mathbb{R}^3 \to \mathbb{R}^2$ defines the lengths of the actuators as a function of q via kinematic relationships (cf. Fig. 7a). For the analytical expressions of M, $D(q)$, $\mathcal{V}(q)$, and $l_{DE}(q)$, the reader may refer to [32].

Model (4) is used to perform a quasi-static analysis, by applying slow-varying sinusoidal input commands u alternating between the two actuators, with amplitude chosen such that the DE voltages range within the physical actuator limits (i.e., $v_{DEi} \in [0,\ 3]\ kV$), and evaluating the resulting q. This study is repeated for different combinations of geometric parameters, and used estimate the robot input-output map. The right-hand side of Fig. 7b shows an example of this study in which the lower mounting points of the rolled DEAs are varied, namely a and b in Fig. 7a. For the ease of visualization, only the angle α is shown, reported as a function of the difference between actuators voltages $\Delta v_{DE} = v_{DE1} - v_{DE2}$. The left-hand side of Fig. 7b further reports the minimum eigenvalue of the Hessian matrix of the potential energy of the system in the vertical configuration (see [29] for details), whose sign determines the stability properties of the voltage-free vertical equilibrium state (i.e., stable if positive, unstable if negative). As it can be seen from Fig. 7b, some

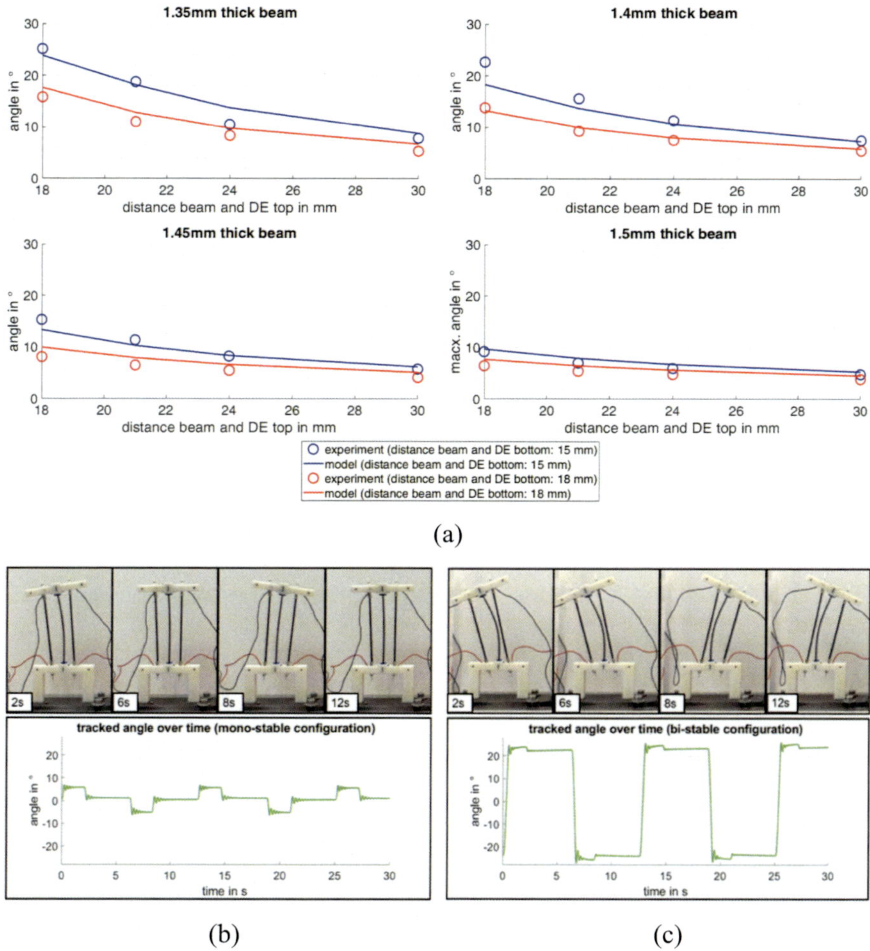

Fig. 8 Planar bending module experimental validation: **a** quasi-static model validation for different system geometry parameters; **b** dynamic response of the mono-stable experimental prototype; **c** dynamic response of the bi-stable experimental prototype. Pictures **a–c** are based on publication [27]

combinations of a and b lead to soft robots whose vertical configuration is mono-stable (depicted in green), while some others are meta-stable (depicted in yellow), and finally some other configurations lead to an unstable vertical equilibrium (depicted in red). It can readily be observed that the unstable design leads to a significant magnification in bending displacement, which is of $\pm 23°$ and about twice as much the bending angle obtained for the best mono-stable configuration. On the other hand, while mono-stable configurations allow to regulate every position in the actuation range, the unstable design results in a hysteretic input-output map. Physically, this hysteresis is explained by considering that the normal force exerted by the pre-tensioned rolled DEAs onto the beam causes the latter to buckle, and if the system parameters are properly tuned the buckling direction can be controlled by switching the actuators voltages. Such hysteresis implies that the robot is not able to operate in a stable way around the vertical configuration, as it remains bent towards the left-hand side or the right-hand side when no voltage is applied to the DEAs. The proposed energy-based method successfully extends well-known bi-stable design principles for single-DoF DEAs [17], which are based on graphical methods and thus only work in one dimension, to general multi-DoF DEA soft robots.

Various experimental prototypes are then built based on the simulations [27], by changing parameters such as rolled DEAs upper mounting point distance w, lower mounting point distance a, and beam thickness t_b (cf. Fig. 7a). A camera-based experimental setup is then built to measure the resulting bending performance. As it can be observed from Fig. 8a, the experimental data are well matched by the model. The measured bending angle is reported alongside snapshots of the recorded video for two selected system configurations, i.e., a mono-stable one shown corresponding to $w = 30\,\text{mm}$, $a = 18\,\text{mm}$, and $t_b = 1.4\,\text{mm}$ and shown in Fig. 8b, and a bi-stable one corresponding to $w = 18\,\text{mm}$, $a = 15\,\text{mm}$, and $t_b = 1.35\,\text{mm}$ and reported in Fig. 8c. The resulting bending angle range is of $\pm 6°$ for the mono-stable case and of $\pm 25°$ for the bi-stable case. The mono-stable design always leads the robot to recover the vertical position when unactuated, while the bi-stable design causes the module to remain displaced even when the voltage is removed. Those experiments validate the theoretical predictions of the model, and highlight the advantages of bi-stability for the design of DE-based soft robots. As a demonstrative example, a tentacle prototype consisting of a stack of three bending modules is shown in Fig. 9.

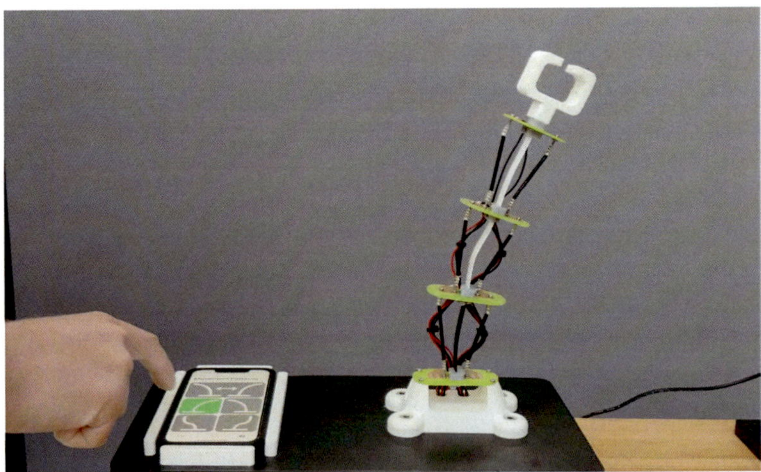

Fig. 9 Prototype of a planar soft tentacle arm made of three single bending modules

5.3 Motion Control

To retain all the advantages of the bi-stable design (i.e., large bending displacement) without losing proportional regulation (because of bi-stability), closed-loop feedback control strategies are developed. In a first step, we addressed the problem of designing feedback controllers which stabilize an arbitrary configuration within the robot workspace in spite of the several challenges posed by model (4), most notably nonlinearities and underactuation. More specifically, the problem of underactuation with configuration-dependent actuation matrix, represented by the rectangular $J^{\mathsf{T}}(q)$ in (4), is known to be highly challenging in soft robotics. While state of the art methods are based on coordinate transformations to address this control problem [45], we developed a new approach which does not require any change of coordinates, but leverages instead an energy-based control framework [31, 32]. The goal is to develop a feedback law u which regulates the robot configuration q to a desired target $q^\star$ starting from an arbitrary condition in the reachable workspace, and explicitly account for system nonlinearities, open-loop instability, underactuation with configuration-dependent actuation matrix, control input saturation, and presence of constant external disturbances. The final form of the controller is reported in the following (for details, please refer to [32]):

$$
\begin{aligned}
\dot{x}_c &= -\dot{\tilde{l}}_{DE} - J(q)\, W_1 \tilde{q} - K_S \mathrm{DZ}_{\rho_I}(K_I x_c) \\
u &= \underbrace{u^\star - \sum_{i=1}^{m} \sum_{j=1}^{h} \theta_{ij} \frac{\partial \phi_{ij}\left(\tilde{l}_{DEi}(q)\right)}{\partial \tilde{l}_{DEi}(q)}}_{\text{stabilizing term } u_s} \\
&\quad + \underbrace{\mathrm{SAT}_{\rho_I}(K_I x_c) + \mathrm{SAT}_{\rho_P}\left(-K_P \dot{\tilde{l}}_{DE} - K_P J(q)\, W_1 \tilde{q}\right)}_{\text{robustifying term } u_r}
\end{aligned}
\tag{5}
$$

Control law (5) consists of the sum of two contributions also illustrated in Fig. 10a, i.e., a stabilizing term u_s which ensures regional stability of the target $q^\star$ in the robot workspace, and a robustifying term u_r that enables the stabilization goal to be satisfied in spite of unmeasurable and constant external disturbances. As for the other quantities appearing in (5), x_c is an internal integral state of the controller, $\tilde{q} = q - q^\star$ and $\tilde{l}_{DE}(q) = l_{DE}(q) - l_{DE}(q^\star)$ denote the control error expressed in configuration and actuation coordinates respectively, W_1 is a constant matrix tuned to ensure that passivity holds true, $\phi_{ij}\left(\tilde{l}_{DEi}(q)\right)$ are freely-chosen regressors, θ_{ij} are stabilizing gains, $\mathrm{SAT}_{\rho_I}(x)$ and $\mathrm{DZ}_{\rho_I}(x)$ denote vector-valued saturation and deadzone functions, respectively, while K_S, K_I, and K_P are free tuning parameters (for more details, please refer to [32]). In case input u does not saturate, control law (5) can be rewritten as:

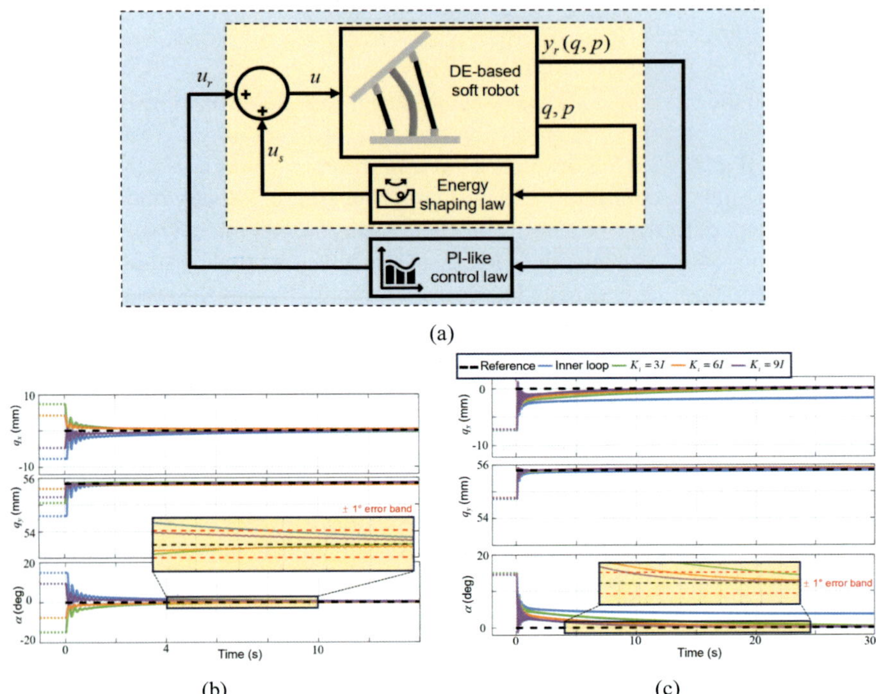

Fig. 10 Planar bending module set-point regulation: **a** control system architecture; **b** stabilization of the vertical configuration from different initial conditions; **c** effect of integral gain on the closed-loop system. Pictures **a–c** are based on [32]

$$u = \underbrace{u^{\star} - (A + K_I)\,\tilde{l}_{DE}(q) - K_P\dot{\tilde{l}}_{DE}}_{\text{stabilizing term } u_s} + \underbrace{\left(-K_P J\,(q)\,W_1\tilde{q} - K_I \int_0^t J\,(q)\,W_1\tilde{q}d\tau\right)}_{\text{robustifying term } u_r}, \tag{6}$$

with A representing the local action of the terms proportional to $\theta_{ij}\frac{\partial\phi_{ij}(\tilde{l}_{DEi}(q))}{\partial\tilde{l}_{DEi}(q)}$. Control law (6) can be intuitively interpreted as the sum of three actions, i.e., a feedforward $u^{\star}$, a PD law in actuator length error $\tilde{l}_{DE}(q)$ where $A + K_I$ is the proportional gain and K_P is the derivative gain, and a PI law in configuration error $\tilde{q}$ projected into actuator space via matrix $J(q)W_1$ where K_P is the proportional gain and K_I is the integral gain. This analogy with a PID law can be used for a more intuitive tuning of free parameters K_P and K_I. Figure 10b shows the effectiveness of the proposed controller in stabilizing the vertical configuration of the system when starting from different initial conditions, while Fig. 10c shows the controller performance for different choices of gain K_I. Further works also investigated the position regulation of the same system via optimal control techniques, the reader may refer to [30] for details.

Next, we investigated the tracking control of smooth arbitrary time-varying trajectories $q^{\star}(t)$. As discussed in [45], control of underactuated robots with configuration-dependent actuation matrix can be achieved provided that a coordinate transformation is found which brings the system in the so-called collocated form, where the actuation matrix becomes $J^{\mathsf{T}} = \begin{bmatrix} I & 0 \end{bmatrix}^{\mathsf{T}}$. Once in collocated form, robot control can be achieved straightforwardly using classic robot control theory via partial feedback linearization [46]. In [33], we provided a closed-form and analytically-invertible solution for a nonlinear coordinate transformation which brings model (4) into collocated form. The proposed choice for the new coordinates is visually illustrated in Fig. 11a. After converting the system in collocated form, the tracking control is successfully achieved in simulation via partial feedback linearization. Figure 11b shows the effectiveness of this approach when tracking a polynomial trajectory, while Fig. 11c shows the same result in case of a sinusoidal trajectory. Aspects related to the planning of feasible trajectories which satisfy constraints due to underactuation and control input saturation are discussed in [34].

5.4 Self-sensing Proprioception

To implement the feedback position control law in the experimental setup, accurate real-time measurements of the robot own state are required. In the experiments conducted in Fig. 10, the feedback is obtained via a camera operating at 65 frames per second. In principle, DE technology allows to implement closed-loop control by means of self-sensing feedback. Self-sensing is enabled by the dependency between actuator elongation and electrical capacitance, see Fig. 5e. By measuring the actuator electrical voltage and current during high-voltage activation, and using such data to estimate the DEA capacitance, the resulting actuator displacement can be estimated in real time. In the case of single-DoF DEA systems, we have successfully used self-sensing to estimate at the same time the actuator displacement and force (the latter reconstructed by means of an additional state observer), and used the resulting displacement-force information to achieve sensorless control of the DEA-environment interaction [35]. In the context of multi-DoF DE soft robots, the system-level self-sensing architecture shown in Fig. 12a is proposed to achieve sensorless proprioception of the entire robot configuration [36, 37]. The architecture in Fig. 12a consists of two functional blocks, i.e., a Capacitance-based Actuator Length Reconstruction block which uses real-time measurements of the DEA voltage and current signals to estimate their lenghts, and a System-Level Self-Sensing which uses combined information from DEA lengths and input voltage to reconstruct the configuration q. The second stage is especially important because the DEA lengths are only 2 while the system has 3 DoF, therefore the kinematic underdetermination is solved by incorporating information on the dynamic model through an extended Kalman filter (EKF). The resulting estimation performance are shown for 2 different versions of the EKF in Fig. 12b for a motion caused by the voltage applied to the DEAs, and Fig. 12c for a motion caused by an external mechanical load. In both

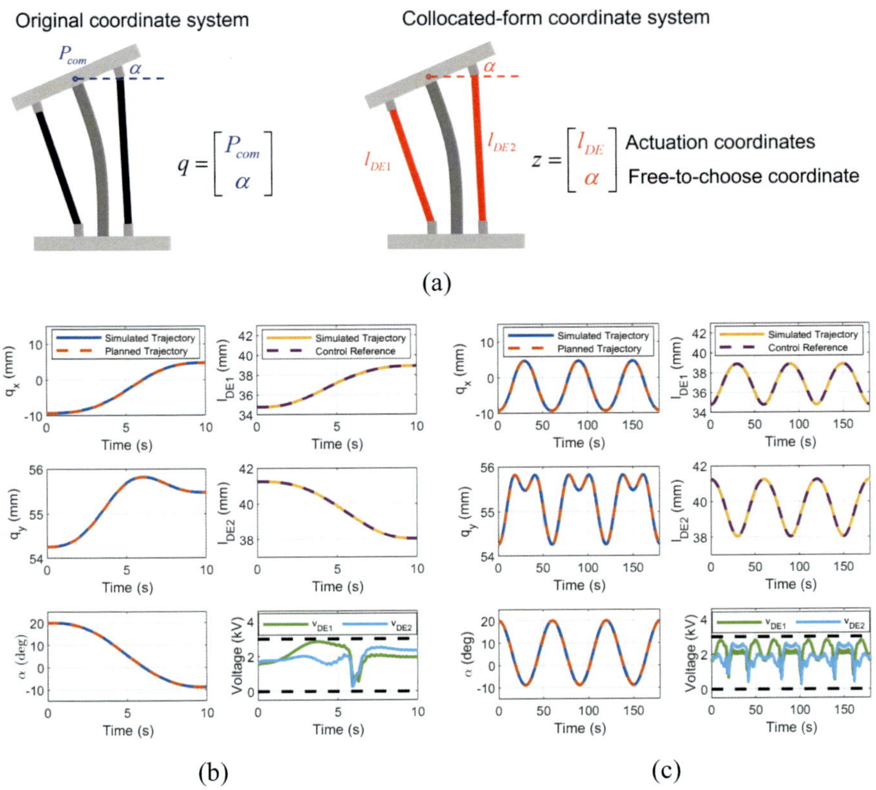

Fig. 11 Planar bending module trajectory tracking: **a** graphical representation of the coordinate transformation which brings the system in collocated form; **b** tracking of a polynomial trajectory; **c** tracking of a sinusoidal trajectory

cases, the self-sensing proprioception algorithm provides estimations very close to the reference camera measurements.

Next, we combined the self-sensing proprioception algorithm with the position feedback law to achieve sensorless closed-loop control. The corresponding sensorless architecture is illustrated in the block diagram in Fig. 13a. In the conducted experiment, we closed the position loop via self-sensing feedback generated by the EKF at 5 kHz, and used control law (5) to stabilize the vertical configuration while simultaneously applying several manual disturbances. The sensorless control architecture successfully stabilizes the vertical configuration after each manual perturbation. Moreover, the resulting self-sensing estimations always remain tightly close to the camera measurements, used here as a means of validation. This result shows that it is possible to compensate for open-loop instability without relying on a complex and expensive camera system, leveraging instead the intrinsic intelligence of DE technology.

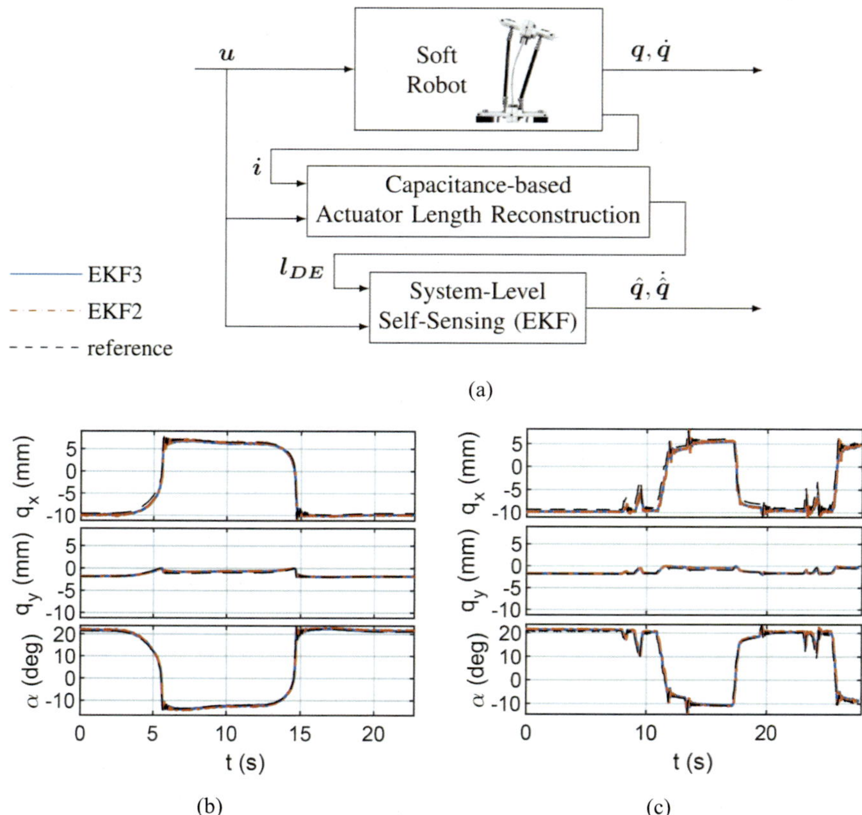

Fig. 12 Planar bending module sensorless proprioception: **a** block diagram of the estimation scheme; **b** self-sensing estimation when the motion is caused by the actuation voltage; **c** self-sensing estimation when the motion is caused by an external mechanical load. Pictures **a–c** are based on [37] licensed under CC-BY 4.0

6 Towards a Rolled DEA-Based 3D Soft Tentacle Arm

This section discusses the first steps towards the development of a rolled DEA-driven soft robotic module capable to bend freely in three dimensions. A preliminary 3D module concept was developed based on a rigid robot design [38]. Although lacking the soft features, this system is used to develop useful insights towards the actuation kinematics of DEA-based manipulators in three-dimensional spaces, and eliminates fabrication uncertainties due to the flexible element. The system, shown in Fig. 14a, consists of two rigid circular plates connected by a ball joint in the center. Nine pre-tensioned rolled DEAs are arranged around the circumference of the two rigid disks, and electrically controlled in groups of three. When a group of rolled DEA is activated, it causes the structure to rotate towards the opposite direction.

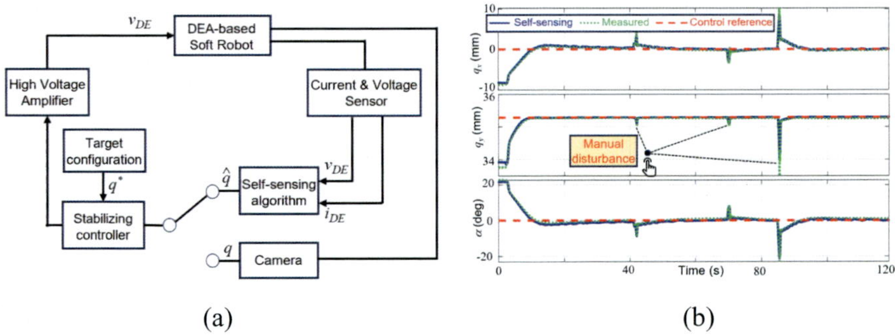

Fig. 13 Planar bending module sensorless position control: **a** block diagram of the sensorless position control scheme; **b** sensorless stabilization of the vertical equilibrium position when external impulsive disturbances are manually applied to the system

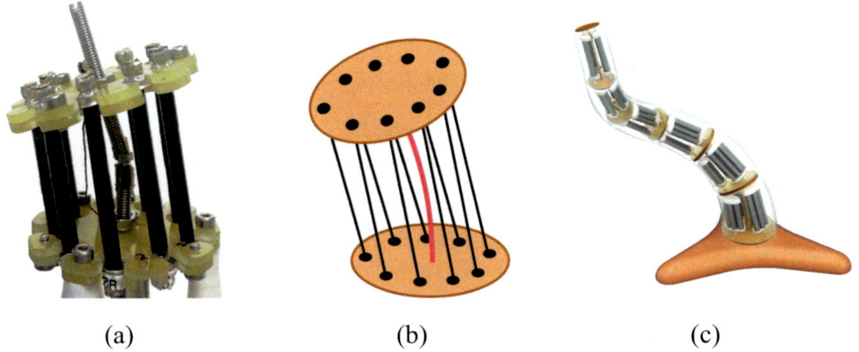

Fig. 14 Towards a three-dimensional soft robotic DEA bending module: **a** initial concept based on ball joint; **b** soft version of the three-dimensional bending module; **c** envisioned full tentacle arm prototype. Picture **a** is based on [38]

The difference between this concept and the one from Fig. 6 is the fact that the 3D structure tends to rotate following a precession-like motion, rather than jumping in-plane. Based on this concept, rotations are achieved with fixed bending angles of about 25°.

Next, we conducted a model-based feasibility study on a soft version of the three-dimensional bending module [39]. This is obtained by simply replacing the ball joint in Fig. 14a with a cylindrical flexible beam capable of bending in all direction, as shown in Fig. 14b. By means of numerical simulations based on a detailed physical model it is shown that, upon a proper tuning of the geometry parameters, the system can be designed to exploit beam instability as a means to expand the motion range also in 3D. The simulations show that a theoretical bending larger than 50° is virtually achievable without exceeding the physical limits of the materials and structure. Moreover, by properly controlling the driving strategy, this version of the soft robot can undergo both a precession (similarly to the ball joint concept in Fig. 14a) as well as an in-plane bending, thus ensuring a much wider reachable workspace. Experimental validation of these simulation findings will be object of future research. Once developed, the soft version of the three-dimensional bending module will be used to develop a soft tentacle arm as the one envisioned in Fig. 14c.

7 Conclusions and Outlook

This chapter has discussed various design, manufacturing, modeling, control, and estimation solutions which led to the realization of an intelligent soft tentacle arm driven by dielectric elastomer transducers. A new type of DEA, namely a coreless rolled DEA, has been used as bio-inspired artificial muscle to develop a bi-dimensional bending module, which in turn serves as a building block for a soft tentacle arm. By means a model-assisted bi-stable design, significant magnification in bending performance is achieved. Although this design solution results in a loss of proportional regulation of the robot motion, it is demonstrated how proportional regulation can be effectively recovered by means of stabilizing feedback control laws, which can be also implemented in an integrated and camera-free fashion by exploiting the self-sensing feature of DE transducers. Finally, the first steps towards the development of a three-dimensional tentacle arm driven by rolled DEAs have been shown.

The obtained results have demonstrated for the first time the benefits of a synergistic model-based approach to design, control, and estimation of multi-DoF soft robots based on DE technology. The obtained results will pave the way for future autonomous and intelligent DE-based soft robots, which fully exploit the high potential of DE transducers to achieve functionalities not possible only via mechanical design or electronic control. In the future, it will be essential to investigate the role of artificial intelligence to further enhance the design, control, and estimation capabilities of DE-driven soft robots. Moreover, potential applications of the developed rolled DEAs in applications beyond soft manipulators will be investigated, e.g., in tensegrity-based soft robots as preliminarily studied in [40, 41].

Acknowledgements Funded by the deutsche forschungsgemeinschaft (DFG, German Research Foundation) under grant no. 405032227.

References

1. Zhang, Z., Wang, X., Wang, S., Meng, D., Liang, B.: Design and modeling of a parallel-pipe-crawling pneumatic soft robot. IEEE access **7**, 134301–134317 (2019)
2. Wilhelm R Wockenfuß, V., Brandt, L., Weisheit, W.-G., Drossel: Design, modeling and validation of a tendon-driven soft continuum robot for planar motion based on variable stiffness structures. IEEE Robotics and Automation Letters **7**(2), 3985–3991 (2022)
3. Federico Carpi, Danilo De Rossi, Roy Kornbluh, Ronald Edward Pelrine, and Peter Sommer-Larsen. Dielectric elastomers as electromechanical transducers: Fundamentals, materials, devices, models and applications of an emerging electroactive polymer technology. Elsevier, 2011
4. Hajiesmaili, E., David R Clarke: Dielectric elastomer actuators. J. Appl. Phys. **129**(15), 2021
5. Holger Böse and Johannes Ehrlich. Dielectric elastomer sensors with advanced designs and their applications. In Actuators, volume 12, page 115. MDPI, 2023
6. Samuel Rosset, Benjamin M O'Brien, Todd Gisby, Daniel Xu, Herbert R Shea, and Iain A Anderson. Self-sensing dielectric elastomer actuators in closed-loop operation. Smart Materials and Structures, 22(10):104018, 2013
7. Guo-Ying, G., Zhu, J., Zhu, L.-M., Zhu, X.: A survey on dielectric elastomer actuators for soft robots. Bioinspiration & biomimetics **12**(1), 011003 (2017)
8. Gupta, U., Qin, L., Wang, Y., Godaba, H., Zhu, J.: Soft robots based on dielectric elastomer actuators: A review. Smart Mater. Struct. **28**(10), 103002 (2019)
9. Oluwaseun A Araromi, Irina Gavrilovich, Jun Shintake, Samuel Rosset, Muriel Richard, Volker Gass, and Herbert R Shea. Rollable multisegment dielectric elastomer minimum energy structures for a deployable microsatellite gripper. IEEE/ASME Transactions on mechatronics, 20(1):438–446, 2014
10. Kovacs, G., Lochmatter, P., Wissler, M.: An arm wrestling robot driven by dielectric elastomer actuators. Smart Mater. Struct. **16**(2), S306 (2007)
11. Maffli, L., Rosset, S., Ghilardi, M., Carpi, F., Shea, H.: Ultrafast all-polymer electrically tunable silicone lenses. Adv. Func. Mater. **25**(11), 1656–1665 (2015)
12. Sun, W., Liang, H., Zhang, F., Wang, H., Yanjun, L., Li, B., Chen, G.: Dielectric elastomer minimum energy structure with a unidirectional actuation for a soft crawling robot: Design, modeling, and kinematic study. Int. J. Mech. Sci. **238**, 107837 (2023)
13. Luo, B., Li, B., Yuan, Yu., Meng, Yu., Ma, J., Yang, W., Wang, P., Jiao, Z.: A jumping robot driven by a dielectric elastomer actuator. Appl. Sci. **10**(7), 2241 (2020)
14. Shintake, J., Cacucciolo, V., Shea, H., Floreano, D.: Soft biomimetic fish robot made of dielectric elastomer actuators. Soft Rob. **5**(4), 466–474 (2018)
15. Ren, Z., Kim, S., Ji, X., Zhu, W., Niroui, F., Kong, J., Chen, Y.: A high-lift micro-aerial-robot powered by low-voltage and long-endurance dielectric elastomer actuators. Adv. Mater. **34**(7), 2106757 (2022)
16. Canh Toan Nguyen, H., Phung, T.D., Nguyen, H., Jung, H.R., Choi: Multiple-degrees-of-freedom dielectric elastomer actuators for soft printable hexapod robot. Sens. Actuators, A **267**, 505–516 (2017)
17. Hodgins, M., York, A., Seelecke, S.: Experimental comparison of bias elements for out-of-plane DEAP actuator system. Smart Mater. Struct. **22**(9), 094016 (2013)
18. Giacomo Moretti, Luca Sarina, Lorenzo Agostini, Rocco Vertechy, Giovanni Berselli, and Marco Fontana. Styrenic-rubber dielectric elastomer actuator with inherent stiffness compensation. In Actuators, volume 9, page 44. MDPI, 2020
19. Nalbach, S., Banda, R.M., Croce, S., Rizzello, G., Naso, D., Seelecke, S.: Modeling and design optimization of a rotational soft robotic system driven by double cone dielectric elastomer actuators. Frontiers in Robotics and AI **6**, 150 (2020)
20. Julian Kunze, Johannes Prechtl, Daniel Bruch, Sophie Nalbach, Paul Motzki, Stefan Seelecke, and Gianluca Rizzello. Design and fabrication of silicone-based dielectric elastomer rolled actuators for soft robotic applications. In Electroactive Polymer Actuators and Devices (EAPAD) XXII, volume 11375, pages 287–294. SPIE, 2020

21. Julian Kunze, Johannes Prechtl, Daniel Bruch, Bettina Fasolt, Sophie Nalbach, Paul Motzki, Stefan Seelecke, and Gianluca Rizzello. Design, manufacturing, and characterization of thin, core-free, rolled dielectric elastomer actuators. In Actuators, volume 10, page 69. MDPI, 2021
22. Johannes Prechtl, Julian Kunze, Daniel Bruch, Stefan Seelecke, and Gianluca Rizzello. Modeling and parameter identification of rolled dielectric elastomer actuators for soft robots. In Electroactive Polymer Actuators and Devices (EAPAD) XXIII, volume 11587, pages 204–212. SPIE, 2021
23. Prechtl, J., Kunze, J., Moretti, G., Bruch, D., Seelecke, S., Rizzello, G.: Modeling and experimental validation of thin, tightly rolled dielectric elastomer actuators. Smart Mater. Struct. **31**(1), 015008 (2021)
24. Johannes Prechtl, Felix Scherf, Julian Kunze, Kathrin Flaßkamp, and Gianluca Rizzello. An energy-based model for both rate-dependent and rate-independent hysteretic effects in uniaxially-loaded dielectric elastomer actuators. In Electroactive Polymer Actuators and Devices (EAPAD) XXV, volume 12482, pages 236–245. SPIE, 2023
25. Rizzello, G.: Energy-based modeling of rate-independent hysteresis and viscoelastic effects in dielectric elastomer actuators. Smart Mater. Struct. **33**(5), 055027 (2024)
26. Matthias Baltes, Julian Kunze, Johannes Prechtl, Paul Motzki, Stefan Seelecke, and Gianluca Rizzello. Soft robotic tentacle arm element actuated by rolled dielectric elastomer artificial muscles. In Electroactive Polymer Actuators and Devices (EAPAD) XXIV, volume 12042, pages 52–61. SPIE, 2022
27. Baltes, M., Kunze, J., Prechtl, J., Seelecke, S., Rizzello, G.: A bi-stable soft robotic bendable module driven by silicone dielectric elastomer actuators: design, characterization, and parameter study. Smart Mater. Struct. **31**(11), 114002 (2022)
28. Johannes Prechtl, Julian Kunze, Sophie Nalbach, Stefan Seelecke, and Gianluca Rizzello. Soft robotic module actuated by silicone-based rolled dielectric elastomer actuators: modeling and simulation. In Electroactive Polymer Actuators and Devices (EAPAD) XXII, volume 11375, pages 276–286. SPIE, 2020
29. Johannes Prechtl, Julian Kunze, Daniel Bruch, Stefan Seelecke, and Gianluca Rizzello. Bistable actuation in multi-DoF soft robotic modules driven by rolled dielectric elastomer actuators. In 2021 IEEE 4th International Conference on Soft Robotics (RoboSoft), pages 82–89. IEEE, 2021
30. Paolo Roberto Massenio, Johannes Prechtl, David Naso, and Gianluca Rizzello. Nonlinear optimal control of a soft robotic structure actuated by dielectric elastomer artificial muscles. In 2022 IEEE/ASME International Conference on Advanced Intelligent Mechatronics (AIM), pages 644–649. IEEE, 2022
31. Soleti, G., Prechtl, J., Massenio, P.R., Baltes, M., Rizzello, G.: Energy based control of a bi-stable and underactuated soft robotic system based on dielectric elastomer actuators. IFAC-PapersOnLine **56**(2), 7796–7801 (2023)
32. Giovanni Soleti, Paolo Roberto Massenio, Julian Kunze, and Gianluca Rizzello. Model-based robust position control of an underactuated dielectric elastomer soft robot. IEEE Transactions on Robotics, 41:1693–1710, 2025
33. Giovanni Soleti, Paolo Roberto Massenio, Julian Kunze, and Gianluca Rizzello. Nonlinear coordinate transformation and trajectory tracking control of an underactuated soft robot driven by dielectric elastomers. In 2024 IEEE 7th International Conference on Soft Robotics (RoboSoft), pages 228–234. IEEE, 2024
34. Giovanni Soleti, Lorenzo Cicali, Michael Franci, Luca Pugi, and Gianluca Rizzello. Task space trajectory planning for an articulated dielectric elastomer soft robot with input saturation and underactuation. In 2024 20th IEEE/ASME International Conference on Mechatronic and Embedded Systems and Applications (MESA), pages 1–6. IEEE, 2024
35. Rizzello, G., Serafino, P., Naso, D., Seelecke, S.: Towards sensorless soft robotics: Self-sensing stiffness control of dielectric elastomer actuators. IEEE Trans. Rob. **36**(1), 174–188 (2019)
36. Johannes Prechtl, Matthias Baltes, Julian Kunze, Stefan Seelecke, and Gianluca Rizzello. Towards sensorless configuration estimation in multi-DoF soft robotic structures driven by rolled dielectric elastomer actuators. In 2022 IEEE/ASME International Conference on Advanced Intelligent Mechatronics (AIM), pages 1152–1158. IEEE, 2022

37. Prechtl, J., Baltes, M., Flaßkamp, K., Rizzello, G.: Sensorless proprioception in multi-DoF dielectric elastomer soft robots via system-level self-sensing. IEEE/ASME Trans. Mechatron. **29**(6), 4365–4376 (2024)
38. J Kunze, G Soleti, D Bruch, P Motzki, S Nalbach, S Seelecke, and G Rizzello. Development and experimental evaluation of a compact 3D bending module actuated by rolled dielectric elastomer actuators (RDEAs). In Electroactive Polymer Actuators and Devices (EAPAD) XXVI, volume 12945, pages 70–78. SPIE, 2024
39. Giovanni Soleti, Julian Kunze, Paolo Roberto Massenio, and Gianluca Rizzello. Model-based design of multi-stable 3D soft manipulators: a dielectric elastomer case study. In 2025 IEEE 8th International Conference on Soft Robotics (RoboSoft)
40. David Herrmann, Julian Kunze, Julian Kobes, Stefan Seelecke, Paul Motzki, Gianluca Rizzello, and Valter Böhm. A mobile tensegrity robot driven by rolled dielectric elastomer actuators. In 2025 IEEE 8th International Conference on Soft Robotics (RoboSoft)
41. Julian Kunze, David Herrmann, Julian Kobes, Paul Motzki, Stefan Seelecke, Gianluca Rizzello, and Valter Böhm. A tensegrity-based locomoting soft robot actuated by rolled dielectric elastomer transducers. In Electroactive Polymer Actuators, Sensors, and Devices (EAPAD) 2025, volume 13431, pages 13–17. SPIE, 2025
42. Rizzello, G., Naso, D., York, A., Seelecke, S.: Closed loop control of dielectric elastomer actuators based on self-sensing displacement feedback. Smart Mater. Struct. **25**(3), 035034 (2016)
43. Pei, Q., Pelrine, R., Stanford, S., Kornbluh, R., Rosenthal, M.: Electroelastomer rolls and their application for biomimetic walking robots. Synth. Met. **135**, 129–131 (2003)
44. Rui Zhang, Patrick Lochmatter, Andreas Kunz, and Gabor M Kovacs. Spring roll dielectric elastomer actuators for a portable force feedback glove. In Smart Structures and Materials 2006: Electroactive Polymer Actuators and Devices (EAPAD), volume 6168, pages 505–516. SPIE, 2006
45. General characterization and its application to soft robotics: Pietro Pustina, Cosimo Della Santina, Frédéric Boyer, Alessandro De Luca, and Federico Renda. Input decoupling of lagrangian systems via coordinate transformation. IEEE Trans. Rob. 49, 2098–2110 (2024)
46. Mark W Spong. Partial feedback linearization of underactuated mechanical systems. In Proceedings of IEEE/RSJ International Conference on Intelligent Robots and Systems (IROS'94), volume 1, pages 314–321. IEEE, 1994

Stiffness Modelling and Control for Soft Material Continuum Robotic Manipulators

Jialei Shi, Wenlong Gaozhang, Sara-Adela Abad, and Helge Wurdemann

Abstract Soft robots, made from compliant materials, offer high flexibility and enable continuous deformation, resulting in inherently safe and adaptive interactions with their environments. Unlike rigid-link robots where compliance is typically localized at discrete joints, soft robots exhibit distributed compliance throughout their structure. This fundamental difference underscores a deep understanding of their intrinsic compliance, or its inverse, stiffness, which is essential for developing other static models such as forward kinematics and contact force models. This work provides a comprehensive review of existing stiffness-related modelling approaches for soft continuum robots, such as techniques based on analytical formulations, numerical methods, and data-driven strategies. Furthermore, it highlights how robot stiffness critically influences the accuracy and effectiveness of static models and contact control methods, which are essential for enabling robust and precise interaction with the environment.

Keywords Soft robots · Compliance and stiffness · Static models

---

J. Shi (✉)
The Hamlyn Centre for Robotic Surgery, Department of Mechanical Engineering, Imperial College London, Exhibition Road, London, UK
e-mail: j.shi@imperial.ac.uk

W. Gaozhang · S.-A. Abad · H. Wurdemann
Department of Mechanical Engineering, University College London, London, UK
e-mail: wenlong.gaozhang.20@ucl.ac.uk

S.-A. Abad
e-mail: s.abad-guaman@ucl.ac.uk

H. Wurdemann
e-mail: h.wurdemann@ucl.ac.uk

S.-A. Abad
Faculty of Agriculture and Renewable Natural Resources, Universidad Nacional de Loja, Loja, Ecuador

A. Raatz et al. (eds.), *Soft Material Robotic Systems*,
https://doi.org/10.1007/978-3-032-22453-8_15

1 Introduction

The mechanical behaviour of soft robots, in particular, their ability to deform under external loads, depends on their stiffness (or its inverse, i.e., the compliance). This property can be characterised by the Young's modulus, which defines the relationship between the stress and resulting strain. Unlike conventional robots constructed from rigid materials, such as metals or hard polymers, which typically exhibit Young's moduli on the order of $10^9 - 10^{10}$ Pa, soft robots are fabricated from significantly more compliant materials with Young's moduli ranging from approximately $10^4 - 10^6$ Pa [1, 2]. This inherent compliance can enable safer interactions with the environment and living tissues but can also limit the robot's ability to apply large forces when needed [3].

Tuning and analysing stiffness properties is central to the design, modelling, and control of soft robotic systems [4–6]. A variety of stiffening mechanisms have been proposed to modulate compliance, broadly categorised into semi-active and active methods [7]. Semi-active techniques operate by modulating the intrinsic mechanical properties of the materials from which they are made. Examples include the use of field-responsive materials, phase-change alloys [8], and granular jamming systems [9]. In addition, switching between actuation media, such as pneumatic and hydraulic systems, can be employed to alter stiffness [10]. On the other hand, active stiffening strategies often utilise antagonistic configurations, where opposing actuators work against each other. Examples include combined pneumatic and tendon-driven systems [11–13], dual pneumatic actuators [14, 15], and hybrid setups where active components are constrained by passive elements. For instance, pneumatic actuators may be mechanically limited using non-extendable fabric [16–18], locked tendons [19], or by leveraging the inherent elasticity of the material itself [20, 21].

This chapter advances the understanding of stiffness-related control of soft material continuum robotic manipulators by investigating how compliance characteristics influence stiffness response and force generation. Developing mathematical models that capture these relationships is essential for both predictive simulation and real-time control of interaction forces between the robot and the environment, for instance. In this context, the work presented here examines the formulation of models that integrate stiffness properties into the kinematic and force control of soft robotic systems.

Section 2 provides a comprehensive overview of existing approaches to modelling compliance and stiffness in soft robotic structures, while sect. 3 reviews established and emerging frameworks for static force modelling in these highly deformable systems. Section 4 transitions from static, passive compliance and stiffness modelling approaches to a review of active, on-demand stiffness control strategies, with a focus on sensing and controlling interaction forces. Section 5 outlines key open challenges and future research directions in modelling and controlling stiffness and interaction forces in soft material continuum robotic manipulators.

2 Compliance/Stiffness Modelling in Soft Material Robotic Systems

In rigid robots, the compliance is provided by variable stiffness joints [22]. For instance, the 3×3 Cartesian stiffness matrix K_X in the task space is mathematically determined based on the Jacobian-based projection in Eq. 1 [23].

$$K_X = J^{+T}\left(\Phi - \frac{\partial J^T}{\partial \theta}F\right)J^{+} \approx J^{+T}\Phi J^{+}. \quad (1)$$

Φ is the stiffness matrix of the joints with a diagonal form, θ is the displacement of joints, J and J^+ are the Jacobian matrix and pseudo-inverse Jacobian matrix under current robot configuration, F is the external force in the Cartesian space. For the static case of unloaded robot configuration, $\frac{\partial J^T}{\partial \theta}F$ can be approximated as a zero item. The robot compliance is the inverse of the stiffness in Eq. 1. Instead, many soft robots are considered continuum manipulators without physical joints, and the compliance is distributed along the robots [5]. Figure 1 demonstrates the difference of compliance generation between rigid-link and soft continuum robots. Moreover, various static modelling techniques of soft robots usually rely on stiffness properties of robots. For instance, the material properties (e.g., Young's modulus), together with bending, elongation and shear stiffness, are usually considered to calculate curvatures and strains from actuation forces. The resulting curvatures and strains complete the statics models, such as the forward kinematics. However, a thorough stiffness description and analysis for soft robots, e.g., the Cartesian stiffness, is not investigated in these kinematic models.

The importance of understanding configuration-dependent stiffness/compliance has been proven for rigid-link and continuum robots [24–27], especially for redundant robotic mechanisms. For instance, the effect of robot configuration on task force capabilities was investigated in [28], force ellipsoid was initially introduced.

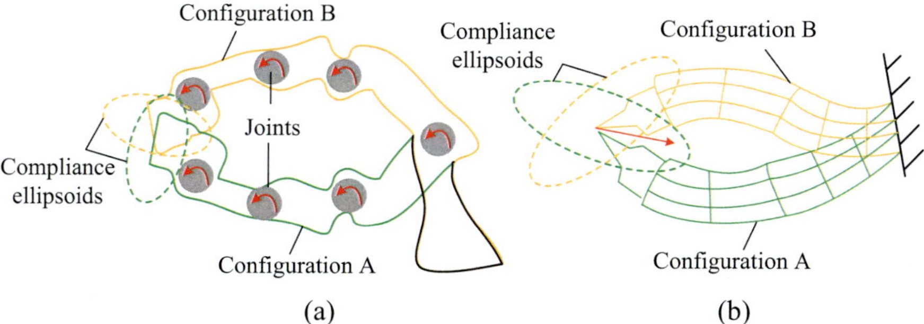

Fig. 1 Illustration of compliance ellipsoids for **a** a rigid-link robot and **b** a soft continuum robot, under two different robot configurations. The compliance of rigid-link robots is typically localized at discrete joints, while soft robots exhibit distributed compliance throughout their structure

To analytically describe the compliance or stiffness properties of soft robots, there are mainly two ways, i.e., Jacobian projection and finite differentiation of forward kinematics. The Jacobian projection approach is motivated by Eq. 1. To this end, the continuum structures of soft robots are discretised into finite elements or joints first to derive the Jacobian matrices, and a pioneer work for achieving compliance modelling of continuum robots using the Jacobian projection approach can be found in [25]. Once the Cartesian compliance matrices are obtained, compliance and force ellipsoids can be employed to achieve analysis, which is analogous to the techniques used in the rigid-link robots. Similarly, the stiffness properties of discretised elements can be written via diagonal stiffness matrices, the Jacobian matrices can be calculated from the Cosserat rod model [29] or piecewise constant curvature (PCC) model [27] for instance. Additionally, the finite differentiation of well-established kinematics models can also deliver the compliance modelling. To achieve this, virtual infinitesimal forces Δf are included in the forward kinematics model to calculate the corresponding variation of robot configuration, the forces and configuration variation are used to derive the compliance or stiffness matrices [30], resulting in Eq. 2.

$$C_X = \left[\frac{p(L) - p(L)^*}{\Delta f_x}, \frac{p(L) - p(L)^*}{\Delta f_y}, \frac{p(L) - p(L)^*}{\Delta f_z} \right] \tag{2}$$

where $p(L)$ and $p(L)^*$ are the 3×1 tip position vectors with and without force increments of Δf_x, Δf_y and Δf_z. For instance, the incremented variables are the tip forces, and the forward kinematics is established based on the Cosserat rod theory [31]. Such a concept was then adopted and developed by [32] to model the stiffness response of a parallel continuum robot under different configurations. Inspired by [31], a new set of compliance differential equations, which was directly coupled to the forward kinematics, to derive the compliance matrix of continuum robots [33]. The stiffness of pneumatic-driven soft parallel robots were analysed with the displacements resulted from applied loading in different directions [34]. However, the robotic manipulators were approximated to the slender Kirchhoff rod, whilst large elongation observed in soft robots were not considered [35, 36]. The stiffness can also be obtained in combination with the Finite Element Method (FEM) and the pseudo-rigid model, as demonstrated in [37]. To account for material nonlinearity resulting from robot elongation, the non-linear matrix structural analysis (NMSA) method presented in [5] can be useful. The NMSA method addresses large deformations in soft bodies by applying load in predefined steps. This method enables the use of a linear strain-stress relation to consider large deformations, where the stiffness matrices update in each step.

As for the efficient mathematical tool for the compliance modelling in robotic field, Lie theory [43] and, in particular, screw theory have been utilised in robotics for it concisely interprets rigid-body motions, such as kinematics and statics modelling [44], parameter calibration [45, 46] and identification [47], and compliance analysis [48–50]. In addition, an analytical model was presented in [40] to model and analyse the stiffness ellipsoids for a tendon-driven soft-rigid gripper. It is worth mentioning that recent study in [41] demonstrated that the Cartesian compliance

Table 1 Examples of modelling and stiffness/compliance analysis for various soft continuum robots

References	Model	Material behaviour	Cartesian compliance modelling	Compliance analysis	Compliance distribution
[31][i], [33][i]	Cosserat rod	Linear	Finite differentiation	✘	✘
[32][i]	Cosserat rod	Linear	Finite differentiation	Ellipsoid	✘
[38][i]	Cosserat rod	Linear	✘	✘	✔
[39][i]	PCC	Linear	Jacobian projection	✘	✘
[40][i]	Pseudo-rigid model	Linear	Jacobian projection	Ellipsoid	✘
[41, 42]	PCC and Cosserat rod	(Non)linear	Screw theory	Ellipsoid	✔
[37][i]	FEM	Linear	Finite differentiation	✘	✘

✘ and ✔ denotes that the corresponding modelling or analysis is not considered or considered, respectively; The superscript [i] denotes the elongation of the backbone is negligible or inextensible.

along the soft continuum robots can be directly integrated with the forward kinematics using the Lie theory. This approach does not require Jacobian transformations or multiple calculations of forward kinematics, which provides a concise approach to model the configuration-dependent compliance at any positions of the continuum robot. Specifically, the general form of the compliance calculation yields in Eq. 3.

$$C(s)^o = \int_0^L \mathrm{Ad}_{ob}^{-T}(s)c(s)^b\mathrm{Ad}_{ob}^{-1}(s)\mathrm{d}(s) \tag{3}$$

$s \in [0, L]$, which is the arc length along the backbone of the robot. $C(s)^o$ is the Cartesian compliance matrix. $\mathrm{Ad}_{ob}(s)$ is the adjoint matrix calculated from the forward kinematics. $c(s)^b$ is the element compliance matrix.

The compliance properties of soft robots exhibit two characteristics. First, the compliance distributes along the robot structure and depends on both the soft robots' configuration and the choice of coordinates (see Fig. 1b). Secondly, modelling of compliance usually involves the calculation of forward kinematics. Specifically, the Jacobian matrices come from the forward kinematics, while the finite differentiation approach needs to calculate the forward kinematics for a few times. The key advantage of Eq. 3 is that it only requires a single forward kinematics computation. It is worth mentioning that these two compliance characteristics are applied for various continuum robots (e.g., tendon-driven robots), as exemplified in Table 1. It can be observed that the compliance/stiffness models are less investigated compared to the kinematics or dynamics models of soft robots. It is also worth mentioning that the robot compliance relates the loads and deformations, so it is also pivotal for understanding the interaction between the robots and environments, such as the contact force.

3 Modelling Interaction Forces in Soft Continuum Robots

For rigid-link soft robots, the interaction force in the task space results from the torque at revolute or prismatic joints. In contrast, the force generation of soft compliant robots is usually produced from the deformation and determined by the compliance properties of robots. Although soft robots are inherently compliant when contacting with environments, the modelling of the interaction force of soft robots is of paramount importance to understand their force capability and achieve active force control [51].

FEM might be the most straightforward approach to achieve the force prediction [52]. For example. constraints need to be applied to one end of soft robots when predicting the tip contact force. After applying actuation pressures on the internal surfaces of robots, the bending deformation and the tip contact force can be obtained. In [53], Abaqus was used to predict the tip force for soft pneumatic actuators with pleated structures. Similarly, the study in [54] used FEM to evaluate the force generation of a pneumatic-driven soft gripper with two pneumatic networks. A toolbox was presented in [55] to expedite the design of bellow soft robots, where the motions and output forces of soft robots with different parameters are assessed using the FEM. FEM is easy to be implemented; however, it lacks an analytical description of the force generation and often involves intensive computation costs.

FEM falls into the category of numerical approaches. Moreover, analytical models have been developed to achieve the force modelling. An analytical contact force model was proposed for PneuNets actuators, where the tip force was analysed considering the equilibrium moment of the PneuNets at the applied pressure [56]. The force model was validated via both FEM simulations and experiments. Overall, the analytical model predicts the tip force with errors of less than 10%. In [57], an analytical model was proposed for PneuNets actuators based on the Euler-Bernoulli finite strain hyperelastic thin cantilever beam theory. The deformation of the air chambers used finite strain membrane theory. The proposed analytical model was applicable to two states of forward kinematics modelling: free-space motion and tip-contact conditions. The model has similar accuracy with the FEM model, but the computational time is only 1% of the FEM model. Building on the PCC assumption and Euler-Bernoulli beam theory, an analytical blocking force model was proposed in [58]. The model reveals that the tip blocking force is impacted by the change of air volume and bending angle, meanwhile, the generated force is highly proportional to the actuation pressure.

In addition to the force models for aforementioned PneuNets actuators, analytical models for fibre-reinforced actuators were investigated. In [59], an analytical model was developed that captures the explicit relationship between the input pressure, the

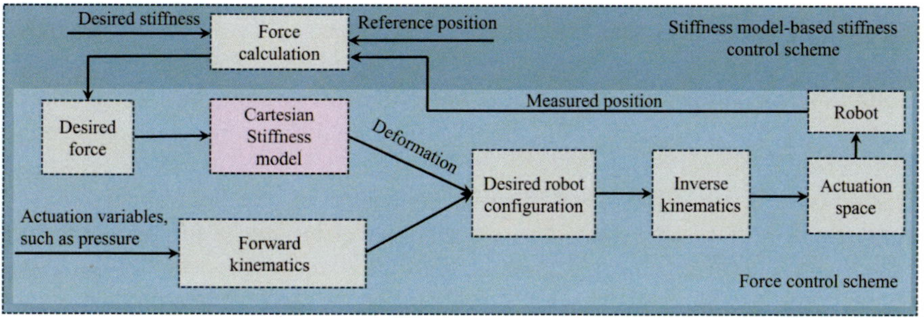

Fig. 2 Diagram of stiffness model-based stiffness and force control, where the stiffness model calculates the robot deformation subject to the desired force

bending angle, and the tip force by considering both the hyperelastic material property of silicone rubber and the geometry of the actuator. The model was built on the Neo-Hookean model and was evaluated on soft robots with different cross-section dimensions and lengths. Results demonstrated that the robot length has minimal influences on the force generation, while the force is primarily impacted by cross-sectional dimensions. The extended work of [59] was reported in [60], where the sensed bending angles of the fibre-reinforced actuator were incorporated to the analytical model. In [61], a comprehensive analytical model was developed for fibre-reinforced bending actuators, the model was also based on the beam mechanics, which can mathematically depict the relation of free-space bending, block force, and deflection shape with respect to the input pressure. Specifically, the block force modelling in [61] was based on the moment equilibrium, and the ODEs of the deflection curve are established. The ODEs were solved using the boundary conditions and resulted in the new force model.

Above force models are usually established on numerical (e.g., FEM) or analytical models, with known robot compliance properties. In particular, the forward kinematics models are often involved to derive the force models. It is worth mentioning that the interaction forces of soft robots can also be predicted via model-free, data-driven approaches, i.e., using neural networks [62]. One prominent example is the soft tactile sensors. The inputs of the neural network come from embedded sensor information, e.g., from camera [20], strain sensors [63], pressure [64]. [65] reviews the recent progress of the flexible tactile sensors. However, these data-driven approaches often overlook the underlying compliance characteristics inherent to soft robots. Section 4.2 further summarizes research on interaction force sensing and control for soft robots.

4 Interaction Compliance and Force Control

Soft robots exhibit passive compliant behaviour resulting from the soft materials. To make the most of their passive compliance properties during robot-environment interaction, active interaction control must be designed. To this end, the robot deformation has to be considered when developing the controllers. Two types of interaction controllers will be discussed in the following sections, including the compliance/stiffness control and the force control. A typical control flow for stiffness model-based interaction control is illustrated in Fig. 2. In this scheme, the stiffness model computes the desired deformation based on the target force. The desired force can be either explicitly specified, as in force control, or derived from the desired stiffness matrix in the case of stiffness control.

4.1 Compliance and Stiffness Control

Models presented in Sect. 2 describe passive compliance/stiffness modelling approaches of soft robots in the Cartesian space. These stiffness properties are inherently determined by the robot design and material selection. Similar to the compliance regulation for rigid-link robots [23], on-demand and active stiffness/compliance control has been investigated for soft robots.

The common approach to regulate robot stiffness is using stiffening mechanisms elaborated in [7]. For instance, once the robots are controlled to desired configurations, the vacuum pressure will be applied to actively increase the rigidity of soft robots, e.g., via layer jamming [66, 67] or granular jamming [68]. To avoid introducing stiffening mechanisms, the antagonistic actuation approach has been developed [69]. For instance, a variable stiffness joint was proposed in [70], where two soft air bellows acted antagonistically on a single joint. In this actuator, various pressure combinations can lead to the same joint angle and allow to additionally adjust the joint stiffness. The same antagonistic stiffening approach was also applied in [71] to vary the stiffness of soft robots made of pneumatic artificial muscles. Thanks to the introduced stiffening mechanisms, the robot configuration and compliance can be controlled independently. However, embedding stiffening mechanisms or applying antagonistic actuation approach in space-restricted applications, for instance, minimally invasive surgery, can be challenging.

As demonstrated in Sect. 2, the robot passive compliance is configuration-dependent and can be analytically or numerically described. In principle, it is possible to control robot configurations to regulate compliance/stiffness of soft robots. The first stiffness control paradigm for soft robots without embedded stiffening mechanisms can be found in [4]. To achieve a desired tip stiffness, the controller actuated the robot to a deflected configuration which produced a required tip force for a measured tip position. And the desired force was calculated from a desired stiffness matrix. The robot kinematics was described using the Cosserat rod model. Similarly, the study in [72] achieved the stiffness control for a parallel continuum robot based on estimated force and robot pose, using the Cosserat rod theory. Results demonstrated that the stiffness controller can vary the natural stiffness of soft robots by about a factor of two along the x and y-axes. Building on the PCC assumption, a Cartesian stiffness control approach was proposed for a multi-segment soft robot [27]. The Cartesian stiffness was described using Eq. 1. A desired compliance ellipsoid was defined to formulate optimization objective functions, and the robot configuration was then derived from solving the optimization problems. The compliance of the robot was regulated by changing the internal pressure of the tendon-driven robot. A tendon-tensioning method was proposed to control the stiffness of a dual-segment tendon-driven soft robot based on depth vision. A closed-loop controller was designed for stiffness compensation [73]. In line with Fig. 2, a compliance model-based control approach was presented in [74] to regulate the Cartesian compliance in desired directions for pneumatically driven soft robots. It is noteworthy that the compliance regulation in [74] exclusively relies on varying the actuation pressure without additional structures.

4.2 Interaction Force Sensing and Control

To regulate the Cartesian forces of soft continuum robots, force control techniques from rigid-link robots could be referred to. A prominent force control approach is discretising soft robots into finite elements and regulating virtual joint forces from each element using Jacobian projection [75], and the Cartesian stiffness matrix is tuned as a control gain. For instance, an intrinsic force sensing approach was proposed to estimate the wrench exerted at a tendon-driven continuum robot tip by monitoring the actuation forces [76]. Consequently, this estimated wrench is further advanced as feedback information to implement closed-loop force control [77]. In particular, a hybrid motion and force control approach was proposed for tendon-driven continuum robots to achieve force regulation or contact surface estimation from sensed forces in the task space. In aforementioned cases, actuation forces usually can be obtained from force transducers in tendon-driven or parallel continuum robots [32, 78]. In contrast, additional consideration is required to convert fluidic pressure or volume to generalised actuation forces or torques for fluidic-driven robots [79]. For hydraulic-driven soft robots, the combination of fluid volume and pressure can be utilised for intrinsic force sensing and control of soft parallel robotic systems [80], where the Cartesian stiffness matrix was estimated by solving the linear least-squares problem. Likewise, a compliance model-based contact force control approach for soft robots was presented in [81], where the compliance model was analytically derived.

In addition to model-based force sensing and control, technological advances enable deployment of force sensing devices on soft robots, e.g., forces exerted on robots can be measured by attached force/torque (F/T) sensors directly [82, 83]. The F/T sensors are usually rigid and limited to its feasibility of being embedded in soft robots. Fibre optics exhibit naturally flexibility [84] and are capable of sensing forces from measuring strains, e.g., stretchable sensors were developed for prosthetic hand based on optical waveguides [85]. Moreover, material sciences have advanced the development of flexible force sensors, e.g., using piezoelectric polymers [86], liquid metals [87] or pneumatic fluids [88]. In particular, fusion of sensed information and models provides new insights on interaction force estimation and regulation of soft robots [89, 90]. A vision-based external forces sensing approach was proposed in combination with finite element method, where the intensities and the locations of the external forces can be estimated [91]. Similarly, the deformation of soft robots derived from vision or motion tracking information can be utilised to achieve force and motion control [92], estimate or attenuate external force disturbances [73, 93].

Another force sensing and control paradigm for soft robots is data-driven approach [94]. A sensorless force and displacement estimation approach was introduced by exclusively utilising fluidic pressure and volume in pneumatically powered soft actuators [64]. A learning-based closed-loop force control approach was presented combining embedded soft sensors and recurrent neural networks, which can cope with significant drift and hysteresis in feedback signals [63]. It is worth mentioning that the efficacy of data-driven approaches rely on and is determined by the training set. Please note that robots' stiffness properties may affect the generalizability of data-driven approaches, as high robot compliance can introduce more uncertainties.

5 Conclusions

Robot compliance/stiffness and force capabilities are exhibited characteristics when soft robots interact with environments. The interaction compliance and force behaviour of soft robots result from the continuous deformability of soft materials. The compliance/stiffness properties essentially connect applied wrenches (including forces and moments) and corresponding twists, i.e., linear or angular deformations. Building on various models developed to describe the statics of soft robots, this review reported on the interrelationship between robot kinematics, Cartesian compliance/stiffness, and force generation capability. Understanding these connections is essential for effectively modelling and controlling interactions between soft robots and their environments, as illustrated in Fig. 3. In particular, these static models can be utilised for achieving various model-based control, e.g., the inverse kinematics, compliance and force regulation. Modelling the stiffness of soft robots with hybrid actuation mechanisms remains a challenging task. In such systems, overall stiffness is influenced not only by the material properties but also by factors like actuation pressure and tendon tension. Accurately modelling and controlling the stiffness of these systems is an interesting research direction.

Whereas notable work has been achieved for static cases, modelling the dynamic behaviour of soft robots remains a significant challenge due to the highly non-linear, time-varying, and often distributed nature of their deformation. Unlike rigid-body systems, where mass and inertia are well defined, soft robots require the incorporation

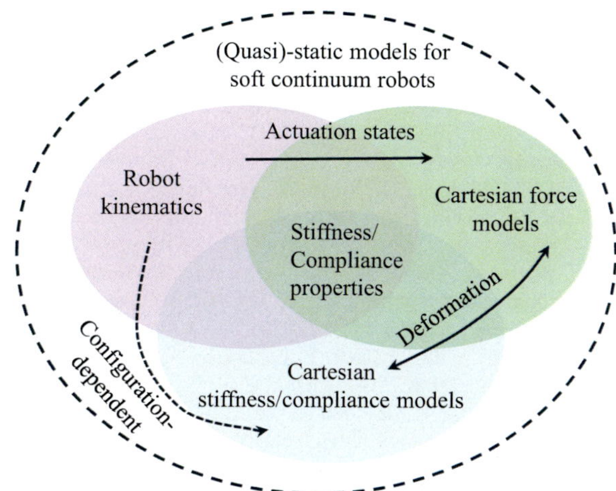

Fig. 3 Summary of the statics models for soft robots. The stiffness properties of soft materials govern the kinematics, as well as Cartesian stiffness and force models of soft robots

of continuum mechanics and viscoelastic properties to accurately capture transient responses. Dynamic modelling is essential for tasks involving high-speed motion, impact mitigation, or interaction with dynamic environments. Coupling dynamic models with stiffness representations enables advanced control strategies, such as variable impedance control, which can regulate both motion and interaction forces in a time-varying manner. However, developing integrated frameworks that jointly model dynamics and compliance remains an challenging problem. Such models must account for actuation principles, non-linear material behaviour, and environmental contact, representing a critical direction for advancing real-time, adaptive control of soft robots.

References

1. Majidi, C.: Soft robotics: a perspective–current trends and prospects for the future. Soft Rob. **1**(1), 5–11 (2014)
2. Shi, J., Abad, S.-A., Dai, J.S., Wurdemann, H.A.: Position and orientation control for hyper-elastic multisegment continuum robots. IEEE/ASME Trans. Mechatron. **29**(2), 995–1006 (2024)
3. Shi, J., Borvorntanajanya, K., Chen, K., Franco, E., Baena, F.R.y.: Design, control, and evaluation of a novel soft everting robot for colonoscopy. IEEE Trans. Robot. **41**, 4843–4859 (2025)
4. Mahvash, M., Dupont, P.E.: Stiffness control of surgical continuum manipulators. IEEE Trans. Rob. **27**(2), 334–345 (2011)
5. Naselli, G.A., Mazzolai, B.: The softness distribution index: towards the creation of guidelines for the modeling of soft-bodied robots. Int. J. Robot. Res. **40**(1), 197–223 (2021)
6. Komatsu, M., Yaguchi, T., Nakajima, K.: Algebraic approach towards the exploitation of softness: the input-output equation for morphological computation. Int. J. Robot. Res. **40**(1), 99–118 (2021)

7. Manti, M., Cacucciolo, V., Cianchetti, M.: Stiffening in soft robotics: a review of the state of the art. IEEE Robot. Automat, Magaz. **23**(3), 93–106 (2016)
8. Peters, J., Nolan, E., Wiese, M., Miodownik, M., Spurgeon, M., Arezzo, A., Raatz, A., Wurdemann, H.: Actuation and stiffening in fluid-driven soft robots using low-melting-point material. In: IEEE/RSJ International Conference on Intelligent Robots and Systems, pp. 4692–4698 (2019)
9. Li, M., Ranzani, T., Sareh, S., Seneviratne, L.D., Dasgupta, P., Wurdemann, H.A., Althoefer, K.: Multi-fingered haptic palpation utilizing granular jamming stiffness feedback actuators. Smart Mater. Struct. **23**(9), 095007 (2014)
10. Peters, J., Anvari, B., Chen, C., Lim, Z., Wurdemann, H.A.: Hybrid fluidic actuation for a foam-based soft actuator. In: IEEE/RSJ International Conference on Intelligent Robots and Systems, pp. 8701–8708 (2020)
11. Shiva, A., Stilli, A., Noh, Y., Faragasso, A., De Falco, I., Gerboni, G., Cianchetti, M., Menciassi, A., Althoefer, K., Wurdemann, H.A.: Tendon-based stiffening for a pneumatically actuated soft manipulator. IEEE Robotics and Automation Letters **1**(2), 632–637 (2016)
12. Stilli, A., Cremoni, A., Bianchi, M., Ridolfi, A., Gerii, F., Vannetti, F., Wurdemann, H., Allotta, B., Althoefer, K.: A novel pneumatic exoskeleton glove for adaptive hand rehabilitation in post-stroke patients. In: IEEE International Conference on Soft Robotics, pp. 579–584 (2018)
13. Stilli, A., Kolokotronis, E., Fraś, J., Ataka, A., Althoefer, K., Wurdemann, H.A.: Static kinematics for an antagonistically actuated robot based on a beam-mechanics-based model. In: IEEE/RSJ International Conference on Intelligent Robots and Systems, pp. 6959–6964 (2018)
14. Gaozhang, W., Shi, J., Li, Y., Stilli, A., Wurdemann, H.: Characterisation of antagonistically actuated, stiffness-controllable joint-link units for cobots. In: IEEE International Conference on Robotics and Automation, pp. 655–661 (2023)
15. Gaozhang, W., Li, Y., Shi, J., Wang, Y., Stilli, A., Wurdemann, H.: A novel stiffness-controllable joint using antagonistic actuation principles. Mech. Mach. Theory **196**, 105614 (2024)
16. Stilli, A., Wurdemann, H.A., Althoefer, K.: A novel concept for safe, stiffness-controllable robot links. Soft Rob. **4**(1), 16–22 (2017)
17. Gandarias, J.M., Wang, Y., Stilli, A., García-Cerezo, A.J., Gómez-de-Gabriel, J.M., Wurdemann, H.A.: Open-loop position control in collaborative, modular variable-stiffness-link (VSL) robots. IEEE Robot. Automat. Lett. **5**(2), 1772–1779 (2020)
18. Shi, J., Shi, G., Wu, Y., Wurdemann, H.A.: A multi-cavity touch interface for a flexible soft laparoscopy device: design and evaluation. IEEE Trans. Med. Robot. Bion. **6**(4), 1309–1321 (2024)
19. Licher, J., Peters, J., Raatz, A., Wurdemann, H.: Tendon locking for antagonistic configuration- and stiffness-control in soft robots. In: IEEE International Conference on Robotics and Automation, pp. 15322–15328 (2025)
20. Raitt, D.G., Abad, S.-A., Homer-Vanniasinkam, S., Wurdemann, H.A.: Soft, stiffness-controllable sensing tip for on-demand force range adjustment with angled force direction identification. IEEE Sens. J. **22**(9), 8418–8427 (2022)
21. Raitt, D.G., Huseynov, M., Homer-Vanniasinkam, S., Wurdemann, H.A., Abad, S.-A.: Soft-tipped sensor with compliance control for elasticity sensing and palpation. IEEE Trans. Rob. **40**, 2430–2441 (2024)
22. Wolf, S., Grioli, G., Eiberger, O., Friedl, W., Grebenstein, M., Höppner, H., Burdet, E., Caldwell, D.G., Carloni, R., Catalano, M.G., Lefeber, D., Stramigioli, S., Tsagarakis, N., Van Damme, M., Van Ham, R., Vanderborght, B., Visser, L.C., Bicchi, A., Albu-Schäffer, A.: Variable stiffness actuators: Review on design and components. IEEE/ASME Trans. Mechatron. **21**(5), 2418–2430 (2015)
23. Alici, G., Shirinzadeh, B.: Enhanced stiffness modeling, identification and characterization for robot manipulators. IEEE Trans. Rob. **21**(4), 554–564 (2005)
24. Ajoudani, A., Tsagarakis, N.G., Bicchi, A.: Choosing poses for force and stiffness control. IEEE Trans. Rob. **33**(6), 1483–1490 (2017)
25. Gravagne, I.A., Walker, I.D.: Manipulability, force, and compliance analysis for planar continuum manipulators. IEEE Trans. Robot. Autom. **18**(3), 263–273 (2002)

26. Ajoudani, A., Tsagarakis, N.G., Bicchi, A.: On the role of robot configuration in Cartesian stiffness control. In: IEEE International Conference on Robotics and Automation, pp. 1010–1016 (2015)
27. Stella, F., Hughes, J., Rus, D., Della Santina, C.: Prescribing cartesian stiffness of soft robots by co-optimization of shape and segment-level stiffness. Soft Rob. **10**(4), 701–712 (2023)
28. Yoshikawa, T.: Manipulability of robotic mechanisms. Int. J. Robot. Res. **4**(2), 3–9 (1985)
29. Renda, F., Boyer, F., Dias, J., Seneviratne, L.: Discrete Cosserat approach for multisection soft manipulator dynamics. IEEE Trans. Rob. **34**(6), 1518–1533 (2018)
30. Nwafor, C., Laurent, G.J., Rabenorosoa, K.: Miniature parallel continuum robot made of glass: analysis, design, and proof-of-concept. IEEE/ASME Trans. Mechatron. **28**(4), 2038–2046 (2023)
31. Rucker, D.C., Webster, R.J.: Computing Jacobians and compliance matrices for externally loaded continuum robots. In: IEEE International Conference on Robotics and Automation, pp. 945–950 (2011)
32. Black, C.B., Till, J., Rucker, D.C.: Parallel continuum robots: Modeling, analysis, and actuation-based force sensing. IEEE Trans. Rob. **34**(1), 29–47 (2017)
33. Smoljkic, G., Reynaerts, D., Vander Sloten, J., Vander Poorten, E.: Compliance computation for continuum types of robots. In: IEEE/RSJ International Conference on Intelligent Robots and Systems, pp. 1066–1073 (2014)
34. Huang, X., Zhu, X., Gu, G.: Kinematic modeling and characterization of soft parallel robots. IEEE Trans. Rob. **38**(6), 3792–3806 (2022)
35. Gong, Z., Fang, X., Chen, X., Cheng, J., Xie, Z., Liu, J., Chen, B., Yang, H., Kong, S., Hao, Y., Wang, T., Yu, J., Wen, L.: A soft manipulator for efficient delicate grasping in shallow water: modeling, control, and real-world experiments. Int. J. Robot. Res. **40**(1), 449–469 (2021)
36. Caasenbrood, B., Pogromsky, A., Nijmeijer, H.: Control-oriented models for hyperelastic soft robots through differential geometry of curves. Soft Rob. **10**(1), 129–148 (2023)
37. Yang, C., Kang, R., Branson, D.T., Chen, L., Dai, J.S.: Kinematics and statics of eccentric soft bending actuators with external payloads. Mech. Mach. Theory **139**, 526–541 (2019)
38. Oliver-Butler, K., Till, J., Rucker, C.: Continuum robot stiffness under external loads and prescribed tendon displacements. IEEE Trans. Rob. **35**(2), 403–419 (2019)
39. Della Santina, C., Katzschmann, R.K., Bicchi, A., Rus, D.: Model-based dynamic feedback control of a planar soft robot: trajectory tracking and interaction with the environment. Int. J. Robot. Res. **39**(4), 490–513 (2020)
40. Hussain, I., Malvezzi, M., Gan, D., Iqbal, Z., Seneviratne, L., Prattichizzo, D., Renda, F.: Compliant gripper design, prototyping, and modeling using screw theory formulation. Int. J. Robot. Res. **40**(1), 55–71 (2021)
41. Shi, J., Shariati, A., Abad, S.-A., Liu, Y., Dai, J.S., Wurdemann, H.A.: Stiffness modelling and analysis of soft fluidic-driven robots using Lie theory. Int. J. Robot. Res. **43**(3), 354–384 (2024)
42. Shi, J., Frantz, J.C., Shariati, A., Shiva, A., Dai, J.S., Martins, D., Wurdemann, H.A.: Screw theory-based stiffness analysis for a fluidic-driven soft robotic manipulator. In: 2021 IEEE International Conference on Robotics and Automation (ICRA), pp. 11938–11944 (2021)
43. Sonneville, V., Cardona, A., Brüls, O.: Geometrically exact beam finite element formulated on the special Euclidean group se (3). Comput. Methods Appl. Mech. Eng. **268**, 451–474 (2014)
44. Renda, F., Cianchetti, M., Abidi, H., Dias, J., Seneviratne, L.: Screw-based modeling of soft manipulators with tendon and fluidic actuation. J. Mech. Robot. **9**(4) (2017)
45. Sun, T., Lian, B., Yang, S., Song, Y.: Kinematic calibration of serial and parallel robots based on finite and instantaneous screw theory. IEEE Trans. Rob. **36**(3), 816–834 (2020)
46. Cibicik, A., Egeland, O.: Kinematics and dynamics of flexible robotic manipulators using dual screws. IEEE Trans. Rob. **37**(1), 206–224 (2021)
47. Fu, Z., Pan, J., Spyrakos Papastavridis, E., Lin, Y., Zhou, X., Chen, X., Dai, J.S.: A Lie theory based dynamic parameter identification methodology for serial manipulators. IEEE/ASME Trans. Mechatron. **26**(5), 2688–2699 (2020)
48. Qi, P., Qiu, C., Liu, H., Dai, J.S., Seneviratne, L.D., Althoefer, K.: A novel continuum manipulator design using serially connected double-layer planar springs. IEEE/ASME Trans. Mechatron. **21**(3), 1281–1292 (2016)

49. Selig, J., Ding, X.: A screw theory of static beams. In: IEEE/RSJ International Conference on Intelligent Robots and Systems, pp. 312–317 (2001)
50. Ding, X., Dai, J.: Compliance analysis of mechanisms with spatial continuous compliance in the context of Screw theory and Lie groups. Proc. Inst. Mech. Eng. C J. Mech. Eng. Sci. **224**(11), 2493–2504 (2010)
51. Shi, J., Jin, H., Abad, S.-A., Gaozhang, W., Shi, G., Wurdemann, H.A.: A static modeling and evaluation framework for soft continuum robots with reinforced chambers. IEEE Trans. Rob. **41**, 6419–6439 (2025)
52. Moseley, P., Florez, J.M., Sonar, H.A., Agarwal, G., Curtin, W., Paik, J.: Modeling, design, and development of soft pneumatic actuators with finite element method. Adv. Eng. Mater. **18**(6), 978–988 (2016)
53. Zhong, G., Dou, W., Zhang, X., Yi, H.: Bending analysis and contact force modeling of soft pneumatic actuators with pleated structures. Int. J. Mech. Sci. **193**, 106150 (2021)
54. Zhang, H., Liu, W., Yu, M., Hou, Y.: Design, fabrication, and performance test of a new type of soft-robotic gripper for grasping. Sensors **22**(14) (2022)
55. Yao, Y., He, L., Maiolino, P.: A simulation-based toolbox to expedite the digital design of bellow soft pneumatic actuators. In: IEEE International Conference on Soft Robotics, pp. 29–34 (2022)
56. Liu, Z., Wang, F., Liu, S., Tian, Y., Zhang, D.: Modeling and analysis of soft pneumatic network bending actuators. IEEE/ASME Trans. Mechatron. **26**(4), 2195–2203 (2020)
57. Sachin, Wang, Z., Hirai, S.: Analytical modeling of a soft pneu-net actuator subjected to planar tip contact. IEEE Trans. Robot. **38**(5), 2720–2733 (2022)
58. Alici, G., Canty, T., Mutlu, R., Hu, W., Sencadas, V.: Modeling and experimental evaluation of bending behavior of soft pneumatic actuators made of discrete actuation chambers. Soft Rob. **5**(1), 24–35 (2018)
59. Polygerinos, P., Wang, Z., Overvelde, J.T.B., Galloway, K.C., Wood, R.J., Bertoldi, K., Walsh, C.J.: Modeling of soft fiber-reinforced bending actuators. IEEE Trans. Rob. **31**(3), 778–789 (2015)
60. Wang, Z., Polygerinos, P., Overvelde, J.T.B., Galloway, K.C., Bertoldi, K., Walsh, C.J.: Interaction forces of soft fiber reinforced bending actuators. IEEE/ASME Trans. Mechatron. **22**(2), 717–727 (2016)
61. Sun, Y., Feng, H., Manchester, I.R., Yeow, R.C.H., Qi, P.: Static modeling of the fiber-reinforced soft pneumatic actuators including inner compression: Bending in free space, block force, and deflection upon block force. Soft Rob. **9**(3), 451–472 (2022)
62. Chen, X., Shi, J., Wurdemann, H., Thuruthel, T.G.: Vision-based tip force estimation on a soft continuum robot. In: IEEE International Conference on Robotics and Automation (ICRA), pp. 7621–7627 (2024)
63. George Thuruthel, T., Gardner, P., Iida, F.: Closing the control loop with time-variant embedded soft sensors and recurrent neural networks. Soft Rob. **9**(6), 1167–1176 (2022)
64. Joshi, S., Paik, J.: Sensorless force and displacement estimation in soft actuators. Soft Matter **19**(14), 2554–2563 (2023)
65. Pyo, S., Lee, J., Bae, K., Sim, S., Kim, J.: Recent progress in flexible tactile sensors for human-interactive systems: from sensors to advanced applications. Adv. Mater. **33**(47), 2005902 (2021)
66. Langer, M., Amanov, E., Burgner-Kahrs, J.: Stiffening sheaths for continuum robots. Soft Rob. **5**(3), 291–303 (2018)
67. Zhang, J., Liu, L., Chen, Y., Zhu, M., Tang, L., Tang, C., Shintake, J., Zhao, J., He, J., Ren, X., Li, P., Huang, Q., Zhao, H., Lu, J., Li, D.: Fiber-reinforced soft polymeric manipulator with smart motion scaling and stiffness tunability. Cell Rep. Phys. Sci. **2**(10), 100600 (2021)
68. Ranzani, T., Cianchetti, M., Gerboni, G., De Falco, I., Menciassi, A.: A soft modular manipulator for minimally invasive surgery: design and characterization of a single module. IEEE Trans. Rob. **32**(1), 187–200 (2016)
69. Best, C.M., Rupert, L., Killpack, M.D.: Comparing model-based control methods for simultaneous stiffness and position control of inflatable soft robots. Int. J. Robot. Res. **40**(1), 470–493 (2021)

70. Habich, T.-L., Kleinjohann, S., Schappler, M.: Learning-based position and stiffness feed-forward control of antagonistic soft pneumatic actuators using gaussian processes. In: IEEE International Conference on Soft Robotics, pp. 1–7 (2023)
71. Bruder, D., Graule, M.A., Teeple, C.B., Wood, R.J.: Increasing the payload capacity of soft robot arms by localized stiffening. Sci. Robot. **8**(81), 9001 (2023)
72. Aloi, V., Black, C., Rucker, C.: Stiffness control of parallel continuum robots. In: Dynamic Systems and Control Conference, vol. 51890, pp. 001–04012 (2018)
73. Lai, J., Lu, B., K. Chu, H.: Variable-stiffness control of a dual-segment soft robot using depth vision. IEEE/ASME Trans. Mech. **27**(2), 1034–1045 (2022)
74. Shi, J., Abad, S.-A., Shi, G., Gaozhang, W., Dai, J.S., Wurdemann, H.A.: Model-based static compliance analysis and control for pneumatic-driven soft robots. IEEE/ASME Trans. Mechatron. **30**(6), 5567–5578 (2025)
75. Yasin, R., Simaan, N.: Joint-level force sensing for indirect hybrid force/position control of continuum robots with friction. Int. J. Robot. Res. **40**(4–5), 764–781 (2021)
76. Xu, K., Simaan, N.: An investigation of the intrinsic force sensing capabilities of continuum robots. IEEE Trans. Rob. **24**(3), 576–587 (2008)
77. Bajo, A., Simaan, N.: Hybrid motion/force control of multi-backbone continuum robots. Int. J. Robot. Res. **35**(4), 422–434 (2016)
78. Li, W., Huang, X., Yan, L., Cheng, H., Liang, B., Xu, W.: Force sensing and compliance control for a cable-driven redundant manipulator. IEEE/ASME Trans. Mechatron. **29**(1), 1–12 (2024)
79. Katzschmann, R.K., Santina, C.D., Toshimitsu, Y., Bicchi, A., Rus, D.: Dynamic motion control of multi-segment soft robots using piecewise constant curvature matched with an augmented rigid body model. In: IEEE International Conference on Soft Robotics, pp. 454–461 (2019)
80. Lindenroth, L., Stoyanov, D., Rhode, K., Liu, H.: Toward intrinsic force sensing and control in parallel soft robots. IEEE/ASME Trans. Mechatron. **28**(1), 80–91 (2022)
81. Shi, J., Abad Guaman, S., Dai, J., Wurdemann, H.: Compliance model-based contact force control for soft continuum robots. Soft Robot. (2026)
82. Sadati, S.M.H., Shiva, A., Herzig, N., Rucker, C.D., Hauser, H., Walker, I.D., Bergeles, C., Althoefer, K., Nanayakkara, T.: Stiffness imaging with a continuum appendage: real-time shape and tip force estimation from base load readings. IEEE Robot. Automat. Lett. **5**(2), 2824–2831 (2020)
83. Shi, J., Gaozhang, W., Wurdemann, H.A.: Design and characterisation of cross-sectional geometries for soft robotic manipulators with fibre-reinforced chambers. In: International Conference on Soft Robotics, pp. 125–131 (2022)
84. Searle, T.C., Althoefer, K., Seneviratne, L., Liu, H.: An optical curvature sensor for flexible manipulators. In: IEEE International Conference on Robotics and Automation, pp. 4415–4420 (2013)
85. Zhao, H., O'brien, K., Li, S., Shepherd, R.F.: Optoelectronically innervated soft prosthetic hand via stretchable optical waveguides. Sci. Robot. **1**(1), 7529 (2016)
86. Cheng, X., Gong, Y., Liu, Y., Wu, Z., Hu, X.: Flexible tactile sensors for dynamic triaxial force measurement based on piezoelectric elastomer. Smart Mater. Struct. **29**(7), 075007 (2020)
87. Chen, S., Wang, H.-Z., Liu, T.-Y., Liu, J.: Liquid metal smart materials toward soft robotics. Adv. Intell. Syst. **5**(8), 2200375 (2023)
88. Tawk, C., Sariyildiz, E., Alici, G.: Force control of a 3D printed soft gripper with built-in pneumatic touch sensing chambers. Soft Rob. **9**(5), 970–980 (2022)
89. Navarro, S.E., et al.: A model-based sensor fusion approach for force and shape estimation in soft robotics. IEEE Robot. Automat. Lett. **5**(4), 5621–5628 (2020)
90. Abbasi, P., Nekoui, M.A., Zareinejad, M., Abbasi, P., Azhang, Z.: Position and force control of a soft pneumatic actuator. Soft Rob. **7**(5), 550–563 (2020)
91. Zhang, Z., Dequidt, J., Duriez, C.: Vision-based sensing of external forces acting on soft robots using finite element method. IEEE Robot. Automat. Lett. **3**(3), 1529–1536 (2018)
92. Wang, H., Ni, H., Wang, J., Chen, W.: Hybrid vision/force control of soft robot based on a deformation model. IEEE Trans. Control Syst. Technol. **29**(2), 661–671 (2019)

93. Cangan, B.G., Navarro, S.E., Yang, B., Zhang, Y., Duriez, C., Katzschmann, R.K.: Model-based disturbance estimation for a fiber-reinforced soft manipulator using orientation sensing. In: IEEE/RSJ International Conference on Intelligent Robots and Systems, pp. 9424–9430 (2022)
94. Thuruthel, T.G., Shih, B., Laschi, C., Tolley, M.T.: Soft robot perception using embedded soft sensors and recurrent neural networks. Sci. Robot. **4**(26), 1488 (2019)

Tendon-Driven Continuum Mechanisms: Modeling, Workspace and Force Control

Tianxiang Dai, Jonas Breuling, Remco I. Leine, and Simon R. Eugster

Abstract Tendon-driven continuum mechanisms (TDCMs) offer high flexibility and adaptability, making them ideal for medical robotics, search and rescue, and inspection applications. However, their inherent nonlinear behavior poses challenges for modeling and control. This study presents the design, modeling, and control of a TDCM consisting of a soft silicone cylinder actuated by four tendons. The system is modeled as a shear-deformable, nonlinear elastic rod using the Cosserat rod theory and discretized by a Petrov–Galerkin finite element formulation. An inverse static control strategy is formulated as a constrained optimization problem to achieve motion control. Experimental validation using strain gauges and fiducial markers demonstrates the accuracy of the modeling and control approach. The results highlight the influence of tendon routing on deformation modes and the effectiveness of the control strategy in regulating movement. While open-loop inverse static control achieves reasonable accuracy, residual errors suggest the need for closed-loop feedback. This study contributes to advancing TDCM experimental research and control methodologies.

1 Introduction

Soft robots are characterized by their ability to continuously deform, providing unparalleled flexibility and adaptability in complex environments. Their inherent compli-

---

T. Dai (✉) · J. Breuling · R. I. Leine
Institute for Nonlinear Mechanics, University of Stuttgart, Stuttgart, Germany
e-mail: tianxiang.dai@inm.uni-stuttgart.de

J. Breuling
e-mail: jonas.breuling@inm.uni-stuttgart.de

R. I. Leine
e-mail: leine@inm.uni-stuttgart.de

S. R. Eugster
Department of Mechanical Engineering, Eindhoven University of Technology, Eindhoven, The Netherlands
e-mail: s.r.eugster@tue.nl

A. Raatz et al. (eds.), *Soft Material Robotic Systems*,
https://doi.org/10.1007/978-3-032-22453-8_16

ance allows them to interact safely with unstructured environments, making them particularly well-suited for applications such as minimally invasive surgery, search and rescue, and inspection in confined or hazardous spaces. However, the nonlinear characteristics of these robots, including hysteresis, stress softening, and extreme hyper-redundancy, pose significant modeling and control challenges [1].

Tendon-driven continuum mechanisms (TDCMs) are a class of soft robots actuated by unilateral tendons. One straightforward modeling strategy to simulate soft robots is to use three-dimensional finite element methods for deformable solids (3D-FEM [2]). While this method provides high geometric fidelity, it is often computationally prohibitive. Additionally, the selection of a specific material model and the identification of its parameters is not always straightforward. For TDCMs with slender structures, Cosserat rod theory [3] is widely regarded as the most comprehensive one-dimensional framework capable of capturing all relevant deformation modes, such as bending, twisting, shearing, and stretching.

Since the seminal work of Simo [4], the structural mechanics community has proposed many rod finite element formulations to address both static and dynamic problems. An extensive survey of Cosserat rod finite element formulations is provided by Meier [5]. In the early stages of soft robotics, however, alternative modeling approaches proved more effective for specific applications. To solve the boundary value problem in static situations, Rucker [6, 7] proposed a shooting method for Cosserat rods. While the classical structural dynamics rod finite elements were mostly position- and orientation-based, Renda [8, 9] introduced rod finite element methods that are based on strains as the unknown fields. Such strain-based formulations would be the ideal model with the vision of being capable of measuring strains—at least discretely.

Discrete elastic rod models are closely related to finite element models [10–12]. These models do not approximate the virtual work of the Cosserat rod but rather its kinetic and potential energy. Then, they apply a variational principle, such as Hamilton's principle in dynamics or the principle of minimum of potential energy in statics. With a particular choice of test and trial functions, as well as with the correct quadrature rule for the spatial integration, discrete rods can also be considered as a low-order finite element method.

In our approach, we adopt a Petrov–Galerkin finite element formulation for Cosserat rods, following the framework established in [14–16]. The rod kinematics are discretized using total Lagrangian interpolation strategies, with nodal positions and orientations serving as the primary degrees of freedom. By independently interpolating both the kinematic fields and their variations, the Petrov–Galerkin projection simplifies the expressions for the virtual work functionals, as discussed in [15, 16]. This formulation can also accommodate alternative orientation parameterizations, such as non-unit quaternions, and provides robust performance in both static and dynamic simulations.

This study presents the design and development of our TDCM consisting of a silicone cylinder actuated by four tendons, detailing the mechanical model formulation and the implementation of an inverse static control strategy. Experimental results

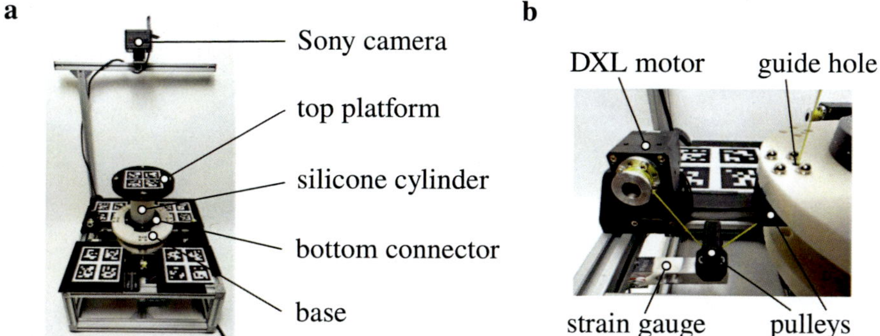

Fig. 1 **a** Tendon-driven continuum mechanism with parallel tendon routing configuration, **b** A close-up view of the left motor and attached strain gauge

validate the proposed approach, demonstrate its feasibility for controlled actuation, and lay the foundation for future advances in tendon-driven continuum robotics.

The remainder of this chapter is structured as follows. First, the experimental test bench used for constructing and actuating the TDCM is introduced in Sect. 2. The forward static model, based on the Petrov–Galerkin finite element formulation for Cosserat rods, is then presented in Sect. 3. Reachable workspace analysis using a brute-force sampling approach is covered in Sect. 4. An inverse static control problem is formulated as a constrained optimization task, and the control strategy is outlined in Sect. 5. Section 6 details the experimental results, including material parameter identification, workspace validation, and control performance assessment. Finally, Sect. 7 summarizes the key findings and discusses potential directions for future research.

2 Test Bench

The TDCM consists of a top platform, a silicone cylinder with two connectors attached to each end, and four tendons (Dyneema) for actuation, as illustrated in Fig. 1a. The soft cylinder was molded from Dragon Skin Shore 20A silicone, following the procedure in [17]. It has a height of 95 mm and a diameter of 60 mm. The connectors and top platform were 3D-printed using polylactic acid (PLA), with the bottom connector fixed to the base, while the top connector is fully embedded into the top platform. The connectors and top platform have heights of 11.5 mm and 14.5 mm, respectively, resulting in a total mechanism height of 121 mm. Excluding the tendons, the top platform and silicone cylinder weigh 185 g and 433 g, respectively.

The tendons are anchored to the top platform and pass through guide holes in the base, with each tendon guided by two pulleys before reaching its respective electric motor (Dynamixel XH430-W350-R), as shown in Fig. 1b. Actuation is achieved by

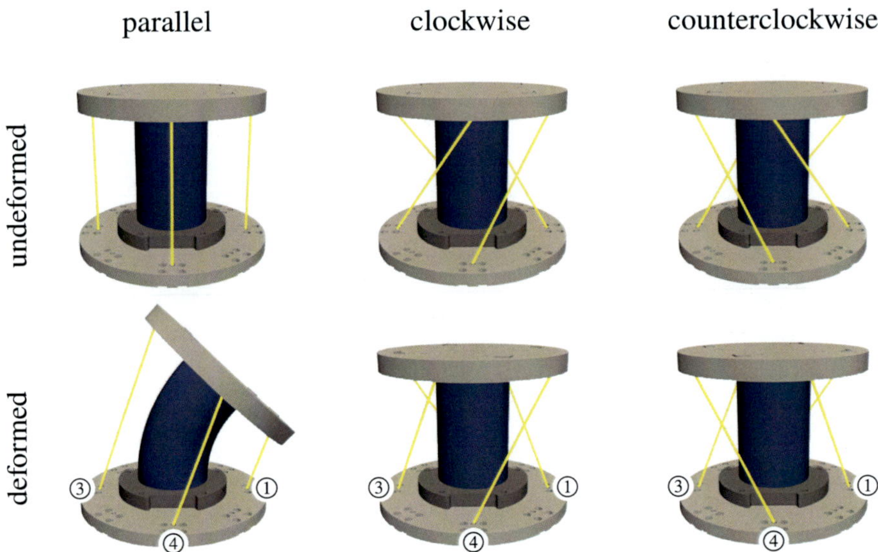

Fig. 2 Comparison of parallel, clockwise, and counterclockwise tendon routings in both undeformed and deformed states. Tendon numbering is indicated by the symbols ①–④. Note that ② is not visible, as it corresponds to the tendon located behind the silicone cylinder

adjusting the tendon lengths, inducing movement in the top platform. For motion tracking, the top platform is equipped with four fiducial ArUco markers [18], which are recorded by a Sony RX0 II camera positioned above the top platform. Sixteen additional markers on the ground enable localization of the camera relative to the inertial frame of reference. Tendon forces are measured using strain gauges mounted on beam-like holders for the second pulley. A PC running ROS 2 nodes [13] serves as the central controller, handling hardware communication. Voltage signals from the strain gauges are digitized using a Texas Instruments ADS1256 analog-to-digital converter and transmitted to the PC via an Arduino. The electric motors are controlled through a USB interface (Dynamixel U2D2). Figure 2 shows three tendon routing configurations employed in our investigation, each inducing different deformations in the soft cylinder. While the parallel tendon routing generates both compression and bending, the clockwise and counterclockwise tendon routings cause torsional deformation.

3 Forward Statics

We model our TDCM as a multi-body system, where the top platform is treated as a rigid body, the soft silicone cylinder as a shear-deformable nonlinear elastic rod, and the tendon actuation and gravity as external forces. Using the non-unit quaternion

Petrov–Galerkin finite element formulation for Cosserat rods [14–16], the elastic rod is discretized into n_{el} elements. Each element contains $p+1$ equally spaced nodes in the reference arc length. The last node of an element and the first node of the next element coincide. Therefore, the generalized coordinates of the rod are given by $\mathbf{q}_{\mathrm{rod}} = (\mathbf{q}_0^0, \ldots, \mathbf{q}_{p-1}^0, \ldots, \mathbf{q}_0^e, \ldots, \mathbf{q}_{p-1}^e, \ldots, \mathbf{q}_0^{n_{\mathrm{el}}-1}, \ldots, \mathbf{q}_{p-1}^{n_{\mathrm{el}}-1}, \mathbf{q}_p^{n_{\mathrm{el}}-1})$, where the $N' = (pn_{\mathrm{el}} + 1)$ nodal generalized coordinates $\mathbf{q}_i^e = (\mathbf{r}_{OP_i^e}, \mathbf{p}_i^e) \in \mathbb{R}^7$ include the nodal centerline positions $\mathbf{r}_{OP_i^e} \in \mathbb{R}^3$ and the non-unit quaternion $\mathbf{p}_i^e \in \mathbb{R}^4$ of the nodal rotations, resulting in $7N'$ degrees of freedom in total. The top platform, modeled as a rigid body, is described by the tuple $\mathbf{q}_{\mathrm{top}} = (\mathbf{r}_{OP}^{\mathrm{top}}, \mathbf{p}^{\mathrm{top}}) \in \mathbb{R}^7$, representing its position and quaternion parameterization for rotation. Thus, the global set of generalized coordinates for the entire system is $\mathbf{q} = (\mathbf{q}_{\mathrm{rod}}, \mathbf{q}_{\mathrm{top}}) \in \mathbb{R}^{7N}$ with $N = N' + 1$. Following the procedure for deriving the virtual work in [14–16], the system satisfies the static equilibrium condition

$$\mathbf{f}(\mathbf{q}, \boldsymbol{\lambda}_{\mathrm{g}}, \boldsymbol{\lambda}_{\mathrm{t}}) = \begin{pmatrix} \mathbf{f}^{\mathrm{int}}(\mathbf{q}) + \mathbf{f}^{\mathrm{ext}}(\mathbf{q}) + \mathbf{W}_{\mathrm{g}}(\mathbf{q})\boldsymbol{\lambda}_{\mathrm{g}} + \mathbf{W}_{\mathrm{t}}(\mathbf{q})\boldsymbol{\lambda}_{\mathrm{t}} \\ \mathbf{g}_{\mathrm{s}}(\mathbf{q}) \\ \mathbf{g}_{\mathrm{g}}(\mathbf{q}) \end{pmatrix} = \begin{pmatrix} \mathbf{0}_{6N} \\ \mathbf{0}_{N} \\ \mathbf{0}_{12} \end{pmatrix}, \tag{1}$$

where $\mathbf{f}^{\mathrm{ext}}$ and $\mathbf{f}^{\mathrm{int}}$ model the generalized external forces and the generalized internal forces of the rod caused by the elasticity, respectively. Since non-unit quaternions do not represent valid rotations, they must be normalized before being used for this purpose. Therefore, we introduce the constraints $\mathbf{g}_{\mathrm{s}}$ to enforce unit length. Additional geometric constraints $\mathbf{g}_{\mathrm{g}} \in \mathbb{R}^{12}$ are employed to ensure the rigid connections between parts, i.e., the first node of the rod is fixed at the base, and the last node of the rod and the top platform are clamped together. This leads to constraint forces $\boldsymbol{\lambda}_{\mathrm{g}} \in \mathbb{R}^{12}$ and their associated generalized force directions $\mathbf{W}_{\mathrm{g}} \in \mathbb{R}^{6N\times 12}$ entering (1). Similarly, the tendon actuation influences the system through tendon forces $\boldsymbol{\lambda}_{\mathrm{t}} \in \mathbb{R}^4$ and force directions $\mathbf{W}_{\mathrm{t}} \in \mathbb{R}^{6N\times 4}$. Currently, only the gravity is considered as the source of external forces. The static properties of the system are characterized by the axial stiffness EA, shear stiffness GA, bending stiffness EI, and torsional stiffness GJ, where the Young's modulus E and shear modulus G must be determined experimentally. These stiffness parameters enter the internal forces $\mathbf{f}^{\mathrm{int}}$ in (1) through the constitutive laws, given by ${}_B\mathbf{n} := \mathbf{C}_\gamma\, \boldsymbol{\varepsilon}_\gamma$ and ${}_B\mathbf{m} := \mathbf{C}_\kappa\, \boldsymbol{\varepsilon}_\kappa$, respectively. Here, $\mathbf{C}_\gamma := \mathrm{diag}(EA, GA, GA)$ and $\mathbf{C}_\kappa := \mathrm{diag}(GJ, EI, EI)$, while $\boldsymbol{\varepsilon}_\gamma$ and $\boldsymbol{\varepsilon}_\kappa$ represent the objective strain variables.

Due to the deformability of the system, a pure kinematic analysis is not suitable. Instead of a forward kinematics, one has to deal with forward and inverse static analyses. In a forward static analysis, the static equilibrium condition (1) is solved for given tendon forces $\boldsymbol{\lambda}_{\mathrm{t}}$. For this purpose, we employ the Newton–Raphson algorithm, an iterative root-finding approach. In each iteration, it solves the linear system

$$\mathbf{J}(\mathbf{x}^k, \boldsymbol{\lambda}_{\mathrm{t}})\Delta\mathbf{x}^k = -\mathbf{f}(\mathbf{x}^k, \boldsymbol{\lambda}_{\mathrm{t}}), \tag{2}$$

where k denotes the iteration index and $\mathbf{J}$ the Jacobian matrix, i.e. the partial derivative of $\mathbf{f}$ with respect to the system state variables $\mathbf{x} = (\mathbf{q}, \boldsymbol{\lambda}_\mathrm{g}) \in \mathbb{R}^{7N+12}$ given by

$$\mathbf{J}(\mathbf{x}, \boldsymbol{\lambda}_\mathrm{t}) = \frac{\partial \mathbf{f}}{\partial \mathbf{x}} = \begin{pmatrix} \frac{\partial \mathbf{f}}{\partial \mathbf{q}} & \frac{\partial \mathbf{f}}{\partial \boldsymbol{\lambda}_\mathrm{g}} \end{pmatrix}. \tag{3}$$

The system states are updated iteratively using $\mathbf{x}^{k+1} = \mathbf{x}^k + \Delta\mathbf{x}^k$ until convergence. This algorithm is initialized with an initial guess, often obtained from a previous solution or the designed configuration without tendon actuation.

4 Reachable Workspace

The reachable workspace of a mechanism consists of all points that one point of moving top platform can access while constrained by its physical limitations [19]. Since tendons can only pull, not push, the tendon forces must be positive. In addition, an upper limit requirement is useful to prevent tendon failure. As a result, the top platform cannot reach all positions in space. Instead, its motion is restricted to the statically reachable workspace [20].

$$\mathcal{W}_\mathcal{U} = \left\{\mathbf{r}_{OQ}(\mathbf{q}_\mathrm{top}) \mid \mathbf{f}(\mathbf{q}, \boldsymbol{\lambda}_\mathrm{g}, \boldsymbol{\lambda}_\mathrm{t}) = \mathbf{0}, \boldsymbol{\lambda}_\mathrm{t} \in \mathcal{U}\right\}, \tag{4}$$

where Q is a reference point on the top platform and $\mathcal{U} = \{\boldsymbol{\lambda}_\mathrm{t} \in \mathbb{R}^4 \mid 0 \le \lambda_{\mathrm{t},i} \le \lambda_\mathrm{max}, i = 1, 2, 3, 4\}$ defines the bounded force space. Without loss of generality, the center of the ArUco-markers on the top platform is defined as Q, which results in

$$\mathbf{r}_{OQ}(\mathbf{q}_\mathrm{top}) = \mathbf{r}_{OP}^\mathrm{top} + \mathbf{A}_{IB}(\mathbf{p}^\mathrm{top})\,{}_B\mathbf{r}_{PQ}, \tag{5}$$

where $\mathbf{A}_{IB}(\mathbf{p}^\mathrm{top})$ is the transformation matrix of the top platform determined by the quaternion $\mathbf{p}^\mathrm{top}$, and ${}_B\mathbf{r}_{PQ}$ is the position vector from point P to point Q expressed in the body-fixed frame B.

A common approach for analyzing the workspace is brute-force sampling. Here, the force space $\mathcal{U}$ is uniformly discretized in all directions, generating a set of grid points $\mathcal{S} = \{\boldsymbol{\lambda}_\mathrm{t}^i \in \mathcal{U} \mid i{=}1, \ldots, M\}$. For every $\boldsymbol{\lambda}_\mathrm{t}^i$, the static equilibrium condition (1) is solved, producing a sampling $\mathcal{W}_\mathcal{S}$ of the reachable workspace. To get its surface $\partial\mathcal{W}_\mathcal{S}$, MATLAB's *alphaShape* function is applied on it. The α-shape algorithm [21] provides a triangulation of the boundary points extracted from $\mathcal{W}_\mathcal{S}$. However, since the displacement of the top platform in the z-direction is significantly smaller than in the x- and y-directions, the raw sample points can lead to an inaccurate shape representation. To improve the algorithm's performance, the samples are first normalized to the interval [0, 1] along each axis before computing the α-shape. This ensures a well-balanced geometric representation and prevents the boundary points from being missed. As the number of samples approaches infinity, the boundary $\partial\mathcal{W}_\mathcal{S}$ converges

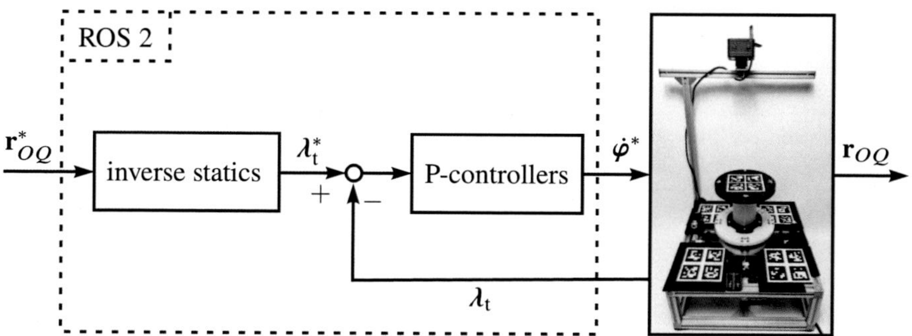

Fig. 3 Inverse static control of the tendon-driven continuum mechanism

to the real workspace boundary $\partial \mathcal{W}_{\mathcal{U}}$. The reachable workspace is visualized and discussed further in Sect. 6.2.

5 Inverse Static Control

For controlling the movement of the top platform, we use cascade control and model inversion techniques. The electric motors are operated in velocity control mode, with each motor regulated by an embedded proportional-integral controller (PI controller). As Fig. 3 shows, the motors receive desired angular velocities $\dot{\boldsymbol{\varphi}}^*$ from the outside. To realize tendon forces as system inputs as required by the mechanical model (1), four decoupled proportional controllers (P controllers) are implemented externally in ROS 2 to command $\dot{\boldsymbol{\varphi}}^*$, based on the difference between the desired tendon forces $\boldsymbol{\lambda}^*_t$ and the force measurements $\boldsymbol{\lambda}_t$. Contrary to the forward statics, the "inverse statics" block computes $\boldsymbol{\lambda}^*_t$ for a desired position $\mathbf{r}^*_{OQ}$. It is not possible to invert the static model (1) directly. Since the number of tendons is more than three, the dimensions of the top platform's position, the model inversion is also not unique. Hence, we formulate the inverse statics as a constrained optimization problem

$$\boldsymbol{\lambda}^*_t = \underset{\boldsymbol{\lambda}_t \in \mathcal{U}}{\arg\min}\, \|\boldsymbol{\lambda}_t\|^2, \quad \text{s.t.} \quad \mathbf{f}(\mathbf{q}, \boldsymbol{\lambda}_g, \boldsymbol{\lambda}_t) = 0 \quad \text{and} \quad \mathbf{r}_{OQ}(\mathbf{q}_{\text{top}}) = \mathbf{r}^*_{OQ}. \tag{6}$$

This problem is solved using the sequential least squares quadratic programming (SLSQP) provided by SciPy [22], where the desired position $\mathbf{r}^*_{OQ}$ is incorporated into the equality constraints as

$$\mathbf{c}(\boldsymbol{\lambda}_t) = \mathbf{r}_{OQ}(\mathbf{q}_{\text{top}}(\boldsymbol{\lambda}_t)) - \mathbf{r}^*_{OQ}. \tag{7}$$

The static equilibrium condition (1) is treated as a subproblem, solved by the Newton–Raphson solver when evaluating the equality constraints and their derivatives with respect to the optimization variables $\boldsymbol{\lambda}_t$ given by

Table 1 Summary of experimental configurations. The tendon routings are illustrated in Fig. 2

Experiment	Relevant stiffness	Tendon routing	Active tendon	Slack tendon
Compression 1	E	Parallel	①, ③	②, ④
Compression 2	E	Parallel	②, ④	①, ③
Torsion 1	G	Clockwise	①, ③	②, ④
Torsion 2	G	Clockwise	②, ④	①, ③
Torsion 3	G	Counterclockwise	①, ③	②, ④
Torsion 4	G	Counterclockwise	②, ④	①, ③
Bending 1	E, G	Parallel	①	②, ③, ④
Bending 2	E, G	Parallel	②	①, ③, ④
Bending 3	E, G	Parallel	③	①, ②, ④
Bending 4	E, G	Parallel	④	①, ②, ③

$$\frac{\partial \mathbf{c}}{\partial \boldsymbol{\lambda}_{\mathrm{t}}} = \frac{\partial \mathbf{r}_{OQ}}{\partial \boldsymbol{\lambda}_{\mathrm{t}}} = \frac{\partial \mathbf{r}_{OQ}}{\partial \mathbf{q}_{\mathrm{top}}} \frac{\partial \mathbf{q}_{\mathrm{top}}}{\partial \boldsymbol{\lambda}_{\mathrm{t}}} = \frac{\partial \mathbf{r}_{OQ}}{\partial \mathbf{q}_{\mathrm{top}}} \mathbf{C} \frac{\partial \mathbf{x}}{\partial \boldsymbol{\lambda}_{\mathrm{t}}}, \tag{8}$$

where $\mathbf{C} \in \mathbb{R}^{7\times(7N+12)}$ is an appropriate Boolean connectivity matrix extracting the coordinates of the top platform out of the global state variables. Since $\mathbf{x}$ and $\boldsymbol{\lambda}_{\mathrm{t}}$ are both arguments of $\mathbf{f}$, the implicit function theorem [23] allows to compute the partial derivatives $\frac{\partial \mathbf{x}}{\partial \boldsymbol{\lambda}_{\mathrm{t}}}$ as

$$\frac{\partial \mathbf{x}}{\partial \boldsymbol{\lambda}_{\mathrm{t}}} = -\left(\frac{\partial \mathbf{f}}{\partial \mathbf{x}}\right)^{-1} \frac{\partial \mathbf{f}}{\partial \boldsymbol{\lambda}_{\mathrm{t}}} = -\mathbf{J}^{-1} \frac{\partial \mathbf{f}}{\partial \boldsymbol{\lambda}_{\mathrm{t}}}, \tag{9}$$

where $\mathbf{J}$ is assumed to be of full rank and therefore invertible as in (2).

6 Experimental Result

To validate our modeling and control approach, we conducted a series of experiments using the presented test bench. The experiments aimed to identify material parameters, assess the accuracy of the static model, and analyze the performance of the inverse static control strategy.

6.1 Parameter Identification

To identify the Young's modulus E and shear modulus G of the silicone cylinder, we conducted ten experiments with different active and slack tendons, as summarized in Table 1. These experiments included two compression experiments, four torsion experiments, and four bending experiments. As their names suggest, the compres-

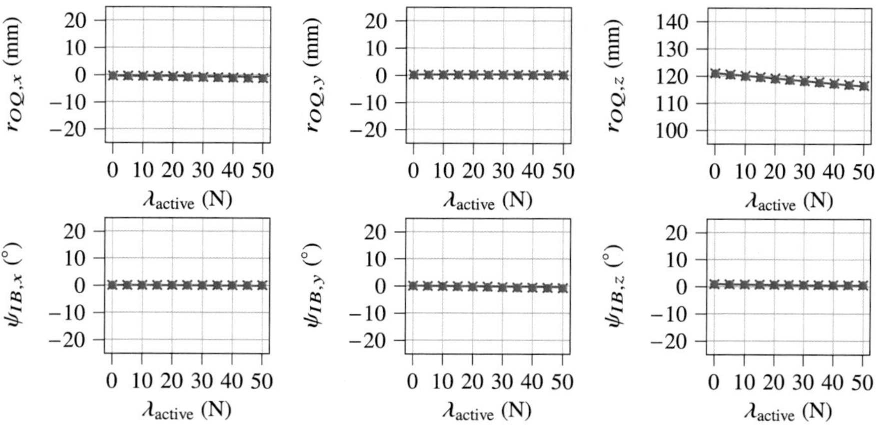

Fig. 4 Compression experiment 1: pose measurements (× and •) compared to model predictions (——). × and • indicate measurements during increasing and decreasing active tendon force λ_{active}, respectively

sion and torsion experiments determined E and G, respectively, while the bending experiments were influenced by both stiffness parameters. During each experiment, the forces in slack tendons were maintained at zero, while the P controllers for active tendons were commanded step-wise from 0 N to $\lambda_{\text{max}} = 50$ N and then back to 0 N with a step size of 5 N. Each step was held for 5 s to ensure that the controller had sufficient time to regulate the tendon forces to their commanded values. To evaluate material isotropy and the consistency of the assembly, experiments of each type—compression, torsion, and bending—were duplicated with the active and slack tendons swapped in the compression and torsion experiments or alternated in the bending experiments.

To exclude dynamic effects of system, only the measurement in the middle of each step was used for the parameter identification. This resulted in 22 measurements for each experiment and 220 measurements in total. These are denoted as $\{(\tilde{\boldsymbol{\lambda}}_{\text{t}}^{i}, \tilde{\mathbf{r}}_{OQ}^{i}, \tilde{\boldsymbol{\psi}}_{IB}^{i}) \mid i=1, \ldots, 220\}$, where the rotation vector $\tilde{\boldsymbol{\psi}}_{IB}^{i}$ represents the measured orientation of the top platform. Then the stiffness parameters E and G were obtained by solving the optimization problem

$$\min_{\substack{E,G \\ \Delta\mathbf{r}_{OP}^{\text{top}}, \Delta\boldsymbol{\psi}_{IB}^{\text{top}}}} \sum_{i=1}^{220} \left\| \mathbf{r}_{OQ}(\mathbf{q}_{\text{top}}^{i}) - \tilde{\mathbf{r}}_{OQ}^{i} \right\|^2 + \left\| \boldsymbol{\psi}_{IB}(\mathbf{q}_{\text{top}}^{i}) - \tilde{\boldsymbol{\psi}}_{IB}^{i} \right\|^2, \tag{10}$$

$$\text{s.t.} \quad \mathbf{f}(\mathbf{q}^{i}, \boldsymbol{\lambda}_{\text{g}}^{i}, \tilde{\boldsymbol{\lambda}}_{\text{t}}^{i}) = 0.$$

Due to assembly tolerances, the top platform was initially displaced in horizontal directions and rotated around the vertical axis relative to the top connector of the silicone cylinder. Therefore, initial deviations $\Delta\mathbf{r}_{OP}^{\text{top}}$ for position and $\Delta\boldsymbol{\psi}_{IB}^{\text{top}}$ for rotation

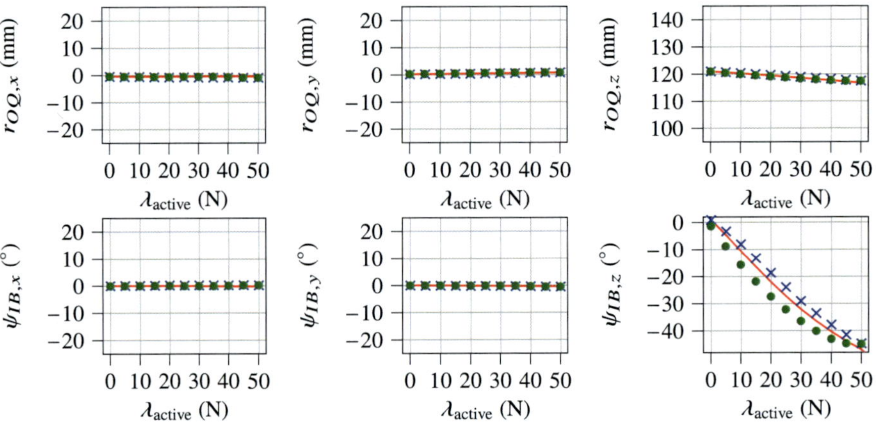

Fig. 5 Torsion experiment 1: pose measurements (× and •) compared to model predictions (——). × and • indicate measurements during increasing and decreasing active tendon force λ_{active}, respectively

were included in the optimization problem beyond E and G. The equilibrium condition (1) was governed by the Newton–Raphson solver for each measurement individually, but the errors between static model and measurement was accumulated globally. The optimization problem was solved with the Levenberg–Marquardt algorithm provided by SciPy [22]. The resulting stiffness parameters, $E = 707.287$ kPa and $G = 228.672$ kPa, were found to be close to the values reported in [17]. The detected initial deviations were sufficiently small, with $\Delta\mathbf{r}_{OP}^{\text{top}} = (-0.443, 0.217, 0)$ mm and $\Delta\boldsymbol{\psi}_{IB}^{\text{top}} = (0, 0, 0.88)$ degrees, such that their impact on the top platform movement was negligible, except for causing slight asymmetry. Due to space limitations, we present only the first experiment from each type. Other experiments of the same type yielded similar results, differing only in displacement direction or rotation axis due to symmetric design.

As shown in Fig. 4, for the compression experiment, the primary deformation in the silicone cylinder is a displacement in the z-direction with maximal 4.5 mm. Displacements in directions other than the z-axis and all rotational components are nearly zero. There is no significant distinction between the loading (crosses) and unloading (dots) cases. The model prediction (red polylines) matches closely the measured data, with an average deviation of 0.16 mm and a maximal deviation of 0.562 mm in position, and an average deviation of 0.136 ° and a maximal deviation of 0.423 ° in rotation.

In the torsion experiment, rotation around the z-axis is dominant, reaching up to 45.8 °, as shown in Fig. 5. A noticeable hysteresis effect is observed. Despite the tendon forces changing, the rotation angles during the first two unloading steps are nearly identical. This hysteresis is attributed to the friction between the tendons and the guide holes. Since our model does not account for friction, the prediction polylines

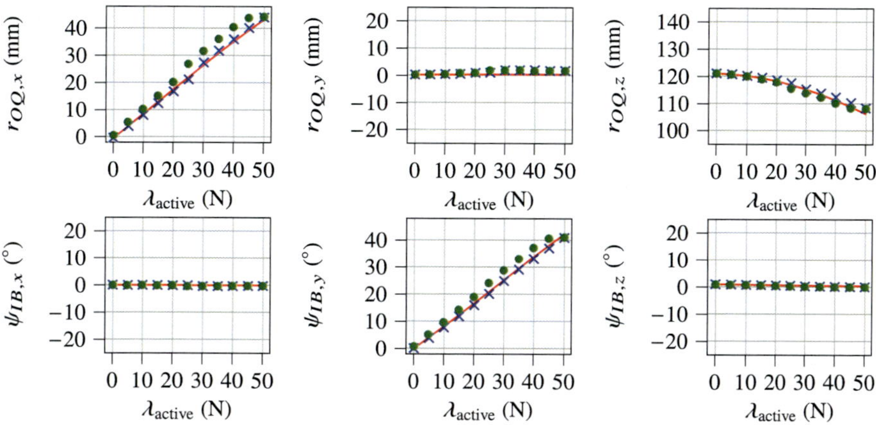

Fig. 6 Bending experiment 1: pose measurements (× and •) compared to model predictions (——). × and • indicate measurements during increasing and decreasing active tendon force λ_{active}, respectively

for loading and unloading overlap and fall in between. The maximum deviations in position and orientation are 0.764 mm and 5.353 °, respectively.

Figure 6 illustrates the bending experiment, where displacements in the x and z directions, along with rotation around the y-axis, are combined. The maximum model prediction errors for these quantities are 5.264 mm, 2.145 mm, and 3.581 °, respectively. While the hysteresis effect in rotation is less pronounced than in the torsion experiment, it introduces hysteresis in the displacements.

6.2 *Workspace Sampling*

Figure 7a illustrates the boundary points and workspace surface of our TDCM with the parallel tendon routing configuration, following the procedure described in Sect. 4. The workspace consists of two umbrella-shaped surfaces sharing a common perimeter, with their boundary points forming polylines define the "ribs" of the umbrellas. The region enclosed by these surfaces approximates the reachable workspace. Due to the nonlinear property of static equilibrium, the shape of the workspace is neither a polyhedron nor a convex hull. To provide further insight, Fig. 7b presents 2D representations of the workspace in a cylindrical coordinate system. The transformation from Cartesian coordinates to cylindrical coordinates, $\Phi : (x, y, z) \mapsto (\rho, \varphi, z)$, is given by

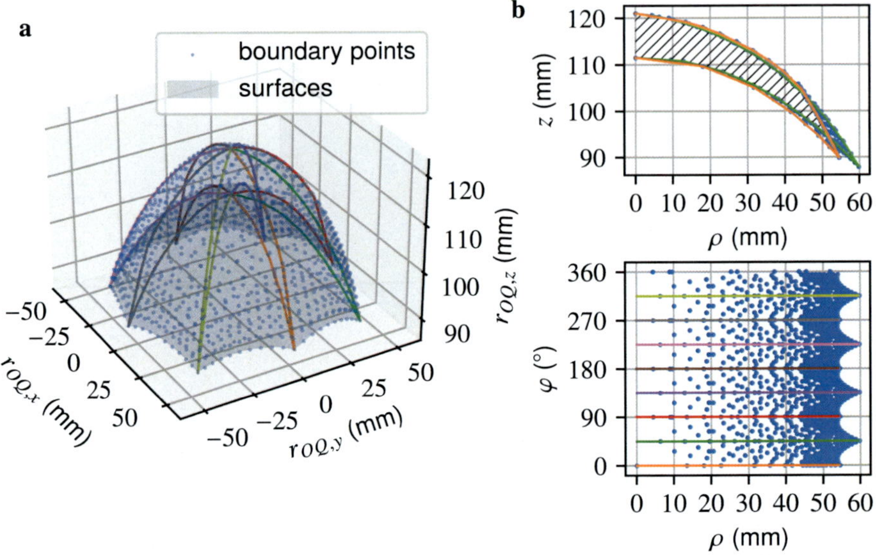

Fig. 7 **a** Boundary points and umbrella-shaped workspace surfaces found by the α-shape method. Lines (e.g. ——) present the "ribs" of the umbrellas. **b** 2D representations of the boundary points and "ribs" in a cylindrical coordinate system

$$
\begin{aligned}
\rho &= \sqrt{x^2 + y^2} \\
\varphi &= \begin{cases} 0 & \text{if } x = 0 \text{ and } y = 0 \\ \arccos\left(\frac{x}{\rho}\right) & \text{if } y \geqslant 0 \\ 2\pi - \arccos\left(\frac{x}{\rho}\right) & \text{if } y < 0 \end{cases} \quad . \\
z &= z
\end{aligned}
\tag{11}
$$

Notably, the workspace exhibits periodicity in φ, repeating every 90°, with the "ribs" appearing at $\varphi = 0°, 45°, \ldots, 315°$. This symmetry arises from the uniform design of the TDCM and the consistent tendon force constraints. By extruding the hatched region—where no boundary points exist—along the φ-direction, we obtain a rotationally uniform workspace, denoted as $\mathcal{W}_S^{\text{unif}}$. While this representation simplifies workspace analysis by avoiding the complexities of the actual geometry, it provides a conservative yet practical estimation of the reachable workspace.

6.3 Inverse Static Control

To validate the inverse static control, ten discrete circular trajectories were generated within the rotationally uniform workspace $\mathcal{W}_S^{\text{unif}}$, as shown in Fig. 8. These trajec-

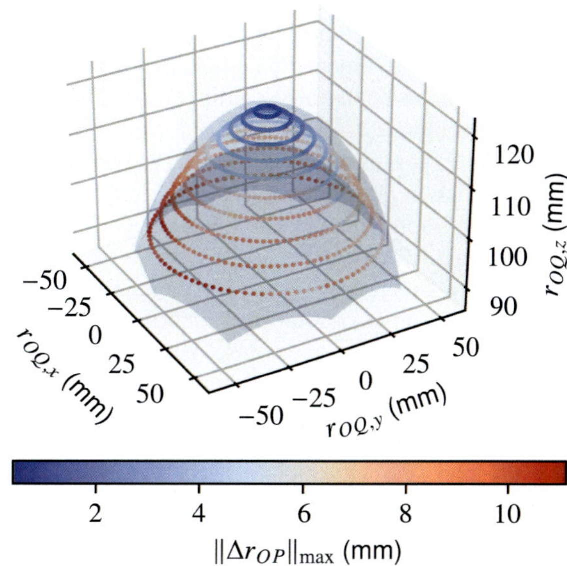

Fig. 8 Ten circular trajectories with varying radii at different heights. From top to bottom, the height decreases from 119.77 to 95.77 mm, while the radius increases from 5 mm to 50 mm. Each trajectory is uniformly discretized into 90 points around the vertical axis

tories were designed to cover a significant portion of the workspace. The discrete positions along each trajectory were sequentially commanded to the "inverse statics" block, with each step held for 5 s. To minimize the influence of dynamic effects, only the measurements taken at the midpoint of each step were considered. The tendon forces were well-regulated, with a mean absolute error of 0.056 N and a maximum absolute error of 0.524 N—relative to a maximum applied force of 50 N—observed in the most challenging trajectory with the largest radius of 50 mm. In contrast to the precise force control, the position errors, color-coded in the trajectory plot, were significantly larger. The maximum position error of 11.1 mm occurred for the trajectory with the largest radius, where the required actuation forces were at their highest. This deviation is not entirely unexpected, as inverse static control operates in an open-loop manner without position feedback.

7 Conclusion

In this study, we conducted a comprehensive investigation into the design, modeling, and control of a tendon-driven continuum mechanism (TDCM). Using strain gauges and fiducial ArUco markers, we measured the partial states of the system and established a robust physical basis for subsequent analysis. Using the finite element formulation of the Cosserat rod model, we successfully analyzed the behavior of the system under various actuation conditions. Our analysis of the reachable workspace of the TDCM revealed a complex, non-convex shape caused by the tendon actuation constraints.

For position control, we formulated the inverse statics as a constrained optimization problem. Combined with a force controller, the effectiveness of this control strategy was experimentally demonstrated. However, the open-loop control approach resulted in non-negligible position errors, highlighting the need for more advanced control techniques.

In our experiments, we also observed hysteresis effects in the system, primarily due to friction between tendons and guide holes. This friction-induced hysteresis affects both force transmission and position accuracy, making precise control more challenging. The current model does not explicitly account for frictional effects, which limits its predictive accuracy under varying loading conditions. Future work should focus on incorporating a friction model [24] into the system to better capture these effects and improve control performance.

Regarding workspace analysis, brute-force sampling, while effective for smaller systems, becomes computationally expensive as the number of tendons increases. To overcome this, alternative methods such as continuation techniques [20, 25] could be explored to more efficiently determine the workspace boundary. However, these methods require refinement to avoid identification of interior boundaries.

In terms of control strategies, the current inverse static control method operates in an open-loop fashion, resulting in residual position errors. Future work should focus on developing nonlinear closed-loop control strategies to improve accuracy. In addition, accelerating the model to enable real-time computation is critical, as the current constrained optimization approach for inverse statics is computationally intensive.

By addressing these challenges, future research can significantly improve the performance and applicability of tendon-driven continuum mechanisms, enabling their broader adoption in complex and dynamic robotic applications.

Acknowledgements This research has been funded by the Deutsche Forschungsgemeinschaft (DFG, German Research Foundation; grant numbers 405032572) as part of the priority program 2100 Soft Material Robotic Systems.

References

1. Alessi, C., et al.: Rod models in continuum and soft robot control: a review. Preprint at https://arxiv.org/abs/2407.05886 (2024)
2. Coevoet, E., et al.: Software toolkit for modeling, simulation, and control of soft robots. Adv. Robot. **31**(22), 1208–1224 (2017)
3. Antman, S.S.: Nonlinear problems of elasticity. Appl. Math. Sci. **107** (2005)
4. Simo, J.C., Vu-Quoc, L.: A three-dimensional finite-strain rod model. Part II: Computational aspects. Comput. Methods Appl. Mech. Eng. **58**(1), 79–116 (1986)
5. Meier, C., Popp, A., Wall, W.A.: Geometrically exact finite element formulations for slender beams: Kirchhoff–Love theory versus Simo–Reissner theory. Arch. Comput. Methods Eng. **26**, 163–243 (2019)
6. Rucker, D.C., Webster, R.J., III.: Statics and dynamics of continuum robots with general tendon routing and external loading. IEEE Trans. Rob. **27**(6), 1033–1044 (2011)

7. Bryson, C.E., Rucker, D.C.: Toward parallel continuum manipulators. In: Proceedings 2014 IEEE International Conference on Robotics and Automation (ICRA), Hong Kong, China (2014)
8. Renda, F., et al.: A geometric variable-strain approach for static modeling of soft manipulators with tendon and fluidic actuation. IEEE Robot. Autom. Lett. **5**(3), 4006–4013 (2020)
9. Mathew, A.T., et al.: Reduced order modeling of hybrid soft-rigid robots using global, local, and state-dependent strain parameterization. Int. J. Robot. Res. **44**(1), 129–154 (2024)
10. Bergou, M. et al.: Discrete elastic rods. In: Proceedings ACM SIGGRAPH 2008 papers (SIGGRAPH'08). Association for Computing Machinery, New York, NY, USA (2008)
11. Gazzola, M., et al.: Forward and inverse problems in the mechanics of soft filaments. Royal Soc. Open Sci. **5**(6), 171628 (2018)
12. Lang, H., Linn, J., Arnold, M.: Multi-body dynamics simulation of geometrically exact Cosserat rods. Multibody Sys.Dyn. **25**, 285–312 (2011)
13. Macenski, S., et al.: Impact of ROS 2 node composition in robotic systems. IEEE Robot. Autonom. Lett. **8**(7), 3996–4003 (2014)
14. Harsch, J., Sailer, S., Eugster, S.R.: A total Lagrangian, objective and intrinsically locking-free Petrov–Galerkin SE(3) Cosserat rod finite element formulation. Int. J. Numer. Meth. Eng. **124**(13), 2965–2994 (2023)
15. Harsch, J., Eugster, S.R.: Nonunit quaternion parametrization of a Petrov–Galerkin Cosserat rod finite element. PAMM **23**(4) (2023)
16. Eugster, S.R., Harsch, J.: A family of total Lagrangian Petrov–Galerkin Cosserat rod finite element formulations. GAMM-Mitteilungen **46**(2) (2023)
17. Deutschmann, B., Reinecke, J., Dietrich, A.: Open source tendon-driven continuum mechanism: a platform for research in soft robotics. In: Paper presented at the 5th International Conference on Soft Robotics (RoboSoft), Edinburgh, 4–8 Apr (2022)
18. Garrido-Jurado, S., et al.: Automatic generation and detection of highly reliable fiducial markers under occlusion. Patt. Recogn. **47**(6), 2280–2292 (2014)
19. Yang, C., Ye, W., Li, Q.: Review of the performance optimization of parallel manipulators. Mech. Mach. Theory **170**, 104725 (2022)
20. Amehri, W., Zheng, G., Kruszewski, A.: Workspace boundary estimation for soft manipulators using a continuation approach. IEEE Robot. Autom. Lett. **6**(4), 7169–7176 (2021)
21. Edelsbrunner, H., Kirkpatrick, D.G., Seidel, R.: On the shape of a set of points in the plane. IEEE Trans. Inf. Theory **29**(4), 551–559 (1983)
22. Virtanen, P., et al.: SciPy 1.0: fundamental algorithms for scientific computing in Python. Nat. Methods **17**, 261–272 (2020)
23. Krantz, S.G., Parks, H.R.: The implicit function theorem. Birkhäuser Boston, MA (2003)
24. Xu, K., Simaan, N.: Actuation compensation for flexible surgical snake-like robots with redundant remote actuation. In: Proceedings 2006 IEEE International Conference on Robotics and Automation, Orlando, FL (2006)
25. Haug, E., et al.: Numerical algorithms for mapping boundaries of manipulator workspaces. ASME J. Mech. Des. **118**(2), 228–234 (1996)

The Soft Material Robotics Toolbox: Coherent Methodology for Modelling and Design of Soft Material Robots

Mats Wiese, Max Niklas Bartholdt, Rebecca Berthold, Jörg Wallaschek, Thomas Seel, and Annika Raatz

Abstract Researchers and developers strive to understand the special behavior of soft material robots (SMR) to leverage their systems' inherent flexibility and adaptability. These robots are revolutionising fields such as industrial automation, healthcare, and exploration due to their flexibility, safety and ability to operate in undefined spaces. To fully exploit their potential and the ability to adapt both actively and passively, a comprehensive set of tools is required for their efficient development for future applications. Effective modeling techniques that address the unique challenges posed by the soft and deformable nature of SMR are crucial. These include modeling kinematics, statics, dynamics, and environmental interactions, which are essential for the design and control of SMR. Such methods enable accurate prediction of robot behavior, enhance motion planning, and improve interactions with humans and unstructured environments. Advanced contact modeling and material characterisation further enhance the development of strategies that leverage environmental contact, significantly extending operational capabilities. The main objective of this

Mats Wiese, Max Niklas Bartholdt, Rebecca Berthold—These authors contributed equally to this work.

M. Wiese (✉) · A. Raatz
Institute of Assembly Technology and Robotics, An der Universität 2, Garbsen, Germany
e-mail: wiese@match.uni-hannover.de

A. Raatz
e-mail: raatz@match.uni-hannover.de

M. N. Bartholdt · T. Seel
Institute of Mechatronic Systems, An der Universität 1, Garbsen, Germany
e-mail: max.bartholdt@imes.uni-hannover.de

T. Seel
e-mail: thomas.seel@imes.uni-hannover.de

R. Berthold · J. Wallaschek
Institute of Dynamics and Vibration Research, An der Universität 1, Garbsen, Germany
e-mail: berthold@ids.uni-hannover.de

J. Wallaschek
e-mail: wallaschek@ids.uni-hannover.de

A. Raatz et al. (eds.), *Soft Material Robotic Systems*,
https://doi.org/10.1007/978-3-032-22453-8_17

proposed project is to develop a coherent methodology for the modeling and design of SMR. These approaches aim to establish a foundation for innovative methodologies that empower engineers and researchers to harness the full potential of soft robotics, paving the way for widespread use and transformative applications outside research institutions.

1 Introduction

The inherent safety of SMR, due to their softness, makes them ideal for interaction with humans or the environment and suitable for operation in narrow or cluttered spaces as well as direct interaction with humans. This potential indicated that SMR could make revolutionary changes in industrial, clinical or exploratory environments. However, from a modeling point of view, the soft material is a challenge itself. For instance, the composition of incompressible material and fiber-reinforced pressure chambers leads to a complex actuation behavior.

Different tasks, ranging from system design over motion planning to control, impose different demands on models and simulations regarding accuracy, time efficiency, and the flexibility to adjust parameters [1].

On the one hand, the system design of SMR is challenging due to the nonintuitive prediction of a soft system's performance [2]. For iterative design finding for a task that involves complex motion and interaction of the robot with its environment, models need to account for contact, be computationally efficient and accurate [3]. On the other hand, the control of SMR becomes difficult as models have to make a trade-off between computational efficiency and accuracy, restricting the choice of methods that are suitable to the demands [4]. In parallel, system parameters change during field tests either due to environmental influences or unmodeled effects [5]. This challenge is heightened by the absence of reliable information on output quantities to implement feedback control, as soft material robots, in theory, inherit an infinite number of degrees of freedom, and commercially available sensors are incapable of directly measuring this complex deformation [1].

So far, existing modeling of SMR rarely includes contact considerations, whether in design, motion planning, or control. But to navigate in environments requires, besides controlled free motion, strategies for contact detection and behavior. Wang et al. [6] provide a comprehensive overview of the contributions to and challenges in the field of perceptive robots, emphasizing particularly the challenge of modeling and predicting contact behavior.

This chapter summarises the outcomes and insights gained into SMRS concerning statics, dynamics, material characterization, contact modeling, and state estimation. As outlined, each topic poses a challenge for engineers and researchers and is difficult to access, especially if approached from scratch. Research in recent years has seen considerable efforts and advancements in developing more general modeling frameworks rather than focusing on specialized models for very specific soft robotic systems. To name a few, we refer the reader to SOFA [7], TMTDyn [8], Sorotoki [9], SoRoSim [10] and PyElastica [11].

Next to these general Frameworks, that have been developed in recent years, our toolbox leans towards widely investigated soft pneumatic systems, offering insights into their unique challenges and opportunities. A strong priority is given to bridging the sim2real gap for these systems.

This chapter highlights our contributions to static and dynamic modelling of these systems, material characterization, contact modeling, and state estimation. For a more comprehensive literature review on these topics we refer the reader to our respective publications.

In the following, we will give a brief overview of the project's objectives and the overall approach. After explaining the basics of the applied models, we will present the basic tools implemented within this project and highlight capabilities and limitations. To demonstrate the toolbox's capabilities in the example of a soft pneumatic continuum robot, we will present key results, foundational to basic tools, findings from these tools, and methods that expand on them.

1.1 Objectives

The overall goal is to leverage the potential of SPA and develop sophisticated methods to facilitate the development of SMR for engineers and researchers. The research provides fundamental methods and implementations for the design, modeling, and control of soft continuum robots. It equips engineers in research with a toolbox to bring soft robots from the lab to practical applications. To summarise, the project's objectives are to

- Acquire a coherent methodology for modeling and designing slender soft material robots.
- Generically determine enhanced models for kinematics, statics, and dynamics under consideration of contact with the environment.
- Transfer methods into easy-to-use design framework (SMaRT).
- Examine soft material robotic systems regarding control, motion planning, and contact.

1.2 Toolbox Concept

The main concept and (software) components of the intended *Soft Material Robotics Toolbox* are visualized in Fig. 1. We set up a basic toolchain starting from commercial FE software. Detailed but computationally costly FEM serve as a reference for model identification of computationally cheaper Cosserat rod models. Set-up rod models are incorporated in subsequent model-based methods for motion planning or model-based control. The toolbox was primarily developed using a soft pneumatic continuum robot with fiber-reinforced air chambers and the ability to move in three-dimensional space.

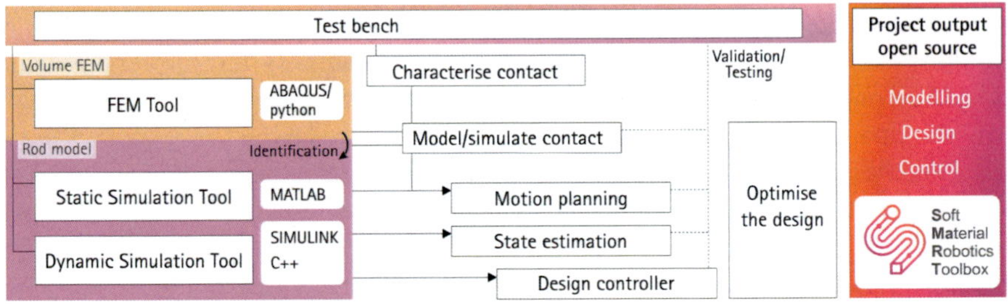

Fig. 1 The diagram visualizes the concept of the Soft Material Robotics Toolbox

For the development and application of the toolbox, extensive material parameter identifications were carried out, detailed FEM were built, computationally efficient rod models were implemented and further developed for the modeling of contact and control purposes. Models and methods were evaluated on a test bench. The interaction of soft manipulators with their environment was considered in FEM as well as rod models. The project contributes estimation strategies that provide system states using the modeling knowledge gained in previous work. Physical parameters are adapted online using a predictor-corrector scheme so that unmodeled effects are effectively compensated, which is an important property for a successful application to real systems. Additionally, this work contributes to closing the loop for the design process for single modules or multi-segmented robots to determine application-specific design parameters to successfully accomplish a given task.

2 Underlying Models

Soft robotic systems are characterized by their low stiffness, which is a great advantage when operating in an unknown environment or in human-robot interaction. For targeted development of SMRS, models with different levels of detail need to be developed as discussed above.

2.1 *Finite Element Model*

In the emerging field of soft material robotics, FEM have become indispensable tools, as demonstrated, for example, in Runge et al. [12] and Moseley et al. [13] for simulating and understanding the complex behaviors of soft robots. These robots employ highly flexible materials, which can, therefore, undergo large deformations, which lead to geometrical nonlinearities. Additional nonlinearities due to material behavior or contact can also occur. All these nonlinearities can be addressed in a finite element model as in Tawk and Alici [14]. Another advantage of finite element modeling is the possibility to use simulation results instead of costly physical experiments [15]. The FEM is set up in ABAQUS in a modular way so that design adjustments can automatically be realized.

2.1.1 Hyperelastic Material Models

Hyperelastic materials show a nonlinear relationship in their stress-strain curves. To describe such a relation, energy density functions are being used. In the following section two of the most commonly used material models are presented.

In the Yeoh model as introduced by Yeoh [16] the strain energy density function U is described based on the strain invariants I of the Cauchy-Green deformation tensor.

$$U_{\text{Yeoh}} = \sum_{i=1}^{n} C_i (I_1 - 3)^i + \sum_{i=1}^{n} \frac{1}{D_i} (J^{\text{el}} - 1)^{2i} \tag{1}$$

The material constants C_i need to be defined for each material that is described by the Yeoh model. Here, J^{el} is the elastic volume strain, that describes the volume change. For incompressible materials $J^{\text{el}} = 1$. In general, the number of terms can be chosen depending on the accuracy of the model. For most cases $n = 3$ leads to a good accuracy.

The Ogden material model, as introduced by Ogden [17], describes the energy density function based on the principle stretches λ.

$$U_{\text{Ogden}} = \sum_{i=1}^{n} \frac{2\mu_i}{\alpha_i^2} (\lambda_1^{-\alpha_i} + \lambda_2^{-\alpha_i} + \lambda_3^{-\alpha_i} - 3) + \sum_{i=1}^{n} \frac{1}{D_i} (J^{\text{el}} - 1)^{2i} \tag{2}$$

In Eq. 2 μ, α and D are material constants that need to be defined for each material. For incompressible materials, the second term of the formula can be neglected for both material models.

2.1.2 Viscoelastic Material Models

Besides the discussed nonlinear hyperelastic material behavior, many materials show frequency—and time-dependent behavior. The deformation of the material depends not only on the applied stress but also on time and frequency. This material behavior needs to be considered for dynamic applications. Regarding their long-time behavior, many polymers used in the field of soft material robotics show viscous behavior. An approach to model this characteristic is a combination of springs and dampers. While the dampers represent the viscous behavior, the springs describe the elastic properties of the material. In the generalized Maxwell model, spring and damper elements are connected in parallel, as shown in Fig. 2. Depending on the material, the complexity and, therefore, the number of spring and damper elements can vary [18].

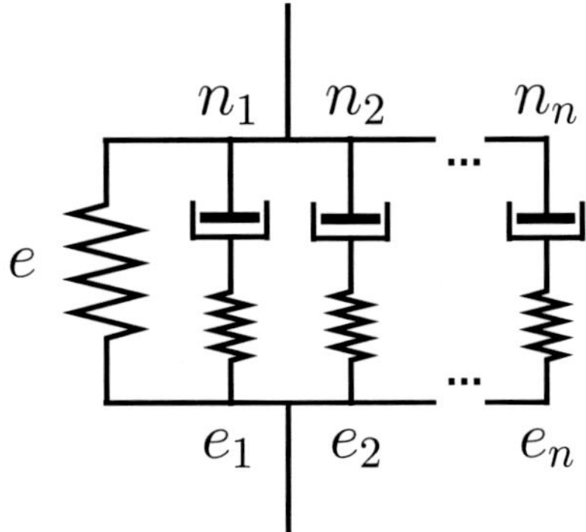

Fig. 2 Schematic figure of the generalized Maxwell model with spring and damper elements

The Prony series, in general, is a method to describe frequency-dependent signals with multiple linear equations. With regards to viscoelastic materials, the Prony series is used to fit the generalized Maxwell model [19]. When used to describe viscous material behavior, the Prony series is also used as a term for the material model itself [20]. In Eq. 3 the storage modulus G' and loss modulus G'' are given, described by a Prony series.

$$
\begin{aligned}
G'(\omega) &= G_0\left(1 - \sum_{i=1}^{N} g_i\right) + G_0 \sum_{i=1}^{N} \frac{g_i \tau_i^2 \omega^2}{1 + \tau_i^2 \omega^2} \\
G''(\omega) &= G_0 \sum_{i=1}^{N} \frac{g_i \tau_i \omega}{1 + \tau_i^2 \omega^2}
\end{aligned}
\tag{3}
$$

While τ is the time relaxation constant, g describes the shear modulus of the material.

2.2 *Cosserat Rod*

A widely used approach to describe the behavior of a (soft) continuum robot in a computationally less expensive manner than a detailed spatial FEM is the Cosserat rod modeling approach. The body of a slender system is reduced to a one-dimensional backbone curve and three directors indicating the orientation of a rigid cross-section. The Cosserat rod model incorporates both translational and rotational deformation (shear, elongation, bending, torsion) to describe the state of soft continuum robots.

A set of differential equations describes the rod's statics. These equations are derived from the rod's kinematics (relation of local strains to the position of a cross-section and an attached orientation frame) and the balance of forces and moments at an infinitesimal element along the rod. Constitutive relations link internal forces to strains based on the rod's cross-sectional and material properties. The dynamics can be formulated as a set of PDE, considering time-dependencies (velocities, accelerations, and related damping and inertia terms).

To solve the differential equations, various numerical solutions have been applied in the area of (soft) continuum robotics in recent years, ranging from approaches like the shooting method to *Ritz* methods. For a comprehensive overview, refer to [4] or [21].

3 Basic Tools

3.1 Static Simulation Tool

The static simulation tool for soft continuum robots is built upon the Cosserat rod modeling approach. It is implemented in MATLAB to simulate multi-module soft robots under various conditions[1] [22, 23]. The simulation relies on the single shooting method, a numerical technique well-suited for static simulations of slender structures [24]. By iteratively refining initial condition estimates (unknown base reaction forces and moments), the method ensures that predicted static configurations of soft robots under given loads achieve equilibrium.

The simulation tool can handle various forces and moments that affect soft continuum robots. It accounts for internal actuating forces, such as those generated by pneumatic pressure, as well as external influences like gravitational loads and additional forces and moments applied along the robot's structure. Incorporating these factors into the simulation is crucial for achieving realistic predictions of how the robot behaves under real-world conditions. Moreover, the tool offers a flexible modular architecture, allowing users to combine multiple robot segments, both active and passive, and thus facilitates the exploration of diverse robotic designs and configurations.

A contact modeling component further enhances the tool's functionality. Utilizing a penalty method, it effectively simulates interactions between the robot and its environment [23]. This includes basic rigid primitives acting as contact partners, providing a simple yet effective way to predict the interplay between soft robots and surrounding objects. This aspect of the tool is particularly relevant in scenarios where the robots must navigate cluttered or narrow populated spaces or manipulate external objects.

A specific focus in developing the tool was laid on SPA. The *lateral compression effect*, a critical phenomenon in SPA, is taken into account [25]. This effect describes the influence of internal actuating pressure on the lateral walls of the cavities. By integrating this effect into the Cosserat model, the simulator can more accurately predict how pneumatic pressure influences the actuator's performance. Material parameter identification further ensures the accuracy of the simulations. By leveraging the basic FE tool, linear elastic material parameters are identified.

[1] https://github.com/match-SoRo/soft-continuum-robot-modeling.

With this tool, researchers and can simulate the behavior of soft pneumatic continuum robots under various conditions, significantly reducing the trial-and-error process traditionally required in soft robotics design and analysis. The simulation tool serves as a basis for higher-level algorithms, such as motion planning and design finding algorithms, which are currently under investigation.

3.2 Dynamic Simulation Tool

The following summarises the findings from our recent publication [26]. The study presents a coherent methodology for simulating soft-material robots, emphasizing integrating Cosserat rod theory into the SIMULINK environment while enhancing computational efficiency. Central to this approach is using neural networks to approximate the inverse of nonlinear material models, a crucial step for achieving faster computation speed. The work outlines a procedure to derive material parameters for these models using FE simulations, thereby eliminating the need for exhaustive experimental analyses.

The methodology, illustrated in Fig. 3, employs FEM to simulate the detailed behavior of soft pneumatic actuators and to derive essential material parameters validated against experimental data. Cosserat rod theory is applied for dynamic simulations, effectively reducing the model to a 1D representation suitable for slender structures like manipulators. This leads to partial differential equations using (differential) kinematics, conservation laws, and constitutive equations describing the material's behavior. Extending the linear Kelvin-Voigt material model, nonlinear effects, including hyperelasticity and damping, are incorporated. To solve the partial differential equations, the BDF-α method is applied to replace gradients w. r. t. time [27]. The spatial discretization is then addressed using the indirect single-shooting method. A feed-forward neural network approximates the inverse function of the nonlinear material model required for the spatial integration scheme, further enhancing computational efficiency. Thus, the simulation time remains comparable to that of linear material models without sacrificing accuracy compared to, e. g., Newton-Raphson methods to solve the nonlinear material equation.

The presented methodology effectively identifies material parameters from FE simulations without requiring labor-intensive experiments and data collection. Validation against dynamic experimental data for a soft pneumatic actuator shows good agreement, with a mean squared error of approximately 7%, demonstrating the model's validity. Additionally, the proposed methodology successfully reconstructs a benchmark dynamics simulation from existing literature, confirming its accuracy.

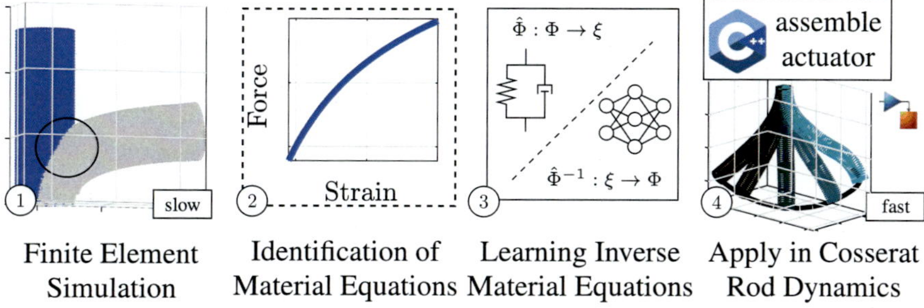

Fig. 3 Methodology and implementation of the proposed parameter identification and dynamics simulation integration into simulink. Taken from [26]

The development includes a modular, object-oriented simulation framework, which is publicly available.[2] This allows users to quickly configure and simulate soft-material robots in SIMULINK for control applications and further research. Overall, this study contributes a systematic approach to designing soft pneumatic actuators in simulation, facilitating rapid development and iteration of soft-material robotic systems.

4 Model Based Methods and Results

To showcase the toolbox's capabilities, we will present selected key results that build the backbone of basic tools, selected findings obtained with basic tools, and selected methods that build on these basic tools. Therefore, we will take a closer look at the finite element model, the contact simulation for soft continuum robots, and adaptive state estimation.

4.1 Study Case: Soft Pneumatic Actuator

All methods discussed in this work are implemented and investigated in the example of SPA. They rely on pneumatic pressure to move. The central body part of the SPA is made of a silicone rubber named Ecoflex 0050 (Shore Hardness 00-50) [28] with a very low stiffness. As shown in Fig. 4, the actuator is supported with a layer of Dragonskin 30 (Shore Hardness 30A) [29] to prevent the cross section from deforming strongly. The SPA has three air chambers circumferential along the longitudinal axis. Two fibers reinforce all chambers to ensure a mere lengthening when pressurized.

[2] https://gitlab.com/soft_material_robotics/cosserat-rod-simulink-sfunction.

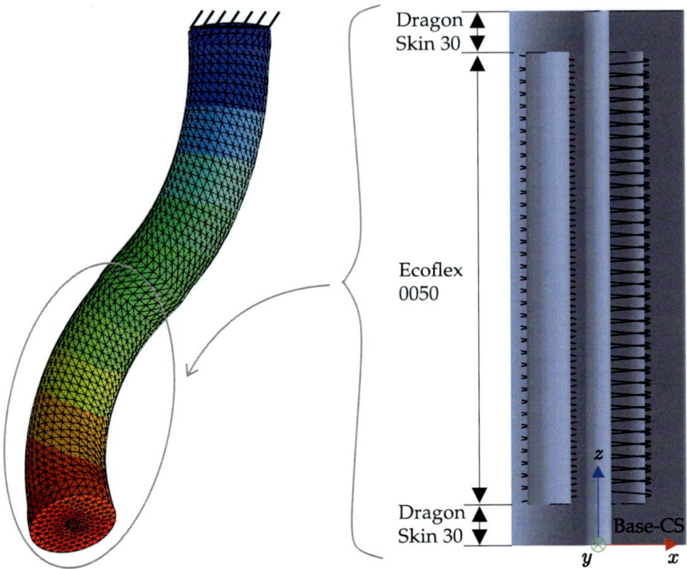

Fig. 4 Pressurised two segment SPA including a schematic sketch of a single segment with three fiber-reinforced chambers

4.2 Material Models

While the finite element method is well known in all engineering fields, the proper modeling of the soft material behavior is a challenge that will be addressed in the following. Both silicone rubbers used for the SPA show hyper-viscoelastic material behavior. The respective material models are implemented in a finite element model in ABAQUS. The results are then compared to physical measurement data to validate the material models. Since various research groups work with the same materials, this paper aims to contribute material models that can be used in a common soft material database like Marechal et al. [30]. For determining the parameters of a hyperelastic material model, a tensile test was conducted with five different samples of 2 mm thickness at room temperature.[3] According to DIN 53504 norm [31], dumbbell shaped specimens were considered for the stress-strain measurement. Uni-axial tensile measurements are performed using a Zwick 1456 tensile tester (ZwickRoell GmbH, Ulm, Germany) equipped with optical strain measurement according to DIN EN ISO 527-2 [32]. A cross-head speed of 200 mm/min is used, and then the samples are stretched to the failure. The resulting stress-strain curves from these measurements are shown in Fig. 5 for the materials Ecoflex (left) and Dragonskin (right). For both materials, the results for all samples show a good agreement except for one sample, which was removed in the following evaluation. Furthermore, it can

[3] Hyperelastic Measurements conducted by the team of Dr. Amit Das, Leibniz-Institut für Polymerforschung Dresden e. V.

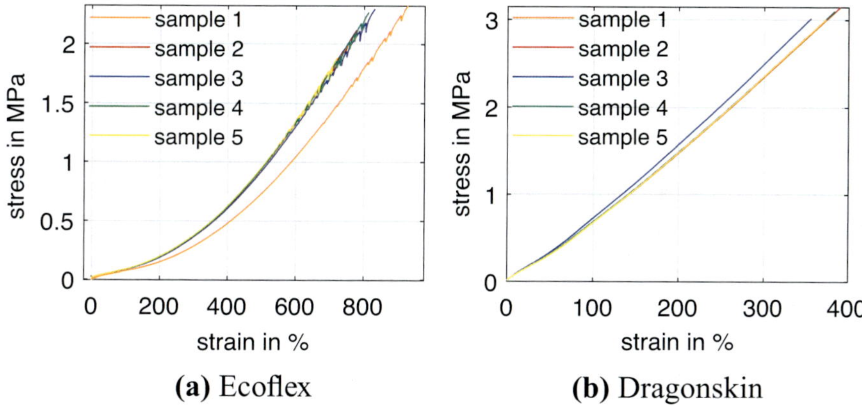

Fig. 5 Measurement data from a tensile test for the material Ecoflex 0050 (*left*) and Dragonskin 30 (*right*) for five different samples

Table 1 Hyperelastic material model for Ecoflex and Dragonskin

Material model	Coefficient	Ecoflex	Dragonskin
Yeoh model	C_1 in Pa	26042.6565	123698.092
	C_2 in Pa	357.697778	11666.9401
	C_3 in Pa	10.9015536	−247.930046
Ogden model	α_1	2.08687140	0.900606064
	α_2	2.25048648	1.07566285
	α_3	1.91473039	0.720953758
	μ_1 in Pa	−7643088.22	−108804905
	μ_1 in Pa	3663693.98	54896236.8
	μ_1 in Pa	4036560.00	54141661.1

be seen that the Ecoflex material is very elastic and can undergo large deformation. For Ecoflex, a clear nonlinear stress-strain relation can be shown, while for Dragonskin, the stress-strain relation is approximately linear, especially for small strains. The tensile test data was evaluated up to a strain of 400 %. The discrete stress-strain values were exported to ABAQUS to fit the hyperelastic material models using a least-squares method. The resulting coefficient for the material models can be seen in Table 1. Both material models were validated using physical experiments to ensure good model accuracy.

To describe viscoelastic behavior of the silicone rubbers, time and temperature depending test data is needed. In common testing setups a sample of the tested material undergoes a tensile test with sinusoidal excitation. The response force is measured and can be evaluated with respect to the excitation. The test is conducted with different frequencies at different ambient temperatures. With the help of the WLF-shift a master curve can be created for a specific temperature [33]. Both mate-

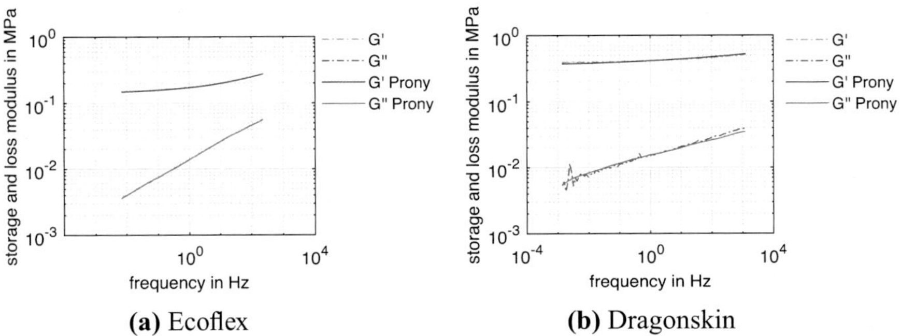

(a) Ecoflex **(b)** Dragonskin

Fig. 6 Ecoflex material master curve (*left*) and Dragonskin master curve (*right*) with fitted Prony parameter

Table 2 Prony parameter for Ecoflex and Dragonskin

	Ecoflex		Dragonskin	
G_0 in MPa	0.424164963		0.56	
i	τ_i in s	g_i	τ_i in s	g_i
1	1.00E+03	2.03E–02	1.00E+03	2.43E–02
2	2.31E+02	4.31E–03	2.31E+02	0.00E+00
3	5.34E+01	4.73E–03	5.34E+01	1.34E–02
4	1.23E+01	8.41E–03	1.23E+01	1.34E–02
5	2.85E+00	1.36E–02	2.85E+00	1.78E–02
6	6.58E–01	1.86E–02	6.58E–01	2.15E–02
7	1.52E–01	2.81E–02	1.52E–01	2.63E–02
8	3.51E–02	4.25E–02	3.51E–02	2.78E–02
9	8.11E–03	6.53E–02	8.11E–03	3.62E–02
10	1.87E–03	9.02E–02	1.87E–03	4.33E–02
11	4.33E–04	1.37E–01	4.33E–04	4.42E–02
12	1.00E–04	2.48E–01	1.00E–04	9.54E–02

rials show very low stiffness. As a result, the well known measurement setup is not suitable. Instead a cubical sample with a torsional excitation was used to investigate shear behavior. The measurement was performed by the DEUTSCHES INSTITUT FÜR KAUTSCHUKTECHNOLOGIE E. V.. Different ambient temperature values were chosen in the range from -85° to 55 °C. The measured frequency range was set from 0.1 $\frac{rad}{s}$ to 100 $\frac{rad}{s}$. In Fig. 6, the calculated master curves for Dragonskin and Ecoflex are shown at a reference temperature of 20 °C. The Prony parameters were fit to the master curve for $i = 12$ Prony parameters as listed in Table 2. These parameters can be implemented in finite element software.

4.3 Mechanical Contact Model for Soft Continuum Robots

Soft robotic systems provide built-in safety and adaptability because of their high material compliance. Consequently, these robots are well-suited for operating in unfamiliar environments and performing tasks that require or leverage interactions with their surroundings. When applying soft robots, integrating contact models into modeling frameworks is essential for simulation tasks for as design finding, motion planning, or control. In [23], the static simulation tool from Sect. 3.1 is enhanced to deal with contact anywhere along its length. The proposed contact implementation for Cosserat rod models of slender soft continuum robots is capable of simulating multiple contact points along the structure. The model considers the robot's radius for contact detection and forces and moments on the structure's backbone curve induced from the point of contact.

The penalty contact method uses virtual springs to apply normal forces based on penetration depth and penalty stiffness to simulate contact constraints. This approach helps separate the soft robot and its contact partner by generating forces when they penetrate each other. A regularised coulomb friction model handles tangential forces. For frictional contact, the coefficient of friction is based on [34].

Cosserat rod simulations are evaluated in different scenarios against the high fidelity FEM from Sect. 2.1. The Cosserat rod model and the FEM exhibit similar behavior, suggesting that the stiffness parameters identified from FE to Cosserat are valid. This validity extends to contact simulations (see Fig. 7), although minor discrepancies arise due to the rod model's simplifications, such as using linear material models and constant cross-sections.

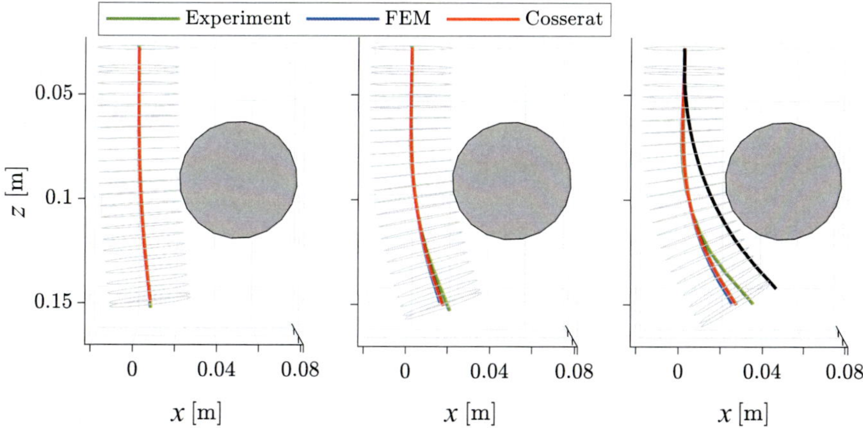

Fig. 7 Experimental, FE simulated backbone curve and Cosserat rod simulated backbone curve when in contact with a cylindrical rigid object. The deformation for three different pressure states is visualized (increasing from *left* to *right*). The dark grey curve in the last tile indicates the Cosserat rod simulated backbone curve with the same actuation but no contact as a reference. For more insights see [23]

Despite these differences, the deviations remain acceptably low for the case studies discussed in the article. However, greater contact forces can lead to significant deformations, and simulations diverge notably from experimental data. This divergence is possibly due to manufacturing inaccuracies and idealizations, like constant friction coefficients not accurately representing real conditions. As a result, the models might not perfectly reflect real-world scenarios.

Further investigations are necessary to better describe contact behavior. Overall, despite some accuracy issues compared to real-world data, the Cosserat rod model, which is computationally efficient, can rival the detailed FEM in terms of accuracy. In future research, we will conduct additional studies and examine the contact area more thoroughly. We also intend to incorporate the contact model into motion planning schemes and a dynamic framework.

4.4 Adaptive State Estimation for Soft Continuum Robots

This section summarises the works in [35, 36]. The papers address the complex challenge of achieving accurate trajectory tracking for SPA by introducing a novel method for state estimation using base-reaction forces and torques. This study uniquely contributes to the field by employing an UKF to estimate the dynamic state of soft robots based on force-torque sensors, avoiding reliance on additional rigid structures or internal sensors that could alter the robot's dynamics. This approach enables real-time state estimation using a special case of the CS model, a low-order spatial discretization of the Cosserat rod model.

The method begins by modeling the system's kinematics and dynamics with a floating-base formulation and a CS kinematics approach [35]. This allows for fast predictions of the dynamics at the cost of model accuracy. The reaction force and torque at the robot's base are the outputs of the model. Mounting the sensors at the robot's base avoids adding additional mass to the end-effector, which would limit the actuator's workspace.

The discretized model is tailored to the UKF's requirements. Furthermore, the state space is augmented with the bending stiffness to mitigate errors introduced by the strong kinematic assumption of constant strains. This dominant parameter within the system is, therefore, identified online.

The method was validated through experiments (Fig. 8) using both experimental data from a real SPA and simulation data generated from the Cosserat rod simulation in Sect. 3.2. The UKF demonstrated real-time capability, achieving satisfactory estimation accuracy across various dynamic conditions. Findings include an average position estimation error of about three to five millimeters, illustrating the method's potential for tackling more complex control tasks.

The paper concludes by highlighting opportunities for future research, such as integrating more complex models like variable curvature or rod models. Future work may also explore the online identification of dynamic parameters to accommodate changing conditions such as load variations or material fatigue. Overall, this approach

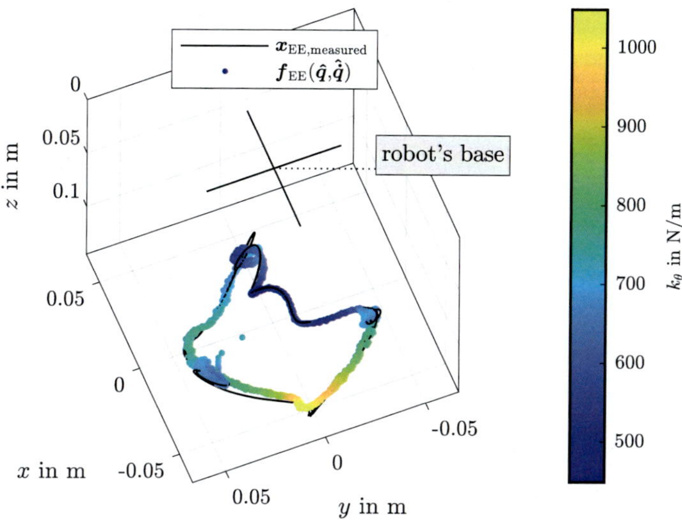

Fig. 8 Results of the proposed state estimation using the constant-curvature models and an unscented Kalman filter. The validation is done using the tip positions from camera tracking. The color indicates the estimated parameter of the bending stiffness. Taken from [36]

offers a reliable solution for SPA state estimation that preserves the robot's inherent compliance and dynamics characteristics.

5 Conclusion

The project outlines a methodology from CAD design over optimal path generation to control applications. The methodology is validated using a soft continuum robot with pneumatic actuation. The Cosserat rod model has become the gold standard in soft continuum robotics and is also fundamental to this toolbox. However, the findings in the project indicate, that depending on design and actuation of the robot as well as the application of model, different depths of modeling are necessary for sufficient results.

The described FEM enables a detailed analysis of the SPA's static as well as dynamic behavior. The level of detail comes at a high computational cost. Nevertheless, the SPA's design can be adapted, and it was shown that the FEM could be used to identify model parameters for rod models and consequently serve as the basis for simplified and, therefore, efficient model approaches.

Design optimization and offline path generation require details like the lateral compression effect that can be derived from the analysis in FE a priori to any experiments. With the help of FEM, it is possible to acquire computationally cheaper yet comparably accurate models that can be used for prediction and higher-level algorithms like path generation and system design. The applicability of a derived Cosserat rod model expands to the simulation of contact scenarios, as evaluations against FE simulations and experiments suggest [23].

In control applications, on the other hand, these details complicate the computation to the point that the requirement of real-time capability is usually not met. If they are left out, it follows that model uncertainty has to be taken into account, and it is required to implement models that are compatible with estimation algorithms such as the unscented Kalman filter or particle filter. It was indicated that this leads to accurate information for the control of soft continuum robots.

A variety of identification strategies of models and parameters have been investigated to bridge the sim2real gap, including the use of hybrid methods (model- and learning-based) [22, 26], online identification [36] and analysis using FEM [25].

5.1 Discussion

The soft material robotics toolbox is, by choice of the model order reduction to 1D-continua, suitable for soft continuum robots. It follows, that it is not applicable for arbitrary designs and deformation. However, a manifold of current soft robot designs is, in fact, slender structures [4, 37]. While a broad range of core challenges for this type of SMR are successfully addressed, open aspects remain.

Thus far, derived simulation tools show satisfactory accuracy and simulation times for the considered soft pneumatic system and its range of motion. They are able to handle moderate deformations and strains, but their accuracy for models with very high deformations and material nonlinearities remains to be verified.

With the toolbox, we are able to simulate contact with the robot's environment, both with the FEM tool and the rod-based static simulation tool. The agreement with experimental data is promising even without explicit identification for contact modeling. However, the rod-based simulation tool operates with a very basic contact model that requires fine-tuning of method-specific parameters (e. g., penalty stiffness) to avoid convergence issues. Besides this, the contact area plays a significant role in more complex contact scenarios, which are not yet considered, neither in the static nor the dynamic Cosserat rod tools.

The claim, that the derived and implemented models are suitable for motion planning, system design, and precise control is still unproven and requires further testing. Moreover, the methods have only been evaluated on a single system for now. In terms of a toolbox this raises questions about the tools' applicability to the totality of different soft robotic systems such as different pneumatic or hydraulic robotic systems or even different actuation systems like e. g. dielectric elastomers.

5.2 Outlook

Building on the static simulation tool, current and future work evolves around motion planning schemes for multi-module soft continuum robots in cluttered environments. Aside from obstacle avoidance, leveraging of contact is a crucial aspect considered. Having a path planning scheme to indicate whether (and to what extent) an intended design is able to fulfil a task at hand, design optimisation methods will be applied to adapt the design for this specific task. Thereby, we will close the loop from the initial design idea over FE and rod modeling to design findings.

This work introduced modeling approaches for incorporating contact in rod models, specifically for static scenarios. Currently, dynamic modeling with contact has only been achieved using FEM, but future work will extend this to dynamic rod models. Furthermore, the contact model will be enhanced to handle more detailed and multi-dimensional scenarios, addressing area contact and deformation within the contact zone. Current and future work in control is focused on hybrid modeling strategies, that can be incorporated with the proposed adaptive estimation method. Especially in the domain of soft continuum robot control, hybrid methods like physics-informed neural networks [38] are of interest. The current focus lies on implementing models suitable for machine-learning frameworks like PyTorch. Moreover, the application in model predictive control using modern optimization algorithms have a high potential since the inference of shallow neural networks is computationally fast compared to a model inference of Cosserat rod models.

6 Material Data

Acknowledgements Funded by the Deutsche Forschungsgemeinschaft (DFG, German Research Foundation) under grant no. 405032969.

References

1. Santina, C.D., Duriez, C., Rus, D.: Model-based control of soft robots: a survey of the state of the art and open challenges. IEEE Control. Syst. Mag. **43**(3):30–65 (2023)
2. Pinskier, J., Howard, D.: From bioinspiration to computer generation: developments in autonomous soft robot design. Adv. Intell. Syst. **4**(1), 2100086 (2022)
3. Chen, F., Wangm M.Y.: Design optimization of soft robots: a review of the state of the art. IEEE Robot. Autom. Mag. **27**(4), 27–43 (2020)
4. Armanini, C., Boyer, F., Mathew, A.T., Duriez, C., Renda, F.: Soft robots modeling: a structured overview. IEEE Trans. Robot. **39**(3), 1728–1748 (2023)
5. Laschi, C., Thuruthel, T.G., Lida, F., Merzouki, R., Falotico, E.: Learning-based control strategies for soft robots: theory, achievements, and future challenges. IEEE Control. Syst. Mag. **43**(3), 100–113 (2023)

6. Wang, H., Totaro, M., Beccai, L.: Toward perceptive soft robots: progress and challenges. Adv. Sci. **5**(9), 1800541 (2018)
7. Coevoet, E., Bieze, T.M., Largilliere, F., Zhang, Z., Thieffry, M., Lopez, M.S., Carrez, B., Marchal, D., Goury, O., Dequidt, J., Duriez, C.: Software toolkit for modeling, simulation and control of soft robots. Adv. Robot. **31**, 1208–1224 (2017)
8. Hadi Sadati, S.M., Elnaz Naghibi, S., Shiva, A., Michael, B., Renson, L., Howard, M., Rucker, C.D., Althoefer, K., Nanayakkara, T., Zschaler, S., Bergeles, C., Hauser, H., Walker, I.D.: Tmtdyn: A matlab package for modeling and control of hybrid rigid—continuum robots based on discretized lumped systems and reduced-order models. Int. J. Robot. Res. **40**(1), 296–347 (2020)
9. Caasenbrood, B.: Sorotoki—a soft robotics toolkit for matlab. https://github.com/BJCaasenbrood/SorotokiCode (2020)
10. Mathew, A.T., Hmida, I.B., Armanini, C., Boyer, F., Renda, F.: Sorosim: a matlab toolbox for hybrid rigid—soft robots based on the geometric variable-strain approach. IEEE Robot. Autom. Mag. **30**(3), 106–122 (2023)
11. Tekinalp, A., Kim, S.H., Bhosale, Y., Parthasarathy, T., Naughton, N., Albazroun, A., Joon, R., Cui, S., Nasiriziba, I., Stölzle, M., Shih, C.H., Gazzola, M.: Gazzolalab/pyelastica: v0.3.2 (2024)
12. Runge-Borchert, G., Wiese, M., Günther, L., Raatz, A.: A framework for the kinematic modeling of soft material robots combining finite element analysis and piecewise constant curvature kinematics. In: 2017 3rd International Conference on Control, Automation and Robotics (ICCAR), pp. 7–14 (2017)
13. Moseley, P., Florez, J.M., Sonar, H.A., Agarwal, G., Curtin, W., Paik, J.: Modeling, design, and development of soft pneumatic actuators with finite element method. Adv. Eng. Mater. **18**(6), 978–988 (2016)
14. Tawk, C., Alici, G.: Finite element modeling in the design process of 3d printed pneumatic soft actuators and sensors. Robotics **9**(3) (2020)
15. Xue, X., Zhan, Z., Cai, Y., Yao, L., Lu, Z.: Design and finite element analysis of fiber-reinforced soft pneumatic actuator. In: Yu, H., Liu, J., Liu, L., Ju, Z., Liu, Y., Zhou, D. (eds.) Intelligent Robotics and Applications. Springer International Publishing, Cham, pp. 641–651 (2019)
16. Oon Hock Yeoh: Some forms of the strain energy function for rubber. Rubber Chem. Technol. **66**, 754–771 (1993)
17. Ogden, R.W.: Recent advances in the phenomenological theory of rubber elasticity. Rubber Chem. Technol. **59**(3), 361–383 (1986)
18. Nasdala, L.: Materialmodelle. Springer Fachmedien Wiesbaden, Wiesbaden, pp. 181–226 (2015)
19. Tzikang, C.: Determining a prony series for a viscoelastic material from time varying strain data. Technical Report (2000)
20. Barrientos, E., Pelayo, F., Noriega, A., Lamela, M.J., Fernández-Canteli, A., Tanaka, E.: Optimal discrete-time prony series fitting method for viscoelastic materials. Mech. Time-Depend. Mater. **23**, 193–206 (2019)
21. Boyer, F., Lebastard, V., Candelier, F., Renda, F., Alamir, M.: Statics and dynamics of continuum robots based on cosserat rods and optimal control theories. IEEE Trans. Rob. **39**(2), 1544–1562 (2023)
22. Bartholdt, M., Wiese, M., Schappler, M., Spindeldreier, S., Raatz, A.: A parameter identification method for static cosserat rod models: Application to soft material actuators with exteroceptive sensors. In: 2021 IEEE/RSJ International Conference on Intelligent Robots and Systems (IROS), pp. 624–631 (2021)
23. Wiese, M., Berthold, R., Wangenheim, M., Raatz, A.: Describing and analyzing mechanical contact for continuum robots using a shooting-based cosserat rod implementation. IEEE Robot. Autom. Lett. **9**(2), 1668–1675 (2024)
24. Till, J.D.: On the statics, dynamics, and stability of continuum robots: model formulations and efficient computational schemes. Ph.d. thesis, University of Tennessee (2019). Available at https://trace.tennessee.edu/utk_graddiss/5379

25. Berthold, R., Wiese, M., Raatz, A.: Investigation of lateral compression effects in fiber reinforced soft pneumatic actuators. In: 2022 International Conference on Electrical, Computer, Communications and Mechatronics Engineering (ICECCME), pp. 1–7 (2022)
26. Bartholdt, M., Berthold, R., Schappler, M.: Towards a modular framework for visco-hyperelastic simulations of soft material manipulators with well-parameterised material. In: 2023 IEEE International Conference on Soft Robotics (RoboSoft), pp. 1–8 (2023)
27. Till, J., Aloi, V., Rucker, C.: Real-time dynamics of soft and continuum robots based on cosserat rod models. Int. J. Robot. Res. **38**(6), 723–746 (2019)
28. Smooth On: *Ecoflex Series Datasheet* (2021)
29. Smooth On: *Dragonskin Series Datasheet* (2021)
30. Marechal, L., Balland, P., Lindenroth, L., Petrou, F., Kontovounisios, C., Bello, F.: Toward a common framework and database of materials for soft robotics. Soft Rob. **8**(3), 284–297 (2021). PMID: 32589507
31. Testing of rubber—determination of tensile strength at break, tensile stress at yield, elongation at break and stress values in a tensile test, March 2017. DIN 53504:2017-03
32. Plastics-determination of tensile properties part 2: test conditions for moulding and extrusion plastics (1993)
33. Williams, M.L., Landel, R., Ferry, J.: The temperature dependence of relaxation mechanisms in amorphous polymers and other glass-forming liquids. J. Am. Chem. Soc. **77**, 3701–3707 (1955)
34. Berthold, R., Burgner-Kahrs, J., Wangenheim, M., Kahms, S.: Investigating frictional contact behavior for soft material robot simulations. Meccanica **58**(11), 2165–2176 (2023)
35. Mehl, M., Bartholdt, M., Schappler, M.: Dynamic modeling of soft-material actuators combining constant curvature kinematics and floating-base approach. In: 2022 IEEE 5th International Conference on Soft Robotics (RoboSoft), pp. 1–8 (2022)
36. Mehl, M., Bartholdt, M., Ehlers, S.F.G., Seel, T., Schappler, M.: Adaptive state estimation with constant-curvature dynamics using force-torque sensors with application to a soft pneumatic actuator. In: 2024 IEEE International Conference on Robotics and Automation (ICRA), pp. 14939–14945 (2024)
37. Alessi, C., Agabiti, C., Caradonna, D., Laschi, C., Renda, F., Falotico, E.: Rod models in continuum and soft robot control: a review (2024)
38. Krauss, H., Habich, T.L., Bartholdt, M., Seel, T., Schappler, M.: Domain-decoupled physics-informed neural networks with closed-form gradients for fast model learning of dynamical systems. Accepted to International Conference on Informatics in Control, Automation and Robotics (ICINCO) (2024)

Applications of Soft Robots

Wireless Miniature Medical Soft Robots Inside Our Body

Mingtong Li, Tianlu Wang, Ziyu Ren, and Metin Sitti

Abstract The complex internal biological environment of the human body presents numerous organ movement, spatial variations and constraints, and fluid flows, posing significant challenges to the mobility, positioning, and functional performance of wireless miniature medical robots within the body. To tackle these challenges, this work proposes shape-programmable and physically adaptable soft miniature magnetic robots with multimodal locomotion and versatile functionality. We specifically propose three types of wireless miniature soft milli-scale robot designs: a sheet-shaped magnetically controlled soft robot, a stent-shaped robot, and a balloon-shaped robot. These robots respectively demonstrate multimodal locomotion in physiologically relevant mockup and ex vivo environments, controllable movement and anchoring capability in vascular models with fluid flow, large output force (~ 70 N), and high work capacity (~ 175.2 J g^{-1}). Thus, they can achieve adaptive locomotion and multiple medical functionality within confined fluid-filled body sites that are risky or impossible to access with current minimally invasive medical devices, such as catheters. These robot designs enable novel miniature medical devices for various clinical applications, including targeted/local drug delivery, embolization, blood clot opening, aneurysm treatment, balloon-angioplasty, in-situ sensing, and reversible stenting.

Keywords Miniature soft robot · Multimodal-locomotion · Adaptability · Work capacity · Biomedical application

M. Li · T. Wang · Z. Ren · M. Sitti
Physical Intelligence Department, Max Planck Institute for Intelligent Systems, Stuttgart, Germany

T. Wang
Department of Mechanical Engineering, University of Hawai'i at Mānoa, Honolulu, HI, USA

M. Sitti (✉)
School of Medicine and College of Engineering, Koç University, Istanbul, Türkiye
e-mail: sitti@is.mpg.de

A. Raatz et al. (eds.), *Soft Material Robotic Systems*,
https://doi.org/10.1007/978-3-032-22453-8_18

1 Introduction

Wireless miniature soft robots that are capable of adaptively, robustly, and minimally invasively reaching risky, challenging or inaccessible body sites, hold significant promise for future healthcare applications, [1–7] including targeted drug delivery, [8, 9] minimally invasive surgery, [10–14] tissue engineering, [15–17] endoscopy, [18–21] diagnostics, [22, 23] and in-situ sensing [24–26]. Inside our bodies, numerous physical constraints and fluidic environments, such as the gastric mucosa in the gastrointestinal tract, food residues, and blood flow in the brain, cardiac or peripheral vasculature, pose significant challenges to the movement, positioning, and functionality of miniature soft robots [27–30]. Consequently, how to design the structure and composition of soft robots, as well as to develop gait control strategies to enable multimodal locomotion, precise positioning, and specialized functionality in complex, confined fluid-filled body sites, remains a critical challenge in the medical robotics field.

The shape- and motion-programming capability and adaptability of wireless miniature soft robots are critical for enabling locomotion, positioning, and functionality within the body [31–33]. The softness of the constituent stimuli-responsive materials provides a foundation for the adaptive deformation capabilities, enabling soft-bodied multimodal locomotion of miniature soft robots [34, 35]. This approach has allowed robots to pass through dry confined regions or be passively transported by fluid flows in bent tubes [36, 37]. However, achieving active locomotion and precise positioning in specific biological environments, such as fluid-filled blood vessels and stomach with food residues, still requires further research and validation. Active soft-bodied locomotion and maneuverability in fluid-filled confined spaces have yet to be fully demonstrated, and the impact of various factors in such environments on soft-bodied locomotion remains poorly understood.

Additionally, while the softness of the materials enables robots to undergo large deformations, it also limits their force output performance [2, 38]. For instance, existing magnetically responsive soft robots generate a small output force of approximately 60 μN and a low power density of around 10^{-3}–10^{2} J kg^{-1}, whereas common in-body medical applications, such as stent deployment or tissue cutting, require a large force of about several Newtons [39]. This limitation constrains the application of wireless miniature soft robots in biological environments. Therefore, generating sufficient thrust to overcome fluid drag and tissue boundary friction is essential for enabling the functionality of miniature soft robots within the body. To enhance the output performance of soft robots, researchers have recently developed various high-performance soft actuators, including dielectric elastomer actuators (DEA), [40, 41] artificial muscles, [39, 42–44], and hydraulically amplified self-healing electrostatic actuators (HASEL) [45]. These actuators, driven by stimuli, such as electricity and heat, can achieve driving forces of up to 30 N and work capacities of 100 J g^{-1}. However, these high-performance soft actuators are typically larger in size, often on the centimeter scale, can require high voltages, and are predominantly driven by wired systems, limiting their feasibility for inside-the-body medical applications.

To address these challenges, this work uses soft magnetic actuation methods that are compatible with medical applications inside the human body and can generate fast and precise actuation in 3D translation and rotation [4, 31, 46]. Using elastomeric soft composites with embedded hard magnetic microparticles, we can shape program, rotate or pull soft magnetic millirobots. As soft millirobot designs, we introduce three design types. First, focusing on fluid-filled confined body sites, we investigate the locomotion capabilities of a sheet-shaped miniature magnetic soft robot. By implementing a series of magnetic programming-based control strategies, the robot can switch between different locomotion modes, such as rolling, undulatory crawling, swimming, and helical surface crawling, adapting to various environmental conditions in confined spaces with either stagnant or low-velocity fluids. Second, targeting cardiovascular applications, we develop a stent-shaped miniature magnetic soft robot. Under magnetic control, this robot demonstrates adaptive locomotion and safe anchoring capabilities in arteries characterized by high-speed pulsatile blood flow, complex curved pathways, bifurcating branches, and dynamic lumen diameters. Finally, to enhance the output performance of wireless miniature soft robots and expand their functional capabilities within the body, we design a balloon-shaped inflatable wireless miniature soft robot composed of a magnetic soft composite. This design significantly improves the robot's output force (up to 70 N) and work capacity (175.2 J g^{-1}). Through remote magnetic radiofrequency (magnetic-RF) heating, the robot achieves controllable functions such as opening, closing, jumping, and stenting. In summary, this work explores the locomotion, functionality, and medical device application capabilities of wireless miniature soft robots, which show great potential for minimally invasive disease diagnosis and treatment in future clinical applications.

2 Soft-Bodied Sheet-Shaped Robot for Adaptive Multimodal Locomotion in Fluid-Filled Confined Spaces

Soft-bodied locomotion in fluid-filled, confined spaces is essential for the next generation of wireless medical robots designed to navigate the vessels, tubes, channels, and cavities of the human body, which are often filled with stagnant or flowing biological fluids. However, achieving soft-bodied active locomotion becomes highly challenging when the size of the robot is comparable to the cross-sectional dimensions of confined spaces [27–29]. Inspired by small-scale soft-bodied organisms, in this part, we introduce a wireless sheet-shaped magnetic soft robot, demonstrating its multimodal locomotion capability inside fluid-filled confined environments (Figs. 1 and 2).

We achieve various adaptive locomotion modes by exploiting the surrounding boundaries, hydrodynamics, frictional forces, and active soft-bodied deformation of the robots. We also develop a series of control strategies for magnetically programmed sheet-shaped soft robots, enabling them to switch locomotion modes to adapt to different environmental conditions in confined spaces filled with stagnant or flowing

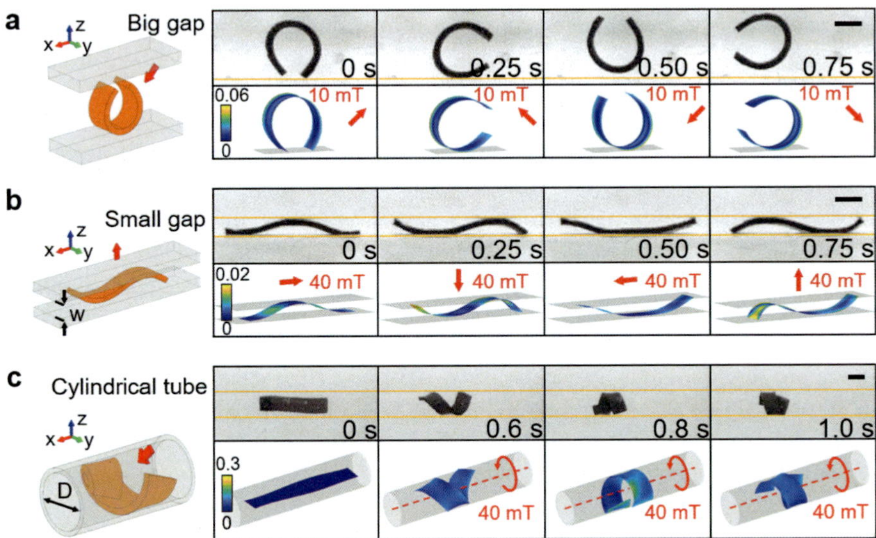

Fig. 1 Miniature sheet-shaped magnetic soft robots for adaptive multimodal locomotion in various confined body sites, from [47], licensed under CC-BY 4.0. **a** In a big gap, the robot curls into a C-shape when the magnetic field ***B*** lies in the x-z plane. **b** In a small gap, the robot deforms into the sinusoidal shape when ***B*** lies in the x-z plane. **c** Inside a cylindrical tube, the robot adopts a helical shape when ***B*** lies in the y-z plane. Finite element simulations predict deformation modes under specific boundary conditions. Red arrows denote the direction of ***B*** and the colormap represents equivalent von Mises strain. Experimental setups are filled with viscous fluid. Scale bars, 1 mm. This figure is adapted with permission from Ref. [47], AAAS

fluids. Moreover, we further computationally simulate and experimentally characterize the locomotion of the sheet-shaped soft robots within fluid-filled channels, by utilizing single- and multi-wave body undulation as well as novel helical locomotion modes to enhance our understanding. In environments with gaps much larger than the robot's size, the robot can roll by curling into a circular shape (Fig. 1a). In narrower gaps, it can perform undulatory crawling or swimming locomotion (Fig. 1b). Within cylindrical tubes, the robot achieves helical surface crawling and can even withstand dynamic fluid flows (Fig. 1c). Each locomotion mode offers distinct advantages. For instance, undulatory crawling excels in navigating highly bent paths with small gaps, while undulatory swimming enables rapid traversal through straight slits. Helical surface crawling provides propulsion both with and against fluid flows and can even resist flow forces when actuation is turned off, making it a versatile and robust solution for challenging in vivo environments.

Furthermore, we investigate the physical mechanisms underlying the locomotion modes of the wireless miniature magnetic sheet-shaped soft robot. The results indicate that undulatory crawling and helical surface crawling rely on friction between the body surface of the robot and the surrounding boundaries to achieve efficient and robust propulsion. In contrast, the undulatory swimming mode generates thrust force

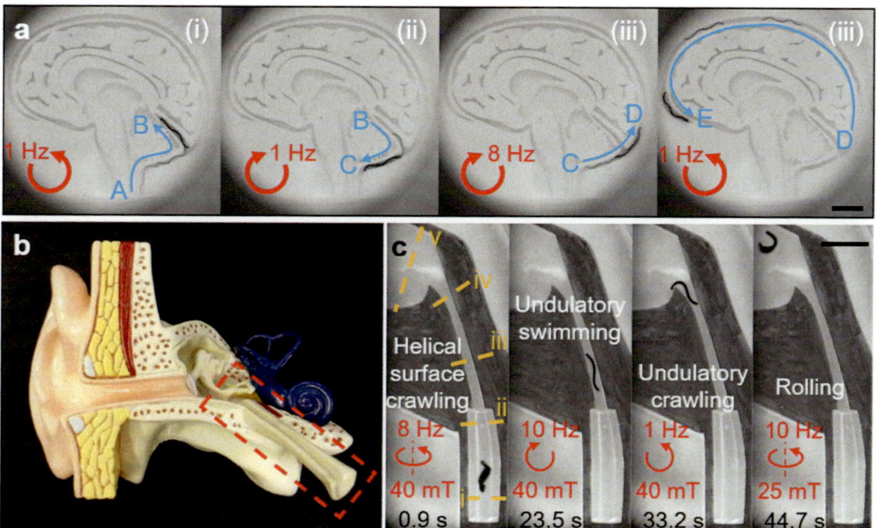

Fig. 2 Demonstration of the sheet-shaped soft robot navigating phantom structures with diverse cross-sectional geometries and sizes using its multimodal locomotion capabilities, from [47], licensed under CC-BY 4.0. **a** The robot is steered in a brain aqueduct-mimicking phantom by adjusting the magnetic field's rotation frequency. Red arrows show rotating magnetic field directions. Scale bar, 5 mm. **b** Anatomical model of the human ear, with the red dashed box highlighting the Eustachian tube, mimicked by the simplified phantom. **c** The robot adapts to local environmental constraints in the phantom by switching locomotion modes, including helical surface crawling, undulatory swimming, undulatory crawling, and rolling. Red arrows indicate the ***B*** direction. This figure is adapted with permission from Ref. [47], AAAS

by transporting the fluid. Building on this understanding, we further propose performance enhancement strategies to improve the robot's locomotion speed, maneuverability, and ability to resist opposing fluid flows. Remarkably, all these locomotion modes can be achieved using a single soft robot design. Finally, we demonstrate the robot can successfully navigate through a brain aqueduct-mimicking phantom and a phantom mimicking the Eustachian tube by employing these multiple locomotion modes (Fig. 2). We believe that the proposed control strategies and performance improvements establish the sheet-shaped robot as a promising tool for future applications within various lumens of the human body.

3 Adaptive Wireless Stent-Shaped Robot Locomotion into Distal Vasculature

We then focus our attention on cardiovascular and cerebrovascular diseases, investigating the locomotion performance of wireless miniature soft robots within vascular environments. In recent clinical practice, miniaturized medical catheterized devices have facilitated various endovascular procedures, including drug delivery, embolization, and thrombectomy within vascular systems. However, deeper vascular regions, including tortuous routes, branches, and distal areas away from the catheter deployment port with complex flow conditions, remain challenging in accessibility [48–51]. Despite burgeoning efforts have been made to develop micrometer and millimeter untethered robots, which could be actively controlled to move into these hard-to-reach regions because of their tether-less performance and possible advantages in dimensions, these state-of-the-art robotic designs lack self-locking capability within the surrounding lumen, making them susceptible to being easily displaced by the fluid flow, especially when the control signals are disrupted and turned off. Additionally, the distributed robots are unpredictable and accumulate in non-target tissues and organs, which might persist for years and lead to prolonged health risks [52–54]. Therefore, in this part, we propose a wireless stent-shaped magnetic soft millirobot that can achieve all of the above requirements to operate in the distal M4 segment of the middle cerebral artery (Fig. 3).

In this work, we design a wireless miniature magnetic soft robot with a cylindrical hollow stent shape featuring a helical surface structure (Fig. 4a). This shape is chosen for its radial deformability and low fluidic drag, enabling effective locomotion and anchoring capability in arteries with high-speed pulsatile flow and dynamic lumen diameters. To facilitate rotational and translational magnetic actuation, we

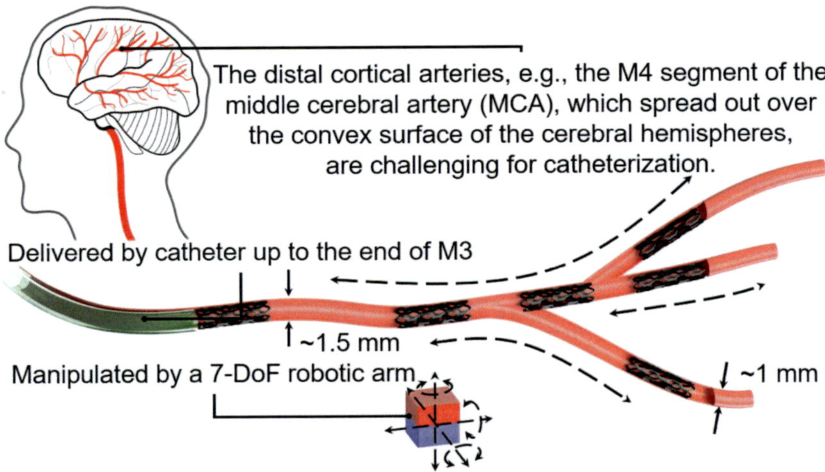

Fig. 3 Overview of application scenarios in the distal vasculature, exemplified by the challenging catheterization case in the M4 segment of the middle cerebral artery, from [55], licensed under CC-BY 4.0. Key locomotion capabilities of the proposed wireless soft robot include forward and backward shape adaptation to varying lumen diameters, passive flow resistance without a magnetic field, and navigation through curved paths and branches. The robot can also serve as a mobile platform for functional tools to treat acute ischemic stroke, aneurysms, and arteriovenous malformations. This figure is adapted with permission from Ref. [55], Springer Nature

incorporate ferromagnetic NdFeB microparticles into the robot's body and magnetized it uniformly. To enhance the adaptive locomotion and anchoring functionality within blood vessels, we optimize the robot's design by adjusting the open area of the stent structure, the density of the helical surface pattern, and the composed material stiffness. After refining the material and structural parameters, we demonstrate that the stent-shaped robot can achieve adaptive locomotion within lumens as small as 1 mm in diameter (Fig. 4b). The performance meets the requirements for forward and backward locomotion in lumens with diameters transitioning from 1.5 mm to 1 mm in the distal M4 segment of middle cerebral artery. Additionally, the robot exhibits reliable locomotion through highly curved routes, bifurcating branches with angles up to 120°, and pulsatile blood flow at speeds up to 26 cm/s with a heart rate of 80 beats per minute (bpm) (Fig. 4c–e). Moreover, it exhibits a safe self-anchoring capability, remaining securely positioned even when the external magnetic actuation input was deactivated.

Beyond controlled and precise navigation, we demonstrate that the robot variants possess several essential medical capabilities. For instance, the robot can deliver tissue plasminogen activator (tPA) on demand to achieve targeted thrombolysis (Fig. 5). Additionally, it can function as a flow diverter to regulate blood flow, preventing it from reaching undesirable sites such as aneurysms or certain branches. These features open new possibilities for minimally invasive, targeted therapies for conditions like acute ischemic stroke (AIS), aneurysms, cerebral arteriovenous malformations (CAVMs), dural arteriovenous fistulas (dAVFs), and brain tumors located in distal and tortuous vascular regions. Given its exceptional locomotion capabilities in distal arterial lumens under pulsatile flow conditions, the miniature stent-shaped magnetic soft robot shows great promise as a wireless medical device. Finally, we demonstrate two proof-of-concept functions: localized on-demand drug delivery and flow diversion. These innovations have the potential to enhance existing catheter-based therapies for AIS, aneurysms, and AVMs in challenging distal artery environments.

4 Wireless Miniature Magnetic Inflatable Balloon-Shaped Soft Robot

Wireless miniature soft robots hold significant promise for impactful applications in medicine, robotic grippers, and artificial muscles [4, 56–58]. However, their potential is currently limited by low output forces and insufficient work capacity. This limitation arises from the softness of their constituent materials and their constrained volume, which make it challenging to store and release high levels of mechanical energy. Most existing miniature magnetic soft robots exhibit a work capacity ranging from approximately 10^{-3}–10^{2} J kg^{-1}, which falls short of the requirements for medical devices, manipulation tasks, and other applications demanding higher energy output. Additionally, the maximum output force of current magnetic soft

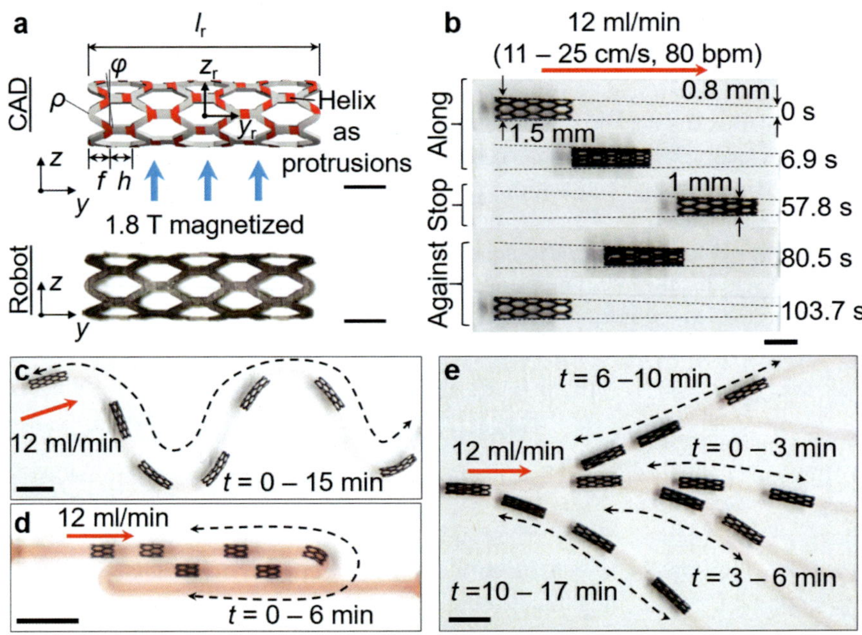

Fig. 4 Adaptive magnetic stent-shaped soft robotic locomotion and anchoring in distal vasculature in the brain, from [55], licensed under CC-BY 4.0. **a** CAD design and prototype of the robot. Key design parameters include strut spacing (h), radius of curvature at the crown junction (ρ), and axial amplitude of each segment (f). The robot is embedded with NdFeB ferromagnetic microparticles, uniformly magnetized using a 1.8 T homogeneous magnetic field. Right-handed helical structures with a helix angle φ were applied. Scale bar: 1 mm. **b** Snapshots of the robot's retrievable radial shape adaptation in a lumen with diameter variations from 1 mm to 1.5 mm. Scale bar: 2 mm. **c** Locomotion of the robot through a tortuous route with radii of curvature ranging from 2.5 mm to 5 mm. **d** Locomotion of the robot through an extremely curved route with a 180° inclination angle and a radius of curvature of 1 mm. **e** Navigation of the robot among branches in a 2D environment. The scale bar from **c** to **e** is 5 mm. This figure is adapted with permission from Ref. [55], Springer Nature

robots is around 60 μN [39, 59]. In contrast, many medical procedures, such as stenting, require devices capable of generating forces exceeding 1 N, which is about 10,000 times greater than the current maximum output force of magnetic soft robots [60].

To address these challenges, we present a balloon-shaped inflatable wireless miniature soft robot made from a magnetic soft composite (Fig. 6). This composite consists of silicone rubber (Ecoflex) embedded with hard-magnetic NdFeB microparticles and iron oxide nanoparticles (Fe_3O_4NPs). The balloon is filled with a low-boiling-point liquid (Novec 7000, with a biocompatible boiling point of $\approx$ 34 °C). The embedded NdFeB microparticles enable actuation via an external magnetic field, while remote magnetic-RF field-based heating of the Fe_3O_4NPs triggers a phase change in the liquid, causing inflation (Fig. 6). This phase change mechanism significantly enhances the robot's output force (up to 70 N) and work capacity (175.2 J g^{-1})

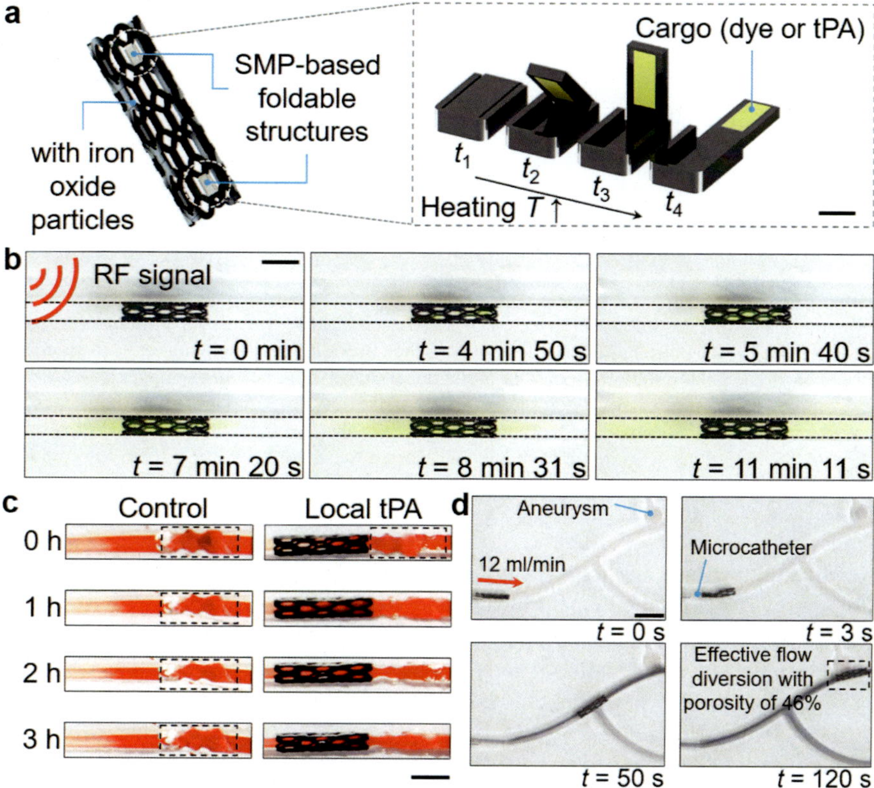

Fig. 5 Demonstration of stent-shaped robot on-demand drug delivery and flow diversion from [55], licensed under CC-BY 4.0. **a** A design variant incorporating shape memory polymer (SMP) based foldable structures for local on-demand delivery of endovascular tPA for acute ischemic stroke therapy. Scale bar: 0.5 mm. **b** RF-based heating demonstration for releasing composite cargo composed of silk fibroin and fluorescein dye. Scale bar: 2 mm. **c** Thrombolysis effect using carried tPA, with thrombi outlined by black dotted lines. Scale bar: 2 mm. **d** Experimental demonstration of flow diversion by the robot, deployed via a microcatheter and magnetically guided to the target lesion in phantom. Scale bar: 5 mm. This figure is adapted with permission from Ref. 55, Springer Nature

(Fig. 6c). The proposed balloon-shaped actuator combines magnetic and pneumatic soft actuation methods, offering a novel approach to overcoming the limitations of existing soft robots.

First, we systematically characterize the inflation behavior of the balloon-shaped soft actuator. Next, we investigate the design principles, including material properties and geometrical parameters, to optimize its output mechanical performance. Building on these insights, we develop a generalized theoretical model to serve as a reference for guiding design and understanding the actuator's nonlinear behavior. This model establishes the relationships among design parameters (composite modulus and thickness), control inputs (internal pressure), and outputs (actuation strain, force,

Fig. 6 Design methodology of wireless miniature magnetic inflatable balloon-shaped soft robot, from [61], licensed under CC-BY 4.0. **a** Schematic and **b** experimental camera video snapshots illustrating the inflation process of the balloon-shaped actuator under magnetic-RF heating. **c** Comparison of various soft actuators in terms of the materials modulus and the output work capacity, indicating the proposed balloon-shaped robot achieves a higher work capacity despite its low material modulus. This figure is adapted with permission from Ref. [61], John Wiley and Sons

and work capacity). We then demonstrate the actuator as a soft device prototype for potential future applications in angioplasty (Fig. 7a–b). To further broaden its functionality for biomedical applications, we integrate the actuator with a thermally responsive shape-memory composite and a bistable metamaterial sleeve (Fig. 7c). This bistable structure enables reversible shape changes and self-locking capabilities. By adjusting the power of the remote magnetic-RF heating input, the device achieves controllable opening and closing functions (Fig. 7d). Additionally, we showcase its ability to navigate and even jump out of granular media (Fig. 7e). Finally, the actuator is encoded with a custom magnetization profile, allowing it to undergo deformation induced by internal magnetic torques under an external magnetic field (Fig. 7f). Beyond enhancing force output, this integration of magnetic soft materials with phase-change materials introduces programmable bending deformation, in addition to programmable inflation and deflation. This expand functionality further enhances the actuator's potential applications in soft robotics and biomedical engineering.

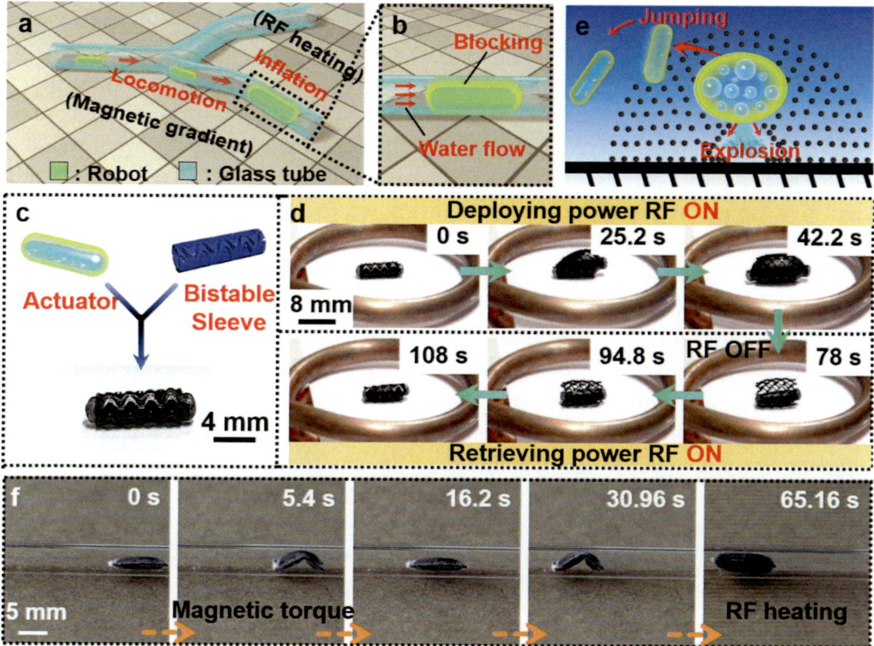

Fig. 7 Wirelessly activated balloon toward angioplasty, locomotion inside granular media, and bimodal actuation, from [61], licensed under CC-BY 4.0. **a–b** Schematic showing the robot's locomotion and inflation behavior under a rotating magnetic field and magnetic-RF heating. **c** Schematic and camera image of the reversible stent design. **d** Camera images showing the reversible stent's operation under varying magnetic-RF heating power inputs. Reversible anchoring is achieved by leveraging the different response temperatures of the magnetic balloon-shaped actuator and the bistable sleeve, controlled by adjusting the Fe_3O_4NPs concentration in the SMP matrix. **e** Schematic illustrating the jumping behavior of the balloon-shaped actuator under magnetic-RF heating inside the granular media. **f** Sequential images of the magnetic balloon-shaped soft robot walking and inflating inside a glass tube under magnetic torques and magnetic-RF heating. This figure is adapted with permission from Ref. [61], John Wiley and Sons

5 Conclusion

Faced with the challenges of enabling wireless miniature soft robots for various disruptive medical applications inside the human body, this project proposes three types of wireless magnetically controlled soft robot designs: sheet-shaped, stent-shaped, and balloon-shaped robots. These robots demonstrate, respectively, multimodal locomotion in confined fluid-filled spaces, adaptive locomotion and anchoring in blood vessels, and enhanced force output and functional capabilities, broadening their potential medical applications. Despite these advancements, challenges remain in applying wireless miniature soft robots within the body. Looking forward, we propose several future directions for their development.

First, although we have demonstrated the soft robotic capabilities for multi-modal deformation and locomotion ex vivo and in physiologically relevant artificial mockups, the highly complex in-vivo environment remains a significant hurdle. For instance, the frictional properties of human tissue differ from those of the materials used to construct the test environments, which could alter the robot's locomotion and function behavior. Additionally, many bodily fluids, such as mucus and blood, are non-Newtonian. While the robot has been shown to perform all locomotion modes, anchoring, and function capability in a shear-thinning fluid, further investigation is needed to understand how the fluid's shear-thinning properties and viscoelasticity affect the robot's locomotion. The non-Newtonian behavior of these fluids can modify the fluid forces acting on the soft body, influencing both the robot's deformation amplitude and the propulsion it generates. Therefore, these robots need to be tested in in-vivo animal models to validate their capabilities and medical functions.

Next, a promising future approach (such as inverse design) would be the development of automated design frameworks. By integrating measured environmental properties, such as boundary conditions and fluid characteristics into dynamic locomotion model-based or data-driven learning-based algorithms, the robot designs and control signals could be automatically generated. Such automation combined with active programmable magnetic materials would optimize the use of adaptive multi-modal locomotion for biomedical applications, ensuring more effective and robust performance in complex internal environments.

Third, since the proposed robot design is not restricted to a specific material, the materials currently used in the prototype might not simultaneously meet the biocompatibility and hemocompatibility requirements of a medical device. Biocompatibility is crucial as it determines whether the robot can be safely used within the human body and for how long. The robot can be fabricated using various FDA-approved materials for medical devices or coated with biocompatible materials, such as polyurethane, polyethylene, or even metals, enabling a wide range of potential applications in the future.

Finally, current high-output soft actuators still face challenges in their application for wireless medical devices inside the body. For instance, thermo-responsive actuators suffer from low cooling efficiency due to their slow thermal diffusion process, particularly in rubber polymer materials on the actuator's outer surface, which have poor thermal diffusion properties, leading to prolonged cooling times [4]. For alternatives, electro-responsive actuators require a high voltage (approximately 10 kV) to achieve high output performance, making them unsuitable for in-body use [45]. Additionally, other high-output flexible actuators, such as pneumatic and hydraulic actuators, face difficulties in achieving wireless remote control and miniaturization [62]. Acoustic waves present a promising energy source for mechanical soft actuation due to their deep penetration into biological tissues and fluidic media. However, ultrasonic actuation mechanisms are limited by material density, making them challenging to implement in flexible materials [63]. Therefore, future work should focus on exploring novel actuation mechanisms or structural designs to amplify the output performance of the wireless miniature soft actuators, enabling them to perform more functions within the body.

Acknowledgments This work was funded by the Max Planck Society, German Research Foundation (DFG) Soft Material Robotic Systems (SPP 2100) Program with grant no: 497562474.

Competing Interests M.S. and T.W. are listed as inventors on pending US patent application US18/133,104, US18/855,998 and European patent application EP22167673.7A, EP23719347.9A submitted by the Max–Planck–Gesellschaft zur Forderung der Wissenschaften e.V. that covers the fundamental design, fabrication, and control principles of the stent-shaped magnetic soft robots included in this work. The other authors declare no competing interests.

References

1. Sitti, M.: Miniature soft robots-road to the clinic. Nat. Rev. Mater. **3**, 74–75 (2018)
2. Cianchetti, M., Laschi, C., Menciassi, A., Dario, P.: Biomedical applications of soft robotics. Nat. Rev. Mater. **3**, 143–153 (2018)
3. Wang, T., Wu, Y., Yildiz, E., Kanyas, S., Sitti, M.: Clinical translation of wireless soft robotic medical devices. Nat. Rev. Bioeng. **2**, 470–485 (2024)
4. Li, M., Pal, A., Aghakhani, A., Pena-Francesch, A., Sitti, M.: Soft actuators for real-world applications. Nat. Rev. Mater. **7**, 235–249 (2022)
5. Kim, H., Lee, K., Go, G.: Wireless hybrid-actuated soft miniature robot for biomedical applications. Actuators. **13**, 341 (2024)
6. Ceylan, H., Yasa, I.C., Kilic, U., Hu, W., Sitti, M.: Translational prospects of untethered medical microrobots. Prog. Biomed. Eng. **1**, 012002 (2019)
7. Ren, Z., Sitti, M.: Design and build of small-scale magnetic soft-bodied robots with multimodal locomotion. Nat. Protoc. **19**, 441–486 (2024)
8. Ng, C.S.X., Tan, M.W.M., Xu, C., Yang, Z., Lee, P.S., Lum, G.Z.: Locomotion of miniature soft robots. Adv. Mater. **33**, 2003558 (2021)
9. Yim, S., Sitti, M.: Shape-programmable soft capsule robots for semi-implantable drug delivery. IEEE Trans. Robot. **28**, 1198–1202 (2012)
10. Runciman, M., Darzi, A., Mylonas, G.P.: Soft robotics in minimally invasive surgery. Soft Robot. **6**, 423–443 (2019)
11. Zhu, J., Lyu, L., Xu, Y., Liang, H., Zhang, X., Ding, H., Wu, Z.G.: Intelligent soft surgical robots for next-generation minimally invasive surgery. Adv. Intell. Syst. **3**, 2100011 (2021)
12. Soon, R.H., Ren, Z., Hu, W., Bozuyuk, U., Yildiz, E., Li, M., Sitti, M.: On-demand anchoring of wireless soft miniature robots on soft surfaces. Proc. Natl. Acad. Sci. USA. **119**, e2207767119 (2022)
13. Son, D., Ugurlu, M.C., Sitti, M.: Permanent magnet array-driven navigation of wireless millirobots inside soft tissues. Sci. Adv. **7**, eabi8932 (2021)
14. Kozielski, K.L., Jahanshahi, A., Gilbert, H.B., Yu, Y., Erin, Ö., Francisco, D., Alosaimi, F., Temel, Y., Sitti, M.: Nonresonant powering of injectable nanoelectrodes enables wireless deep brain stimulation in freely moving mice. Sci. Adv. **7**, eabc4189 (2021)
15. Yasa, I.C., Tabak, A.F., Yasa, O., Ceylan, H., Sitti, M.: 3D-printed microrobotic transporters with recapitulated stem cell niche for programmable and active cell delivery. Adv. Funct. Mater. **29**, 1808992 (2019)
16. Sitti, M., Ceylan, H., Hu, W.Q., Giltinan, J., Turan, M., Yim, S., Diller, E.: Biomedical applications of untethered mobile milli/micro-robots. Proc. IEEE Inst. Electr. Electron. Eng. **103**, 205–224 (2015)
17. Dogan, N.O., Suadiye, E., Wrede, P., Lazovic, J., Dayan, C.B., Soon, R.H., Aghakhani, A., Richter, G., Sitti, M.: Immune cell-based microrobots for remote magnetic actuation, antitumor activity, and medical imaging. Adv. Healthc. Mater. **13**, 2400711 (2024)
18. Son, D., Gilbert, H., Sitti, M.: Magnetically actuated soft capsule endoscope for fine-needle biopsy. Soft Robot. **7**, 10–21 (2019)

19. Wang, B., Chan, K.F., Yuan, K., Wang, Q., Xia, X., Yang, L., Ko, H., Wang, Y.J., Sung, J.J.Y., Chiu, P.W.Y., Zhang, L.: Endoscopy-assisted magnetic navigation of biohybrid soft microrobots with rapid endoluminal delivery and imaging. Sci. Robot. **6**, eabd2813 (2021)
20. Yasa, I.C., Ceylan, H., Bozuyuk, U., Wild, A.M., Sitti, M.: Elucidating the interaction dynamics between microswimmer body and immune system for medical microrobots. Sci. Robot. **5**, eaaz3867 (2020)
21. Yim, S., Sitti, M.: Design and rolling locomotion of a magnetically actuated soft capsule endoscope. IEEE Trans. Robot. **28**, 183–194 (2012)
22. McCandless, M., Perry, A., DiFilippo, N., Carroll, A., Billatos, E., Russo, S.: A soft robot for peripheral lung cancer diagnosis and therapy. Soft Robot. **9**, 754–766 (2021)
23. Qiu, Y., Ashok, A., Nguyen, C.C., Yamauchi, Y., Do, T.N., Phan, H.P.: Integrated sensors for soft medical robotics. Small. **20**, 2308805 (2024)
24. Wang, C., Wu, Y., Dong, X., Armacki, M., Sitti, M.: In situ sensing physiological properties of biological tissues using wireless miniature soft robots. Sci. Adv. **9**, eadg3988 (2023)
25. Tiryaki, M.E., Sitti, M.: Magnetic resonance imaging-based tracking and navigation of submillimeter-scale wireless magnetic robots. Adv. Intell. Syst. **4**, 2100178 (2022)
26. Ceylan, H., Giltinan, J., Kozielski, K., Sitti, M.: Mobile microrobots for bioengineering applications. Lab Chip. **17**, 1705–1724 (2017)
27. Lum, G.Z., Ye, Z., Dong, X.G., Marvi, H., Erin, O., Hu, W.Q., Sitti, M.: Shape-programmable magnetic soft matter. Proc. Natl. Acad. Sci. USA. **113**, E6007–E6015 (2016)
28. Ren, Z., Hu, W., Dong, X., Sitti, M.: Multi-functional soft-bodied jellyfish-like swimming. Nat. Commun. **10**, 2703 (2019)
29. Dong, X., Lum, G.Z., Hu, W., Zhang, R., Ren, Z., Onck, P.R., Sitti, M.: Bioinspired cilia arrays with programmable nonreciprocal motion and metachronal coordination. Sci. Adv. **6**, eabc9323 (2020)
30. Qiu, F., Nelson, B.J.: Magnetic helical micro- and nanorobots: toward their biomedical spplications. Engineering. **1**, 021–026 (2015)
31. Hines, L., Petersen, K., Lum, G.Z., Sitti, M.: Soft actuators for small-scale robotics. Adv. Mater. **29**, 1603483 (2017)
32. Xia, N., Zhu, G., Wang, X., Dong, Y., Zhang, L.: Multicomponent and multifunctional integrated miniature soft robots. Soft Matter. **18**, 7464–7485 (2022)
33. Ashuri, T., Armani, A., Jalilzadeh Hamidi, R., Reasnor, T., Ahmadi, S., Iqbal, K.: Biomedical soft robots: current status and perspective. Biomed. Eng. Lett. **10**, 369–385 (2020)
34. Hu, W.Q., Lum, G.Z., Mastrangeli, M., Sitti, M.: Small-scale soft-bodied robot with multimodal locomotion. Nature. **554**, 81–85 (2018)
35. Ebrahimi, N., Bi, C., Cappelleri, D.J., Ciuti, G., Conn, A.T., Faivre, D., Habibi, N., Hošovský, A., Iacovacci, V., Khalil, I.S.M., Magdanz, V., Misra, S., Pawashe, C., Rashidifar, R., Soto-Rodriguez, P.E.D., Fekete, Z., Jafari, A.: Magnetic actuation methods in bio/soft robotics. Adv. Funct. Mater. **31**, 2005137 (2021)
36. Huang, H.W., Uslu, F.E., Katsamba, P., Lauga, E., Sakar, M.S., Nelson, B.J.: Adaptive locomotion of artificial microswimmers. Sci. Adv. **5**, eaau1532 (2019)
37. Do, T.N., Phan, H.P., Nguyen, T.Q., Visell, Y.: Miniature soft electromagnetic actuators for robotic applications. Adv. Funct. Mater. **28**, 1800244 (2018)
38. Tang, Y., Chi, Y., Sun, J., Huang, T.H., Maghsoudi, O.H., Spence, A., Zhao, J., Su, H., Yin, J.: Leveraging elastic instabilities for amplified performance: spine-inspired high-speed and high-force soft robots. Sci. Adv. **6**, eaaz6912 (2020)
39. Li, M., Tang, Y., Soon, R.H., Dong, B., Hu, W., Sitti, M.: Miniature coiled artificial muscle for wireless soft medical devices. Sci. Adv. **8**, eabm5616 (2022)
40. Gupta, U., Qin, L., Wang, Y., Godaba, H., Zhu, J.: Soft robots based on dielectric elastomer actuators: a review. Smart Mater. Struct. **28**, 103002 (2019)
41. Chi, Y., Zhao, Y., Hong, Y., Li, Y., Yin, J.: A perspective on miniature soft robotics: actuation, fabrication, control, and applications. Adv. Intell. Syst. **6**, 2300063 (2024)
42. Haines, C.S., Lima, M.D., Li, N., Spinks, G.M., Foroughi, J., Madden, J.D.W., Kim, S.H., Fang, S., Jung de Andrade, M., Göktepe, F., Göktepe, Ö., Mirvakili, S.M., Naficy, S., Lepró,

X., Oh, J., Kozlov, M.E., Kim, S.J., Xu, X., Swedlove, B.J., Wallace, G.G., Baughman, R.H.: Artificial muscles from fishing line and sewing thread. Science. **343**, 868–872 (2014)
43. Higueras-Ruiz, D.R., Shafer, M.W., Feigenbaum, H.P.: Cavatappi artificial muscles from drawing, twisting, and coiling polymer tubes. Sci. Robot. **6**, eabd5383 (2021)
44. Nan, M., Go, G., Song, H.W., Darmawan, B.A., Zheng, S., Kim, S., Nguyen, K.T., Lee, K., Kim, H., Park, J.O., Choi, E.: Multistimulus-responsive miniature soft actuator with programmable shape-morphing design for biomimetic and biomedical applications. Adv. Funct. Mater. **34**, 2401776 (2024)
45. Acome, E., Mitchell, S.K., Morrissey, T.G., Emmett, M.B., Benjamin, C., King, M., Radakovitz, M., Keplinger, C.: Hydraulically amplified self-healing electrostatic actuators with muscle-like performance. Science. **359**, 61–65 (2018)
46. Sitti, M.: Mobile Microrobotics. MIT Press, Cambridge, MA (2017)
47. Ren, Z., Zhang, R., Soon, R.H., Liu, Z., Hu, W., Onck, P.R., Sitti, M.: Soft-bodied adaptive multimodal locomotion strategies in fluid-filled confined spaces. Sci. Adv. **7**, eabh2022 (2021)
48. Saver, J.L., Chapot, R., Agid, R., Hassan, A.E., Jadhav, A.P., Liebeskind, D.S., Lobotesis, K., Meila, D., Meyer, L., Raphaeli, G., Gupta, R., Amista, P., Andsberg, G., Cagnazzo, F., Isalberti, M., Karabegovic, S., Kollia, K., Mangiafico, S., Mis, M., Moreno, A., Mudersbach, P.W., Nossek, E., Pero, G., Piasecki, P., Raz, E., Reis, J., Rudnicka, S., Sinisalo, M., Spinetta, M., Stavngaard, T., Undren, P., Zamaro, J.: Thrombectomy for distal, medium vessel occlusions. Stroke. **51**, 2872–2884 (2020)
49. Grossberg, J.A., Rebello, L.C., Haussen, D.C., Bouslama, M., Bowen, M., Barreira, C.M., Belagaje, S.R., Frankel, M.R., Nogueira, R.G.: Beyond large vessel occlusion strokes. Stroke. **49**, 1662–1668 (2018)
50. Rotim, K., Splavski, B., Kalousek, V., Jurilj, M., Sajko, T.: Endovascular management of intracranial aneurysms on distal arterial branches: illustrative case series and literature retrospection. Acta Clin. Croat. **59**, 712–719 (2020)
51. Wang, T., Hu, W., Ren, Z., Sitti, M.: Ultrasound-guided wireless tubular robotic anchoring system. IEEE Robot. Autom. Lett. **5**, 4859–4866 (2020)
52. Alapan, Y., Bozuyuk, U., Erkoc, P., Karacakol, A.C., Sitti, M.: Multifunctional surface micro-rollers for targeted cargo delivery in physiological blood flow. Sci. Robot. **5**, eaba5726 (2020)
53. Zenych, A., Fournier, L., Chauvierre, C.: Nanomedicine progress in thrombolytic therapy. Biomaterials. **258**, 120297 (2020)
54. Zhang, J., Ren, Z., Hu, W., Soon, R.H., Yasa, I.C., Liu, Z., Sitti, M.: Voxelated three-dimensional miniature magnetic soft machines via multimaterial heterogeneous assembly. Sci. Robot. **6**, eabf0112 (2021)
55. Wang, T., Ugurlu, H., Yan, Y., Li, M., Li, M., Wild, A.M., Yildiz, E., Schneider, M., Sheehan, D., Hu, W., Sitti, M.: Adaptive wireless millirobotic locomotion into distal vasculature. Nat. Commun. **13**, 4465 (2022)
56. Tasoglu, S., Diller, E., Guven, S., Sitti, M., Demirci, U.: Untethered micro-robotic coding of three-dimensional material composition. Nat. Commun. **5**, 3124 (2014)
57. Sinatra, N.R., Teeple, C.B., Vogt, D.M., Parker, K.K., Gruber, D.F., Wood, R.J.: Ultragentle manipulation of delicate structures using a soft robotic gripper. Sci. Robot. **4**, eaax5425 (2019)
58. Kim, Y., Parada, G.A., Liu, S., Zhao, X.: Ferromagnetic soft continuum robots. Sci. Robot. **4**, eaax7329 (2019)
59. Rich, S.I., Wood, R.J., Majidi, C.: Untethered soft robotics. Nat. Electron. **1**, 102–112 (2018)
60. Cabrera, M.S., Oomens, C.W.J., Baaijens, F.P.T.: Understanding the requirements of self-expandable stents for heart valve replacement: radial force, hoop force and equilibrium. J. Mech. Behav. Biomed. Mater. **68**, 252–264 (2017)
61. Tang, Y., Li, M., Wang, T., Dong, X., Hu, W., Sitti, M.: Wireless miniature magnetic phase-change soft actuators. Adv. Mater. **34**, 2204185 (2022)
62. Jung, Y., Kwon, K., Lee, J., Ko, S.H.: Untethered soft actuators for soft standalone robotics. Nat. Commun. **15**, 3510 (2024)

63. Katzschmann, R.K., DelPreto, J., MacCurdy, R., Rus, D.: Exploration of underwater life with an acoustically controlled soft robotic fish. Sci. Robot. **3**, eaar3449 (2018)

Multichained Grippers Inspired by Insect Tarsi

Julian Winand, Stanislav Gorb, and Mohsen Jafarpour

Abstract Gripping, holding, and moving objects are fundamental tasks for robots. Since the advent of automation, optimization of these functions has been a key focus in the field. Numerous strategies have been pursued to develop grippers that are cost-effective, reliable, and versatile, while reducing their weight and energy consumption, as well as production and maintenance costs. Although significant advancements in precision and flexibility have been achieved, these solutions often rely on complex support systems, such as advanced sensors and sophisticated control methods, making them unsuitable for applications requiring a balance of efficiency, simplicity, and cost-effectiveness. To address these challenges, this chapter introduces bio-inspired gripper designs that achieve adaptive grasping through morphological encoding and passive compliance: (1) a multi-segmented design inspired by insect tarsal chains, and (2) monolithic designs that leverage compliant double-spirals for passive shape adaptation and stiffness tuning. Fabricating the designed grippers and evaluating their performance confirms the effectiveness of bio-inspired morphological encoding for passive, energy-efficient grasping and secure holding. The findings presented in this chapter advance our understanding of the potential of bio-inspired multi-chain structures in robotic grasping, paving the way for the development of energy-efficient, adaptable gripping solutions for diverse applications.

J. Winand (✉) · S. Gorb · M. Jafarpour
Department of Functional Morphology and Biomechanics, Zoological Institute, Kiel University, Kiel, Germany
e-mail: jwinand@zoologie.uni-kiel.de

S. Gorb
e-mail: sgorb@zoologie.uni-kiel.de

M. Jafarpour
e-mail: mohsen.jafarpour@imtek.uni-freiburg.de

A. Raatz et al. (eds.), *Soft Material Robotic Systems*,
https://doi.org/10.1007/978-3-032-22453-8_19

1 Tritrap: A Soft Gripper Based on Insect Tarsal Chains

TriTrap is a biomimetic gripper inspired by the insect tarsal chain. Insects have a rigid exoskeleton that gains mobility through joints, tendons and muscles. The tarsus, or foot, lacks intrinsic muscles and is mainly moved through a long tendon by an external retractor muscle. Mimicking this biological system, the TriTrap gripper employs a set of underactuated digits that rely on morphological encoding and passive conformation resulting in a robust, versatile, and cost-effective design. The gripper's performance was evaluated using various everyday objects of different sizes, weights, and shapes. The TriTrap successfully lifted, rotated, transported, and released most of the objects. These results demonstrate the viability of the insect tarsal chain-inspired approach. Subsequently, interfacing modules inspired by insect adhesive structures were designed to enhance the performance significantly, benefiting from the additional compliance and/or enhanced adhesive/frictional properties.

1.1 Biological Inspiration

A wide range of solutions has been evolved for locomotion and object interaction/gripping in living organisms [1]. Insects, in particular, exhibit a wide array of such structures due to the specific properties of their exoskeleton, which offers significant strength combined with flexibility, enabling the evolution of numerous structural adaptations for gripping [2, 3]. In certain insect species, the typical locomotory legs are modified to carry out more specialized tasks. Examples include praying mantises using their forelegs to grasp prey [4], cicadas jumping with their hind legs [5], corbiculate bees utilizing complex collecting structures to gather pollen [6], webspinners spinning and maintaining silk with their forefeet [7], and mole crickets using shovel-like legs for digging [8], among others. Evolutionary pressures lead to solutions that combine an optimal balance between energy/material investment and functional benefits, supporting survival [9]. Although insects tend to have relatively short lifespans, they are highly adaptable to various environments, employing a simple and effective "digit paradigm" that operates efficiently in diverse conditions with minimal direct control [10]. These characteristics are primarily associated with the distal segments of insect legs, known as the tarsal chain, a highly underactuated system (see Fig. 1).

In robotics, ongoing research is often focused on improving the precision and adaptability of gripping systems, with many innovative and promising approaches emerging [12]. However, nearly all these methods rely on additional infrastructure (e.g., pumps, heating, magnetic field generation), sophisticated sensor arrays (e.g., spatial, optical, or force sensors), and/or complex control systems (e.g., adaptive movement programs, machine learning, artificial intelligence) [12]. Although these requirements can enhance gripping performance, they may lead to challenges

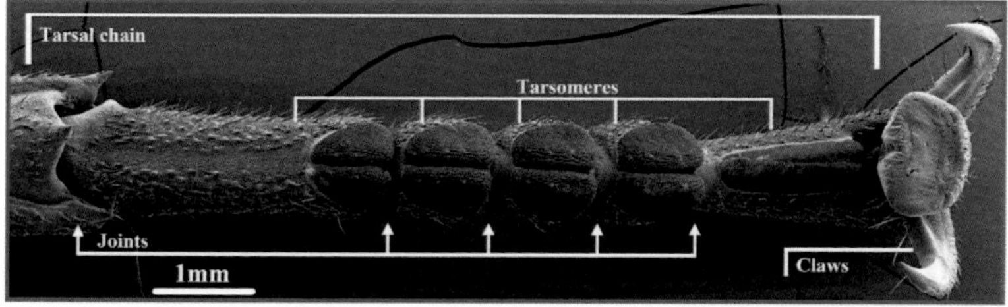

Fig. 1 SEM micrograph of the tarsal chain on one of the legs in the stick insect *Sungaya aeta*. It consists of several, loosely linked elements, called tarsomeres, and is actuated only by a single tendon. This results in a large number of passive degrees of freedom. From [11], licensed under CC-BY 4.0

in certain situations. Environmental factors, such as extreme humidity, temperature variations, or chemical hazards can negatively impact the components of these complex systems. The additional space required by advanced components might be problematic. Moreover, increased complexity often leads to a higher likelihood of malfunctions, requiring more maintenance and greater financial investment. In contrast, the insect tarsal chain operates with a relatively simple internal structure, requiring minimal construction and maintenance when adapted to technical applications [13]. The tarsal chain, despite its simplicity, offers a surprising range of applications, as demonstrated in nature [1–8], and functions with minimal control. In this study, an artificial gripper end effector based on the tarsal chain principle was designed and constructed to leverage these advantages.

As illustrated in Fig. 1, the biological tarsal chain is composed of several loosely connected segments, called tarsomeres [14]. The actuation of the chain is driven by a tendon running along its entire length, offset laterally from the pivot points between the tarsomeres. When the muscle pulls on the tendon, the force applied is off-center relative to the joints' neutral positions, generating a torque toward the tendon side of the chain. Figure 2 illustrates the generalized force interactions in this scenario, using a representative pair of isolated elements connected by a joint. It outlines the key relationships that must be taken into account when designing an artificial folding structure inspired by the tarsal chain. In the schematic, the pulling force along the tendon (F_t) generates a force (F_{Ax}) at the tip of the upper element, which in turn causes a folding motion at the joint. The magnitude of this force can be determined from the total reaction forces resulting from the tendon pull.

1.2 The TriTrap Gripper

The TriTrap gripper prototype, shown in Fig. 3, consists of three parts: (1) the gripper arms for gripping, (2) the actuation cage housing the pulling handle and force distribution disc, and (3) the base plate supporting both the arms and cage. Each arm has

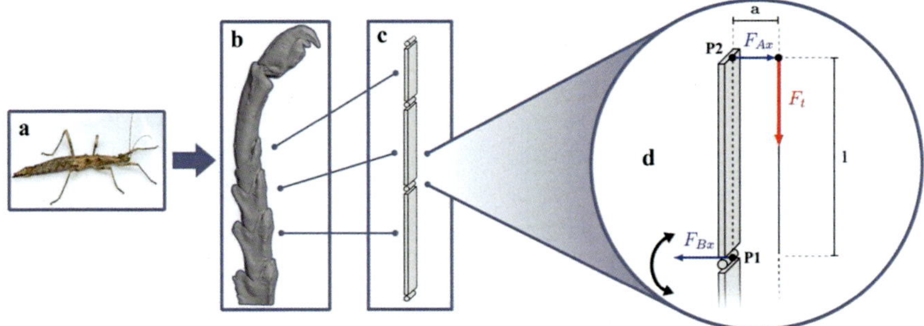

Fig. 2 The generalized working principle of tarsal chain actuation. **a** Model insect species previously studied, *Sungaya aeta* (licensed under Creative Commons (CC BY 4.0)). **b** 3D model of its tarsus created using micro-computed tomography. **c** Simplified mechanical representation as beams linked by joints. **d** Schematic showing the forces at play during the actuation of individual links, using a tarsomere-tarsomere joint as an example. From [11], licensed under CC-BY 4.0

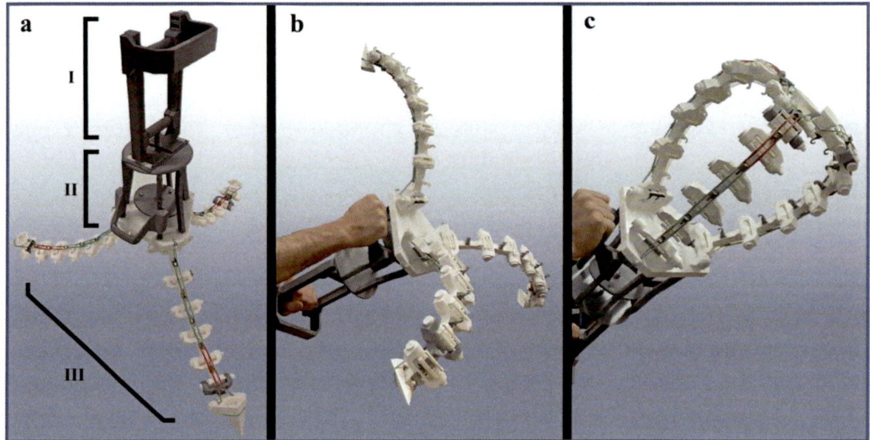

Fig. 3 Overview of the TriTrap gripper. **a** The gripper in its open state, displaying the handle (1), actuation cage (2), and arms (3). **b** The gripper held in hand while fully unflexed. **c** The gripper fully flexed, actuated by hand. From [11], licensed under CC-BY 4.0

seven linked segments, or "links," with the topmost link ending in a pointed wedge resembling an insect leg's claw. The links mimic tarsomeres, and rubber bands act as elastic energy storage units for unfolding, similar to resilin patches on insect legs [15]. The arms' anchor joints are arranged in a triangle on the base plate, converging when fully flexed. The links' lengths decrease from bottom to top, and folding angles are adjusted to imitate natural tarsus-tarsus bending. The arms move in an "embracing" motion, meeting at the base plate's center. The gripper is actuated by pulling a single handle connected to the arms via a force distribution disc, which splits the actuation force across three tendons (Fig. 4).

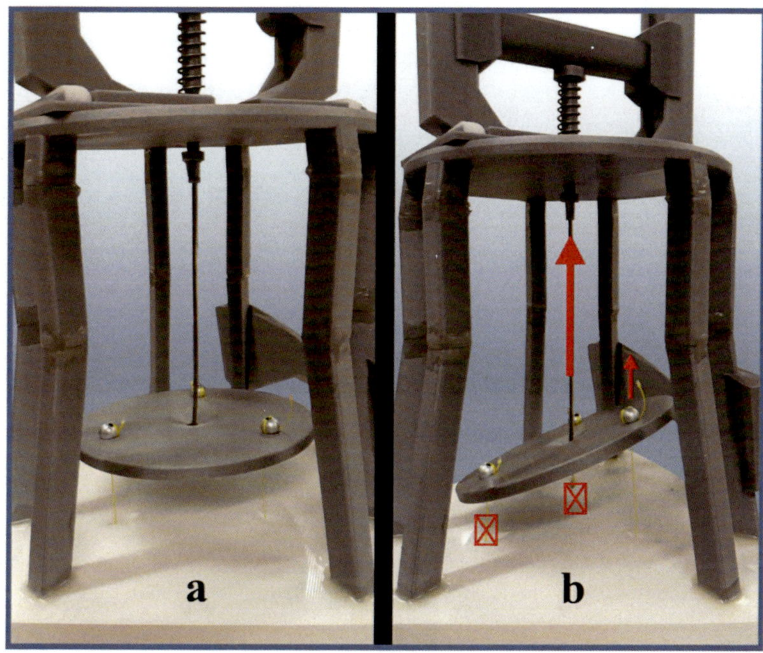

Fig. 4 The TriTrap gripper's actuation cage and its key components, including the tendon and preload spring, handle assembly, and force distribution disc. **a** The disc is in equilibrium, meaning all arms either require the same force to flex or are at rest. **b** The disc is tilted as the main actuation cord is pulled (large arrow). Two of the three arms encounter greater resistance to folding (marked with crossed boxes) or may be jammed due to the object's shape, while the remaining arm is free to continue folding (small arrow). From [11], licensed under CC-BY 4.0

The force distribution disc sits beneath the base plate, within the actuation cage. This structure serves two main functions: housing the force distribution disc and providing an anchor point for the actuation handle. The force distribution disc is a simple 3D-printed component with four holes—one at the center and three positioned equidistantly around it, forming an equilateral triangle (Fig. 4, a). The outer holes serve as attachment points for the tendons controlling the arms, while the central hole secures the handle cord. When the handle is pulled, the cord applies a pulling force at the center of the disc, causing it to move away from the base plate. As a result, the attached arm tendons retract, actuating the arms simultaneously. The disc serves two key purposes. First, it evenly distributes the pulling force among the three arms, ensuring they flex in simultaneously, provided they encounter equal resistance. Second, if the arms experience different levels of resistance—such as when gripping an asymmetrically shaped object—the disc tilts, allowing the free-moving arms to close independently of the obstructed one (Fig. 4, b). This design enables a single linear actuation mechanism while maintaining the gripper's adaptability to various object shapes.

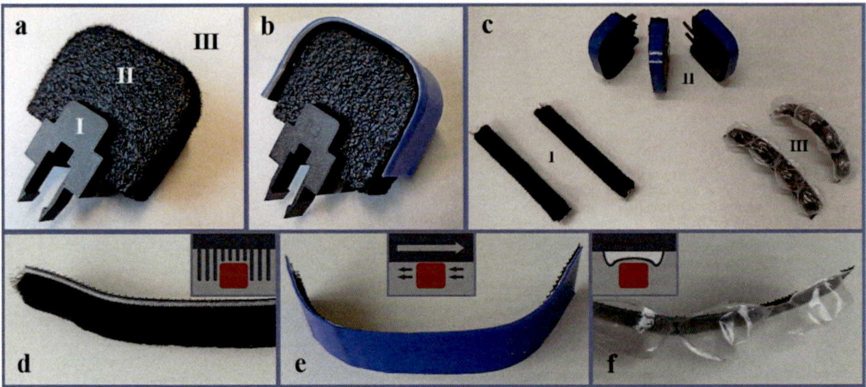

Fig. 5 Interfacing modules created for the TriTrap gripper. **a** snap-on fixture to attach to the TriTrap gripper arms, consisting of the PLA base (1), U-shaped distance pieces made from packaging material (2), and a self-adhering Velcro strip that serves as the attachment point for the interfacing tapes (3). **b** A snap-on fixture fitted with an interface tape (friction tape shown). **c** Hairy tape (1), friction tape (2), and bubble tape (3). The friction tape is already attached to a snap-on fixture. **d–f** Close up images showing the three different tapes (hairy, friction, air bubbles). The insets illustrate their main working principle (interlocking by hairs, friction, compliance). From [16], licensed under CC-BY 4.0

1.3 Interfacing Modules

In order to enhance the TriTrap gripper's compliance, biologically inspired interfacing modules were designed (Fig. 5) [16]. They are inspired by the attachment structures of phasmids (stick insects), and are modeled loosely after their contact-improving pads. Three types of interfacing modules were designed: (1) modules with hairy pads, (2) modules with high-friction tape, and (3) modules with soft bubble-like pads. These interfacing modules were constructed to be easily attached and detached from the TriTrap gripper.

1.4 Gripping Tests and Evaluation

Gripping tests were conducted for two sets of objects. For the first set (Fig. 6, a), a test was performed using just the bare TriTrap gripper without any interfacing modules. It was attempted to pick up, carry, and place back down the objects shown. The TriTrap gripper was deemed viable for an object, if at least one such attempt was successful (Fig. 7).

The TriTrap gripper was able to successfully transport each tested object, though multiple attempts were sometimes required. This test was meant to support the viability of the tarsal chain base concept and to highlight necessary improvements before moving to more quantitative testing. After the interfacing modules had been

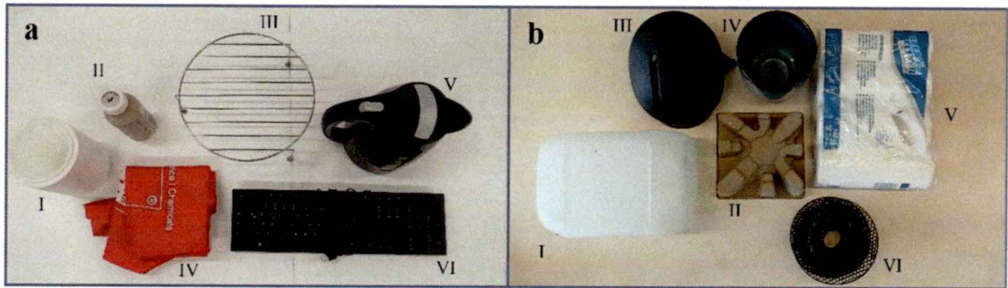

Fig. 6 Test objects used for testing of the TriTrap gripper. **a** First set [11]. 1: Polystyrene vase (20 g). 2: Empty beverage bottle made of plastic (30 g). 3: Steel grill attachment (160 g). 4: Textile bag (40 g). 5: Electric water boiler (600 g). 6: PC keyboard (400 g). **b** New, additional test objects. 1: plastic canister (515 g). 2: cardboard profile (140 g). 3: Bicycle helmet (400 g). 4: Ceramic vase (1500 g). 5: Package of kitchen paper (750 g). 6: Semi-depleted spool of 3D printing filament (285 g). From [16], licensed under CC-BY 4.0

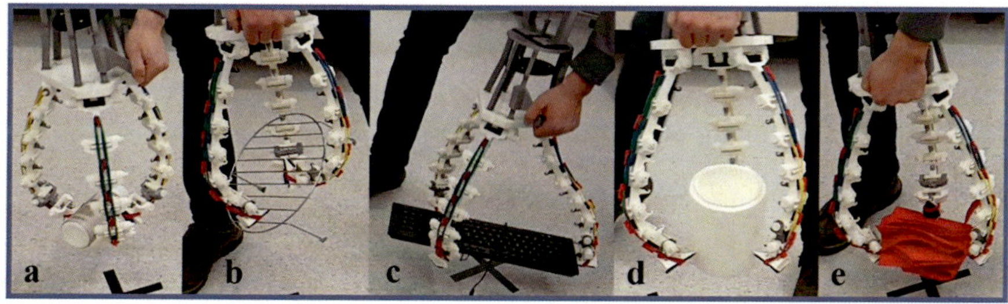

Fig. 7 Snapshots of the TriTrap gripper gripping and transporting objects. **a** Plastic bottle. **b** Grill attachment. **c** Keyboard. **d** Polystyrene vase. **e** Textile bag. From [11], licensed under CC-BY 4.0

designed, a second test was conducted with the same objects using the modules equipped. Each module was able to grip each tested object securely. The two exceptions to this were the combination of the water boiler with hairy and bubble modules, which failed in all cases. Overall, it was shown that the use of interfacing modules drastically improved the TriTrap gripper's performance over its prior capabilities because almost all items could be picked up and held at the first attempt.

Subsequently, a second set of items (Fig. 6, b) was used to test the TriTrap gripper and its interfacing modules quantitatively. Here, each item was attempted to pick up 30 times, with an attempt deemed successful, if the gripper was able to securely hold the item in the air for at least 3 s. The success ratio out of 30 attempts was recorded and is shown in Table 1.

It can be seen that the friction tape was superior in most cases, with the exception of the hairy tape/spool combination, in which the hairy tape slightly outperformed the friction tape. The bubble tape showed itself to be the second-best option out of the three, with the hairy tape coming in last. Overall, the TriTrap gripper successfully proved the viability of the tarsal chain base principle applied to gripping tasks. When

Table 1 Relative gripping success per interface module and object

Module/ Object	Paper package (%)	Cardboard (%)	Spool (%)	Canister (%)	Vase (%)	Helmet (%)
Friction tape	100	100	96.6	96.6	100	100
Bubble tape	100	100	93.0	63.0	100	33.0
Hairy tape	30.0	100	100	26.7	0	0

Values refer to successful attempts out of 30 performed in total per interface module and object [16]

equipped with interfacing modules, such as the friction tape, the gripper demonstrated remarkable performance across objects of varying shape, weight and aspect ratios.

2 Compliant Double-Spirals in the Design of Monolithic Soft Grippers

Following the development of the tarsus-inspired TriTrap gripper, which relies on the rotation of rigid segments interconnected by monoaxial joints [11], compliant double-spirals were used to introduce spiral-based soft grippers with monolithic designs [17]. These grippers aimed to function by leveraging the mechanical compliance of their constituent double-spirals, allowing for adaptive performance without the need for discrete joints. This approach aimed to take one more step towards the world of soft robots, where functionality is achieved through material deformation rather than relying on rigid mechanisms [18, 19].

2.1 Compliant Double-Spirals

Double-spirals are mechanical elements inspired by natural spirals, used in various technical applications ranging from rotational springs and joints [20–24] to stretchable electronics [25, 26] and metastructures [27–29]. Depending on their intended function, double-spirals can have specific geometries and material compositions. Nevertheless, they all feature two free ends, allowing them to serve as joints or modules capable of connecting to other double-spirals and components (Fig. 8) [20].

The planar design and fixed cross-sectional profile of double-spirals facilitate their design, manufacturing, and integration into different mechanical systems. This design results in simple scalability and compatibility with a wide range of manufacturing techniques and materials [30]. Double-spirals fabricated using a single-nozzle fused deposition modeling (FDM) 3D printer and thermoplastic polyurethane (TPU)

Fig. 8 Examples of double-spiral designs with varying geometries. While double-spirals can have different geometries depending on their intended function, all feature two free ends that can be used for connection to other components

filament have been presented as compliant joints with high potential for use in technical applications [17, 23]. TPU, as a material with high elasticity and durability, enables double-spirals to deform in multiple directions without permanent structural changes under repetitive loads, supporting long-term mechanical stability [31–33].

Compliant double-spirals are designed to undergo pre-programmed, reversible deformations in response to mechanical loads, aiming to achieve specific functions. Consisting of well-known spiral curves, the geometry of double-spirals is governed by mathematical equations and can be easily modified using only a few design variables [34]. Properties such as variable stiffness, compressibility, extensibility, and degrees of freedom can be tuned by adjusting the number, thickness, density, and curvature of the coils within a double-spiral [23, 30]. While the mechanical behavior of a single double-spiral is governed by its geometry, arranging multiple rationally designed double-spirals in specific configurations can lead to modular metastructures with a broader range of highly tunable mechanical characteristics, including anisotropic behavior, asymmetric deformation, pre-programmable shape changes, and spatial heterogeneity [28].

All these features make compliant double-spirals valuable mechanical elements for developing simple yet efficient adaptive structures, capable of performing complex tasks through their pre-programmed reversible deformation. Building on the authors' recent research [17], in which various spiral-based structures were developed as examples of engineered designs harnessing the mechanical compliance of their constituent double-spirals to achieve desired functionalities, two gripper prototypes are presented in the following section.

2.2 *Gripper Prototypes*

Freeform Passive Gripper. The first gripper was developed by connecting a series of double-spirals with similar geometries to form a continuous loop (Fig. 9, a).

This design takes advantage of the high rotational deformability of double-spirals, allowing the structure to adapt passively to various objects and surfaces. Each spiral deforms locally, enabling the entire loop to conform to curved or irregular shapes with minimal external force.

To highlight this structural adaptability, a second loop of the same diameter and thickness (1 mm) but without double-spirals was fabricated. The two structures were tested on five substrates with distinct curvatures (Fig. 9, b). When placed on the

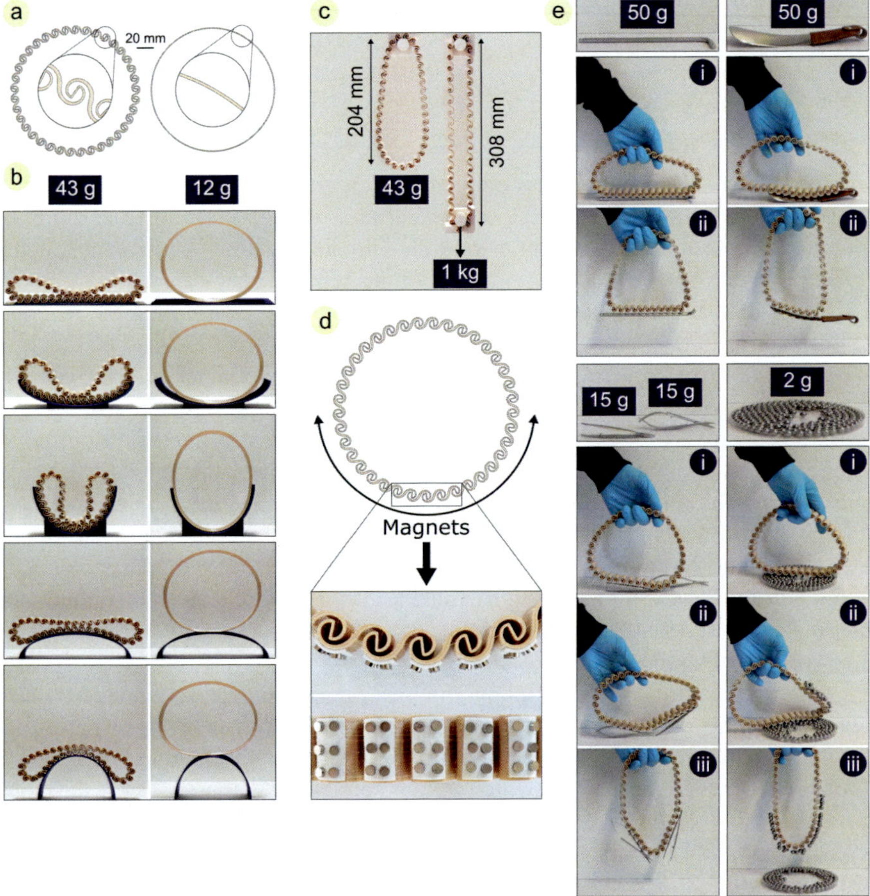

Fig. 9 Freeform passive gripper. **a** Models of double-spiral and circular loops, illustrating their structural differences. **b** Interaction of the double-spiral and circular loops with flat, concave, and convex surfaces, demonstrating deformation under their own weight. **c** Elongation of the double-spiral loop caused by its own weight and an additional 1 kg load. **d** Small disc magnets attached to the lower half of the double-spiral loop using double-sided tape, with close-up views from the top and side perspectives. **e** The spiral-based magnetic gripper used to lift a 50 g Allen wrench, a 50 g shoehorn, two 15 g tweezers, and 2 g steel beads. Reproduced from [17], licensed under CC-BY 4.0

substrates and released, the double-spiral loop exhibited higher compliance and conformed more closely to the surfaces compared to the circular loop, demonstrating the contribution of spiral deformability to overall adaptability.

While the double-spiral loop remains highly compliant under bending and compression, it shows increased stiffness under tensile loading (Fig. 9, c). This characteristic allows the structure to adapt to objects of various shapes while withstanding the forces needed to hold them, making it well-suited for gripping applications. The combination of chain-like deformability and increased tensile stiffness reflects principles observed in biological systems, such as insect tarsi [11]. This balance of flexibility and stiffness is essential for effective attachment and load-bearing in both natural and engineered systems.

To validate the practical performance of this design, the loop was transformed into a magnetic gripper by attaching small neodymium magnets along the outer surface of the lower half, using double-sided tape (Fig. 9, d). This addition did not affect the overall shape or stiffness of the loop. The freeform passive gripper was tested by lifting five objects of various shapes and masses (Fig. 9, e). It successfully conformed to all objects and lifted them, demonstrating its ability to achieve its intended purpose effectively.

Highly Extensible Enveloping Gripper. The second gripper prototype was developed by connecting three pairs of symmetric double-spirals in the form of a triangular structure, creating a three-jaw gripper capable of enveloping objects (Fig. 10, a). This design relies on (1) the high extensibility of double-spirals to significantly increase the area enclosed by the gripper, and (2) their structural compliance to conform to various shapes.

High reversible extensibility of the double-spirals allows the gripper to expand significantly when uncoiled, increasing the enclosed area by up to 100 times (Fig. 10, b). At rest, the triangular area inside the gripper has a side length of 30 mm, which expands to 150 mm upon uncoiling, maximizing the gripping area. After release, the double-spirals return to their original coiled state, demonstrating reversible deformability. The compliance of the double-spirals allows the gripper to adapt to objects of various shapes, generating the normal force required to ensure a stable grip through static friction.

To evaluate the performance of the gripper, tensile tests were conducted on 3D-printed double-spirals (Fig. 10, c). One end of each double-spiral was fixed, and displacement was applied to the other, extending the double-spirals to varying lengths and holding them in place to simulate the gripping function. This process provided insight into the mechanical behavior of the double-spirals in response to repeated loading and unloading. The tests showed that while the double-spirals initially softened during the first few cycles, their performance stabilized in the following cycles. This behavior, known as the Mullins effect [35, 36], indicates that the gripper can adapt to different object sizes while continuing to function reliably over time. The results also demonstrated that the force required to keep the double-spirals extended gradually stabilized during the holding phase of each cycle, confirming their ability

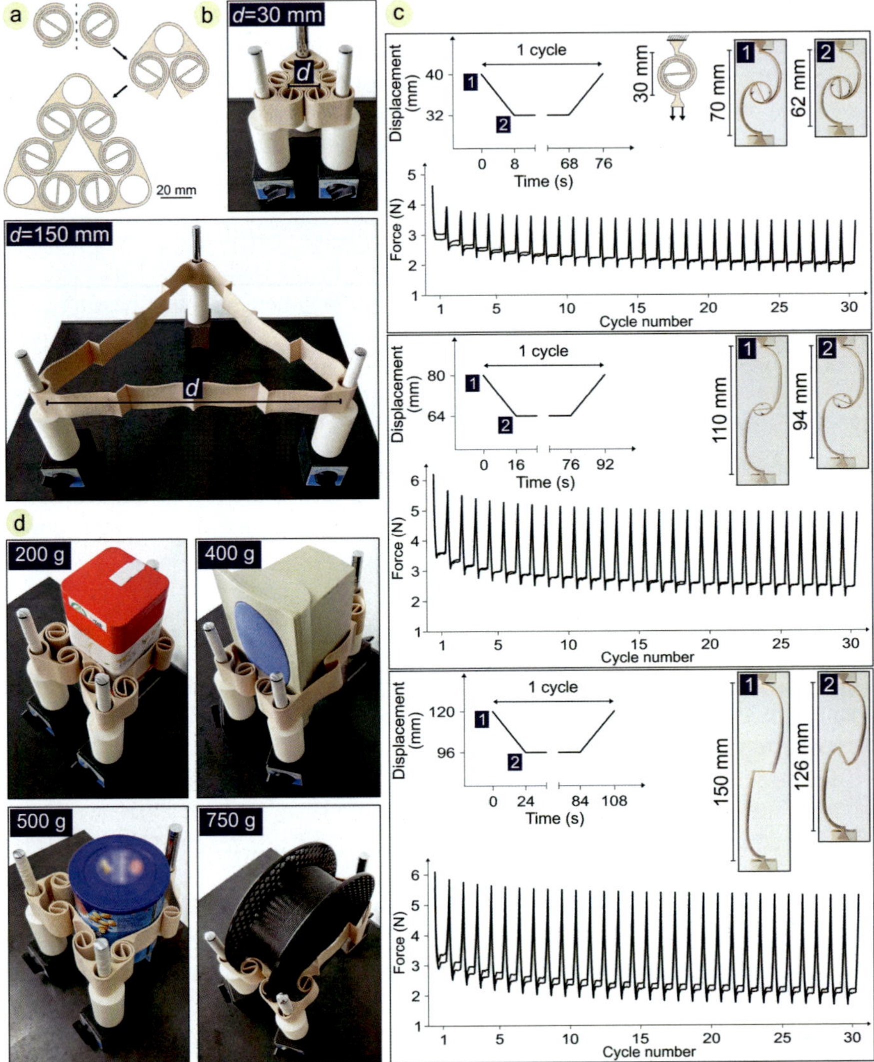

Fig. 10 Highly extensible enveloping gripper. **a** Model of the three-jaw gripper consisting of three pairs of symmetric double-spirals. **b** The fabricated gripper illustrated in its resting and fully extended states. **c** Mechanical characterization of two 3D-printed double-spirals under cyclic loading, reflecting their performance within the gripper. **d** The spiral-based gripper enveloping and holding a 200 g cubic can, a 400 g loudspeaker, a 500 g cylindrical can, and a 750 g filament spool. Reproduced from [17], licensed under CC-BY 4.0

to generate gripping force through static friction. Over multiple cycles, the double-spirals were repeatedly extended, held, and returned to their original state without significant degradation.

The Highly Extensible Enveloping Gripper was tested by lifting objects of various shapes and masses (Fig. 10, d). The extended double-spirals conformed to the surfaces of the objects, held them securely through friction, and maintained a reliable grip. These results show that the double-spiral design provides both adaptability and stability in gripping applications.

3 Summary and Outlook

In this chapter, multiple soft gripping devices were designed and fabricated. In the first part of the chapter, the TriTrap gripper, a gripping device based on insect tarsal chains, was presented. It showcased the potential for passive conformability and morphological computation inherent to insect tarsal chains and was successfully used to pick and place different everyday objects. It furthermore was enhanced with interfacing modules to increase its object contacting, all of which were likewise based on insect attachment modules. The results show that the TriTrap gripper is a viable gripping device that requires very little in terms of active control or additional infrastructure and is simple to manufacture. This way, it represents a very economic gripping principle. However, the discrete nature of its segment–joint–segment structure limits its feature resolution in terms of compliance with target objects. This encouraged the development of a continuous-stiffness alternative.

In the second part, two spiral-based grippers were presented as compliant, monolithic structures capable of passive adaptability through their reversible deformation. Without including discrete joints in their design, these grippers demonstrated the ability to conform to objects of varying shapes and sizes, eliminating the need for external actuators and reducing energy consumption. In contrast, many existing soft grippers rely on active actuation systems, such as vacuum, hydraulic, or electrically driven mechanisms, to perform tasks like tuning stiffness, conforming to surfaces, and holding objects [37–40].

Although the two designs presented here differ in their configuration, both rely on the similar principles of compliance and passive adaptation inspired by insect tarsus. One prototype emphasized surface conformability, while the other combined high reversible extensibility with frictional interaction to envelop and securely hold various objects. Together, they demonstrated the versatility of double-spirals as functional mechanical elements in soft robotics.

The ongoing research is focused on optimizing the performance of spiral-based grippers by exploring the mechanical characteristics of double-spirals with alternative geometries and material compositions, aiming to enhance their functionality and durability. Current efforts also include the development of new gripper designs, such as tendon-driven configurations, that result in grippers with greater efficiency, versatility, and improved gripping precision for more complex tasks. The integration of sensing and actuation elements remains an area of interest, bringing compliant structures closer to fully autonomous soft robotic systems. These advancements are expected to expand the application of multi-chain spiral-based grippers across various engineering fields.

Funding This research was funded by the German Research Foundation (DFG) under grant number 405032442.

References

1. Langowski, J.K.A., Sharma, P., Shoushtari, A.L.: In the soft grip of nature. Sci. Robot. **5**(49), (2020). https://doi.org/10.1126/scirobotics.abd9120 [PMID: 33328299]
2. Winand, J., Büscher, T.H., Gorb, S.N.: Learning from nature: a review on biological gripping principles and their application to robotics. In: Soft Robotics. Bentham Science Publishers, pp. 21–59 (2022)
3. Gorb, S.N.: Attachment Devices of Insect Cuticle. Kluwer Academic Publishers, Dordrecht, Boston (2001)
4. Resh, V.H., Cardé, R.T.: Encyclopedia of Insects, 2nd edn. Academic, Amsterdam (2009)
5. Gorb, S.N.: The jumping mechanism of cicada Cercopis vulnerata (Auchenorrhyncha, Cercopidae): skeleton-muscle organisation, frictional surfaces, and inverse-kinematic model of leg movements. Arthropod Struct. Dev. **33**(3): 201–220 (2004). https://doi.org/10.1016/j.asd.2004.05.008 [PMID: 18089035]
6. Martins AC, Melo GAR, Renner SS: The corbiculate bees arose from New World oil-collecting bees: implications for the origin of pollen baskets. Mol. Phylogenet. Evol. **80**, 88–94 (2014). https://doi.org/10.1016/j.ympev.2014.07.003 [PMID: 25034728]
7. Büsse S, Büscher TH, Kelly ET, Heepe L, Edgerly JS, Gorb SN: Pressure-induced silk spinning mechanism in webspinners (Insecta: Embioptera). Soft Matter **15**(47): 9742–9750 (2019). https://doi.org/10.1039/C9SM01782H [PMID: 31742303]
8. Zhang, Z., Zhang, Y., Zhang, J., Zhu, Y.: Structure, mechanics and material properties of claw cuticle from mole cricket Gryllotalpaorientalis. PLoS One. **14**(9), e0222116 (2019). https://doi.org/10.1371/journal.pone.0222116. [PMID: 31491009]
9. Bell, G.: Selection: the Mechanism of Evolution, 2nd edn. Oxford University Press Incorporated, Oxford (2008)
10. Gorb, S.N.: Design of insect unguitractor apparatus. J. Morphol. **230**(2), 219–230 (1996). https://doi.org/10.1002/(sici)1097-4687(199611)230:2<219::aid-jmor8>3.0.co;2-b
11. Winand, J., Büscher, T.H., Gorb, S.N.: TriTrap: a robotic gripper inspired by insect tarsal chains. Biomimetics. **9**(3), 142 (2024). https://doi.org/10.3390/biomimetics9030142. [PMID: 38534827]
12. Tai, K., El-Sayed, A.-R., Shahriari, M., Biglarbegian, M., Mahmud, S.: State of the art robotic grippers and applications. Robotics. **5**(2), 11 (2016). https://doi.org/10.3390/robotics5020011
13. Shintake, J., Cacucciolo, V., Floreano, D., Shea, H.: Soft robotic grippers. Adv. Mater. **30**(29), e1707035 (2018). https://doi.org/10.1002/adma.201707035. [PMID: 29736928]
14. Betz, O.: Structure of the tarsi in some Stenus species (Coleoptera, Staphylinidae): external morphology, ultrastructure, and tarsal secretion. J. Morphol. **255**(1), 24–43 (2003). https://doi.org/10.1002/jmor.10044. [PMID: 12420319]
15. Michels, J., Appel, E., Gorb, S.N.: Functional diversity of resilin in Arthropoda. Beilstein J. Nanotechnol. **7**(1), 1241–1259 (2016). https://doi.org/10.3762/bjnano.7.115. [PMID: 27826498]
16. Winand, J., Thies, H. B., Stanislav N. G.: Comparison of insect-inspired interfacing mechanics on a macroscopic scale. Submitt. Curr. Rev.

17. Jafarpour, M., Aryayi, M., Gorb, S.N., Rajabi, H.: Double-spiral as a bio-inspired functional element in engineering design. Sci. Rep. **14**(1), 29225 (2024) https://doi.org/10.1038/s41598-024-79630-6[PMID: 39587172]
18. Rus, D., Tolley, M.T.: Design, fabrication and control of soft robots. Nature. **521**(7553), 467–475 (2015). https://doi.org/10.1038/nature14543. [PMID: 26017446]
19. Trivedi, D., Rahn, C.D., Kier, W.M., Walker, I.D.: Soft robotics: biological inspiration, state of the art, and future research. Appl. Bionics Biomech. **5**(3), 99–117 (2008). https://doi.org/10.1155/2008/520417
20. Ahmad, B., Barbot, A., Ulliac, G., Bolopion, A.: Remotely actuated Optothermal robotic micro-joints based on spiral Bimaterial design. IEEE/ASME Trans. Mechatron. **27**(5), 4090–4100 (2022). https://doi.org/10.1109/tmech.2022.3145646
21. Zolfagharian, A., Gharaie, S., Gregory, J., Bodaghi, M., Kaynak, A., Nahavandi, S.: A bioinspired compliant 3D-printed soft gripper. Soft. Robotics. **9**(4), 680–689 (2022). https://doi.org/10.1089/soro.2020.0194. [PMID: 34297904]
22. Tian, Y., Zhou, C., Wang, F., Lu, K., Zhang, D.: A novel compliant mechanism based system to calibrate spring constant of AFM cantilevers. Sens. Actuators Phys. **309**, 112027 (2020). https://doi.org/10.1016/j.sna.2020.112027
23. Jafarpour, M., Gorb, S., Rajabi, H.: Double-spiral: a bioinspired pre-programmable compliant joint with multiple degrees of freedom. J. R. Soc. Interface. **20**(198), 20220757 (2023). https://doi.org/10.1098/rsif.2022.0757. [PMID: 36628530]
24. Georgiev, N., Burdick, J.: Design and analysis of planar rotary springs. In: Design and analysis of planar rotary springs, 4777–84. IEEE, (2017)
25. Yuan, X., Wang, Y.: Nonlinear stretching mechanics of planar Archimedean-spiral interconnects for flexible electronics. Thin-Walled Struct. **185**, 110568 (2023). https://doi.org/10.1016/j.tws.2023.110568
26. Rehman, M.U., Rojas, J.P.: Optimization of compound serpentine–spiral structure for ultra-stretchable electronics. Extrem. Mech. Lett. **15**, 44–50 (2017). https://doi.org/10.1016/j.eml.2017.05.004
27. Zhang, W., Neville, R., Zhang, D., Yuan, J., Scarpa, F., Lakes, R.: Bending of kerf chiral fractal lattice metamaterials. Compos. Struct. **318**, 117068 (2023). https://doi.org/10.1016/j.compstruct.2023.117068
28. Jafarpour, M., Gorb, S.N., Rajabi, H.: Double-spirals offer the development of Preprogrammable modular Metastructures. Adv. Eng. Mater. **25**(13), 2300102 (2023). https://doi.org/10.1002/adem.202300102
29. Zarrinmehr, S., Ettehad, M., Kalantar, N., Borhani, A., Sueda, S., Akleman, E.: Interlocked archimedean spirals for conversion of planar rigid panels into locally flexible panels with stiffness control. Comput. Graph. **66**, 93–102 (2017). https://doi.org/10.1016/j.cag.2017.05.010
30. Jafarpour, M., Gorb, S.N.: PLA double-spirals offering enhanced spatial extensibility. Macromol. Mater. Eng., 2400208 (2024). https://doi.org/10.1002/mame.202400208
31. Desai, S.M., Sonawane, R.Y., More, A.P.: Thermoplastic polyurethane for three-dimensional printing applications: a review. Polym. Adv. Technol. **34**(7), 2061–2082 (2023). https://doi.org/10.1002/pat.6041
32. Arifvianto, B., Iman, T.N., Prayoga, B.T., et al.: Tensile properties of the FFF-processed thermoplastic polyurethane (TPU) elastomer. Int. J. Adv. Manuf. Technol. **117**(5–6), 1709–1719 (2021). https://doi.org/10.1007/s00170-021-07712-0
33. Rigotti, D., Dorigato, A., Pegoretti, A.: Low-cycle fatigue behavior of flexible 3D printed thermoplastic polyurethane blends for thermal energy storage/release applications. J. Appl. Polym. Sci. **138**(3), 49704 (2021). https://doi.org/10.1002/app.49704
34. Tsuji, K., Müller, S.C. (eds.): Spirals and Vortices: in Culture, Nature, and Science. Springer International Publishing, Cham (2019)
35. Cho, H., Mayer, S., Pöselt, E., et al.: Deformation mechanisms of thermoplastic elastomers: stress-strain behavior and constitutive modeling. Polymer. **128**, 87–99 (2017). https://doi.org/10.1016/j.polymer.2017.08.065

36. Qi, H.J., Boyce, M.C.: Stress–strain behavior of thermoplastic polyurethanes. Mech. Mater. **37**(8), 817–839 (2005). https://doi.org/10.1016/j.mechmat.2004.08.001
37. Goh, G.D., Goh, G.L., Lyu, Z., et al.: 3D printing of robotic soft grippers: toward smart actuation and sensing. Adv. Mater. Technol. **7**(11), 2101672 (2022). https://doi.org/10.1002/admt.202101672
38. Qu, J., Yu, Z., Tang, W., Xu, Y., Mao, B., Zhou, K.: Advanced technologies and applications of robotic soft grippers. Adv. Mater. Technol. **9**(11), 2301004 (2024). https://doi.org/10.1002/admt.202301004
39. Zaidi, S., Maselli, M., Laschi, C., Cianchetti, M.: Actuation Technologies for Soft Robot Grippers and Manipulators: a review. Curr. Robot. Rep. **2**(3), 355–369 (2021). https://doi.org/10.1007/s43154-021-00054-5
40. Dou, W., Zhong, G., Cao, J., Shi, Z., Peng, B., Jiang, L.: Soft robotic manipulators: designs, actuation, stiffness tuning, and sensing. Adv. Mater. Technol. **6**(9), 2100018 (2021). https://doi.org/10.1002/admt.202100018

ASDDSA Active Suction Device for Deep Sea Application

Cora Maria Sourkounis, Jan Peters, Oliver Jahns, Tom Kwasnitschka, and Annika Raatz

Abstract Deep-sea research offers invaluable opportunities to uncover hidden ecosystems, unknown biodiversity, insights into Earth's history and climate change impacts. However, exploring these environments traditionally requires costly equipment designed to withstand extreme conditions. Our research addresses this challenge by developing a novel, lightweight suction sampling system, which is a widely used tool for sampling sediments in the deep sea. The system envisioned in our project consists of two primary components: a robotic manipulator for suction sampling, which facilitates precise steering of the sampling tube towards the sediment, and a sample storage system designed to hold the sediment containers. Both components of the suction sampling system depend on a carrier platform, which can be any type of remotely operated vehicle (ROV) rated for operation at depths of up to 6000 m. For the robotic manipulator, our designs take advantage of the benefits offered by soft material actuators and continuum robot concepts. We explored various designs, including soft actuators combined with rigid, bistable mechanisms; spring-loaded continuum robots; and soft actuation systems with stiffening mechanisms. All the designs share a common feature: they are intentionally designed to be modular, scalable, and cost-effective relative to existing solutions for deep-sea suction sampling. Throughout the course of the project, various designs were developed and rigorously

C. M. Sourkounis (✉) · J. Peters · A. Raatz
Institute of Assembly Technology and Robotics, Leibniz University Hannover, Garbsen, Germany
e-mail: sourkounis@match.uni-hannover.de

J. Peters
e-mail: peters@match.uni-hannover.de

A. Raatz
e-mail: raatz@match.uni-hannover.de

O. Jahns
Seagoing Technology Department, GEOMAR Helmholtz Centre for Ocean Research Kiel, Kiel, Germany
e-mail: ojahns@geomar.de

T. Kwasnitschka
RD2 - Deep Sea Monitoring, GEOMAR Helmholtz Centre for Ocean Research Kiel, Kiel, Germany
e-mail: tkwasnitschka@geomar.de

A. Raatz et al. (eds.), *Soft Material Robotic Systems*,
https://doi.org/10.1007/978-3-032-22453-8_20

tested. With its successful implementation, the envisioned system has the potential to greatly enhance deep-sea sampling and research, significantly broadening access and empowering a larger community of researchers to advance this important field .

1 Deep Sea Suction Sampling

The collection of deep-sea sediment is a crucial technique employed in seafloor exploration to gather important information on sedimentation and sediment transport processes [1]. Yet effort and expenses are considerable as the deep ocean is an extreme environment for robotics, characterized by high pressures up to 1100 atmospheres, low temperatures around 4 °C, corrosive seawater, and limited permeability for electromagnetic radiation, resulting in challenges for navigation and communication. Consequently, robotic probes must operate either tethered or autonomously, using limited acoustic communication.

Most scientific research is conducted at depths shallower than 6000 m, while the majority of commercial activities are confined to depths shallower than 4000 m, primarily within a few hundred meters on the continental shelves. However, from an operational standpoint, designing marine scientific equipment to function up to a depth of 6000 m is reasonable and sustainable. This capability allows access to the vast majority of oceanic environments, with the exception of the deepest trenches, which account for less than 1% of the Earth's surface [2]. Attempting to extend capabilities to 100% of marine environments, including these extreme depths, would disproportionately increase the technical complexity and costs involved in the systems development.

There are several types of robots specifically designed to reach the required depths for deep-sea research: Fully or semi-autonomous underwater vehicles (AUVs) are used for inspection, whereas tethered remotely operated vehicles (ROVs) with hydraulic manipulators handle tasks that require more force, like industrial applications. There is a distinction among ROV classes: working class for heavy-duty tasks, inspection class for lighter operations, and miniature ROVs for small-scale (shallow) observation tasks [3].

Currently, great depths of several thousand meters can only be reached by intervention class ROVs that can be equipped with a suction sampler. Thus they are the preferred option used for sampling sediments from deep-sea volcanoes. Traditional methods like dredging are often imprecise and inefficient considering that no spatial information on the sampling location is retained. Push core sampling with hydraulic ROV manipulators aims to obtain stratigraphically controlled samples but is only applicable in flat, soft terrain. Suction samplers, widely used in scientific underwater robotics, provide a solution for precise, localized sediment collection [4]. These devices consist of a large, rigid hose operated by a manipulator, directing samples through a series of interchangeable filters. By generating a vacuum, the sediment is transported through the sampling tube into the containers [5]. Utilizing an intervention class ROV, which can weight up to 5 tons [6], makes the suction sampling process very expensive as an appropriately big research vessel is necessary to deploy this type of ROV.

2 Goals

This interdisciplinary project aims to develop a suction sampling system that is more lightweight, cost-effective, and offers greater sample storage capacity than current systems. The project tasks are divided between two partners: the GEOMAR Helmholtz Centre for Ocean Research Kiel, and the Institute of Assembly Technology and Robotics (match) at Leibniz University Hannover.

Figure 1, illustrates the starting point of the ASDDSA project: the GEOMAR is responsible for providing the carrier platform, which will be outfitted with the new sampling system. To enhance the sampling capacity and prevent cross-contamination, the GEOMAR is designing a sample storage system that holds multiple sampling containers. Meanwhile, the match is tasked with developing the new robotic manipulator for the suction sampler. The envisioned actuation system must enclose the suction sampling tube to ensure precise sediment sampling. The new lightweight design will enhance portability and reduce the operational burden associated with deploying sampling tools. Initially, the intention was to adopt a classic soft robotic design featuring a multi-channel hydraulic silicon actuator.

Over the course of the project, and following initial testing, perspectives shifted slightly to consider new ideas and options for the overall system design. At GEOMAR, the Modular Mobility program introduced the concept of using smaller remotely operated vehicles (ROVs) that are easier to transport and specifically designed for suction sampling. This contrasts with the traditional working-class ROV, which, while versatile for a wide range of applications, becomes bulky and heavy due to its extensive equipment. The change in carrier platform options altered the requirements for the actuation system of the suction sampler. The new suction sampling system needed to be compatible with various sizes of host ROVs, thereby expanding deployment options and reducing the reliance on large, costly research vessels. Given the significance of the Modular Mobility program in shaping these requirements, it will be briefly introduced in the following sections. Subsequently, the developed sample storage system and the different concepts for the robotic manipulator for suction sampling will be presented.

3 Carrier Platform and Sample Storage System

This section introduces the new carrier platform and details the adapted sample storage system. Both designs are informed by GEOMAR's extensive expertise in engineering systems for the exploration of deep-sea volcanoes.

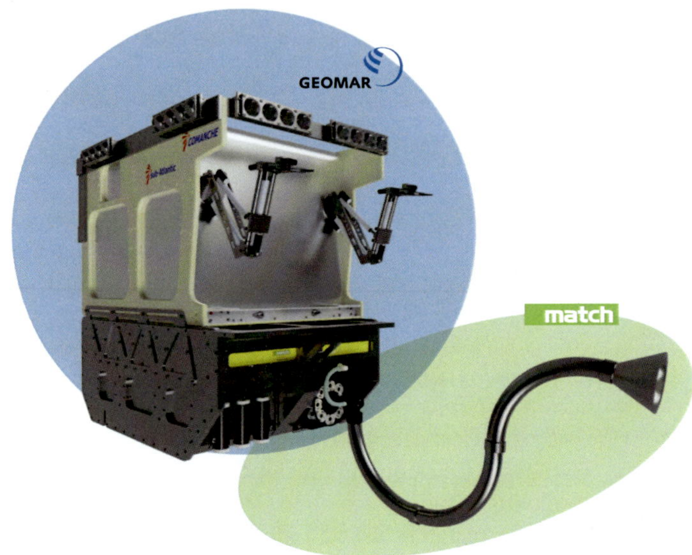

Fig. 1 The ASDDSA project aims to develop an innovative suction sampling system designed to reduce the costs associated with deep-sea sediment sampling. The system comprises two main components: a robotic manipulator, responsible for guiding the sampling tube, developed by the match team (highlighted in green), and a sample storage system, developed by GEOMAR. Additionally, GEOMAR will provide the carrier system (highlighted in blue), which houses the sample storage system and serves as the mounting platform for the robotic manipulator. In this early concept sketch, the envisioned carrier platform was the working-class ROV Kiel 6000 [7]

3.1 Modular Mobility: A Small-Scale Robotic Carrier Platform

Within the GEOMAR internal development program Modular Mobility, we currently develop a small to miniature ROV of less than 250 kg mass that will have a maximum operating depth of up to 6000 m. This is achieved by tethering the vehicle to a bigger towsled which is in turn connected to the host vessel. Thus, the ROV is only affected by the dynamics of a 30 m long neutrally buoyant tether. This platform is ideally suited for inexpensive access to actual, deep seafloor outcrops. At the same time, it provides an ideal showcase of the scaleable approach we take in the design of our actuated suction sampler technology that will not just serve big vehicles but especially small-scale platforms such as Modular Mobility (Fig. 2). The challenges for the ASDDSA system are to minimize power consumption, volume, mass, and buoyancy anomaly of the system, to keep the mode of operation as simple (and fault-tolerant) as possible, while, at the same time, maximizing the number of samples taken. Ultimately, a large number of samples translates to finer detail in the geochemical assessment of seafloor deposits, especially in the case of vertically stratified volcanic outcrops.

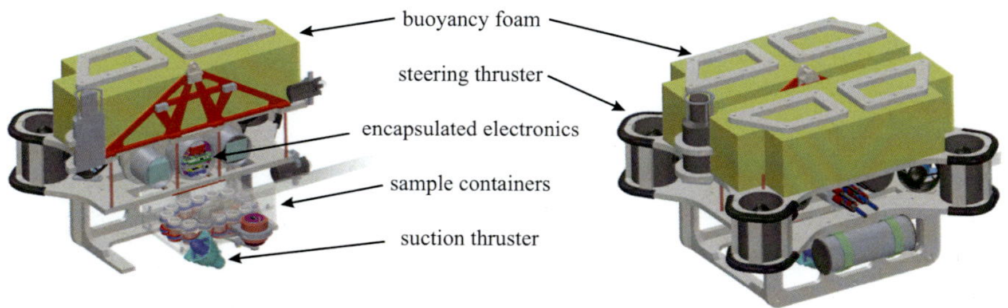

Fig. 2 Functional sketch of the Modular Mobility ROV prototype including the suction and sample storage system prototype underneath. The conduit is depicted schematically without the manipulator

3.2 Sample Storage System

The biggest challenge in the design of the entire conduit and sample container assembly is the minimization of the risk for sample cross contamination, meaning there must be no grooves or dead volumes in the conduit where sediment may get trapped. At their rear end in the direction of suction, the segments contain a sieve of various mesh size (1 mm down to 65 μm) to contain the sample while every other segment is left open without a sieve to facilitate flushing, but also to establish a steady flow of sediment through the conduit before the revolver is advanced towards the next sieve container. When not aligned with the conduit, the segments are hermetically sealed by a top and a bottom plate that also act as a railing for the chain drive. The actuation of the chain is realized by a single brushless DC motor which components have been encapsulated with polyurethane resin to withstand the open exposure to the ocean water. Both positioning, pump control and monitoring of the flushing and sampling process is realized using miniaturized deep-sea cameras, i.e. through personal, visual control. The motor control is relies on industrial subsea robotic components (e.g., Blue Robotics).

To generate the negative pressure necessary for the sampling system to collect sediments, we implement the suction pump using a standard ROV thruster, a common practice in the design of many suction sampler systems. The thruster is mounted at the end of the conduit assembly in a way that its exhaust acts as thrust towards the sample location, i.e. the vehicle is not deflected by the operation of the pump. Critically, this design allows for the reversal of the flow and thus the flushing with fresh water in between samples, to avoid cross contamination. The sampler itself is realized as a set of tube segments that are contained in a revolver chain assembly. By varying length and geometry of the chain drive assembly, the number of sample containers is easily scalable and adapts to a wide range of vehicle geometries (Figs. 3 and 4). The dimensions of 40 mm diameter by 60 mm length for the individual segments are the result of an iterative practical experimentation campaign in which we attempted to contain a volcaniclastic sediment sample suspended in fresh water using a series of 3D printed sample container prototypes.

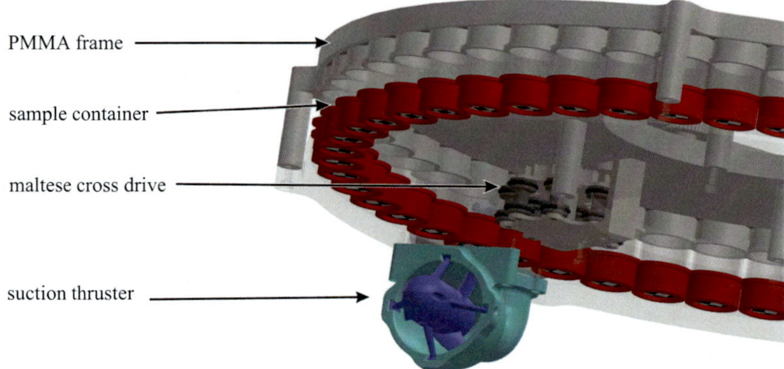

Fig. 3 An early design for integration into a working-class ROV featured a carousel revolver approach, where individual sample containers were advanced using a discrete maltese cross drive, inspired by a design from the Monterey Bay Aquarium Research Institute [8]. The drawback of this design is its large diameter, requiring significant space on the host vehicle

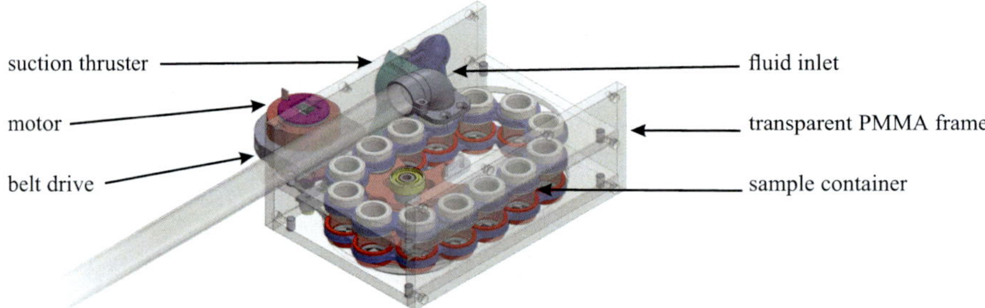

Fig. 4 Schematic close-up of the revolver chain drive sample storage system, designed for integration into the Modular Mobility carrier platform, including the suction sampler pump and drive motor assemblies

4 Robotic Manipulator for Suction Sampling

The design process of the manipulator for suction sampling began with the definition of specific requirements. Following this, an extensive investigation into the current state of the art regarding deep-sea soft and continuum robots was conducted. Based on the findings from this research, multiple design concepts were subsequently developed.

4.1 Requirements

The requirements for the actuation system were developed in collaboration with expert users of the existing suction sampling system, currently integrated into the Kiel 6000 equipment at the GEOMAR research center. The following section is focused specifically on the mechanical design requirements, aligned with the current status of research: Firstly, the system must be capable of operating at a depth of 6000 m, necessitating a pressure-neutral design. Achieving this involves selecting appropriate materials and manufacturing techniques to eliminate air pockets within components. The system components must be adaptable in terms of dimensions, configuration, and fabrication processes. External elements, such as motors, should be selected from a range of scalable products to ensure compatibility with various host vehicles. The design's modularity must also facilitate easy access to components susceptible to breakage. The entire system should be designed to achieve near-neutral buoyancy, thus excluding the use of metal parts. To prevent environmental pollution, alternative actuation approaches should be explored in place of traditional hydraulic systems. The final design of the actuation system must match the length of the current sampling tube, approximately 2–3 m. It should also allow for sediment sampling while the host ROV hovers roughly 1 m above the seafloor to minimize sediment disturbance. Furthermore, the workspace must be situated within the camera's field of view, positioned at the top of the ROV, to ensure effective operation.

4.2 State of the Art: Underwater Continuum and Soft Robots

A similar concept of integrating soft materials or continuous robots with a deep-sea remotely operated vehicle (ROV), illustrated in Fig. 1, was initially proposed by Davies et al. in 1998 [9]. The authors envisioned attaching a soft robot akin to an elephant's trunk to a deep-sea ROV and equipping it with a camera to facilitate exploration tasks requiring high dexterity. Even though this concept was not followed up on by the authors since then, several other underwater robots made from soft or flexible materials have since emerged: Phillips et al. introduced a soft robotic arm specifically designed for handling delicate samples in deep-sea environments [10]. Li et al. developed an untethered soft robot inspired by a snailfish, capable of withstanding the pressures of the Mariana Trench, marking a significant advancement in deep-sea exploration technology [11]. The Eelume robot has been tested at depths of up to 150 m. Although developers are working to improve its pressure resistance for potential deep-sea deployment, its thruster system currently limits its ability to effectively collect sediment samples [12]. Xue et al. presented a tethered, snake-like manipulator designed for deep-sea applications. This system consists of seven joints, each providing two degrees of freedom, granting the manipulator a high degree of dexterity. However, the requirement of 14 motor systems results in a substantial size for the actuation system [13].

Despite the advancements in underwater soft robots and snake-like robots, there remains, to our knowledge, no deep-sea suitable design specifically intended for use as a manipulator for suction sampling. Current approaches predominantly focus on grippers or untethered biomimetic robots, which often lack sufficient pressure resistance to function effectively at depths beyond a few meters [14, 15].

4.3 Concepts

To meet the outlined requirements, a diverse array of concepts was derived from existing literature and constructed for testing.

These structures range from soft material actuators and compliant continuous structures to rigid hyper-redundant systems. Each approach offers unique characteristics and potential advantages for specialized applications. Figure 5 illustrates the various concepts and their current stages of development. Each concept begins as a proof of concept. If successfully implemented, it progresses to a first prototype and, with further development, aims to become a minimally viable product in the near future. The different concepts will be described in more detail in the following section.

4.3.1 Single Channel Soft Robotic Actuator

This prototype is a single-channel soft robotic actuator designed to bend in six directions using a low melting point alloy (LMPA) for switchable strain-limiting structures. It employs a single valve to actuate six hydraulic chambers distributed around the circumference of the cylindrical silicone body. Three LMPA chambers allow the actuator to lock in its bent state, enabling energy-efficient operation during deep-sea suction sampling. The major advantage of using LMPA in water is its rapid cooling capability in cold environments [16] (Fig. 6).

In the future, we plan to scale this proof of concept into a full-scale prototype. To achieve this, adjustments will be made to the shape of the LMPA chambers. A potential concern with the current design is that it might be too heavy to develop into a full 2-meter-long actuation system. Alternatively, it may serve effectively as a dexterous tip for the actuation system.

4.3.2 Bistable Mechanism with Torsional Origami-Inspired Actuators

This prototype addresses the concern that soft actuators alone are not sturdy enough to form a long actuation system. Therefore, we paired soft actuators with a supporting skeleton. The prototype consists of a simple bistable mechanism integrated into the outer rigid shell. Each module includes three actuators arranged in a circular formation and is designed for hydraulic actuation.

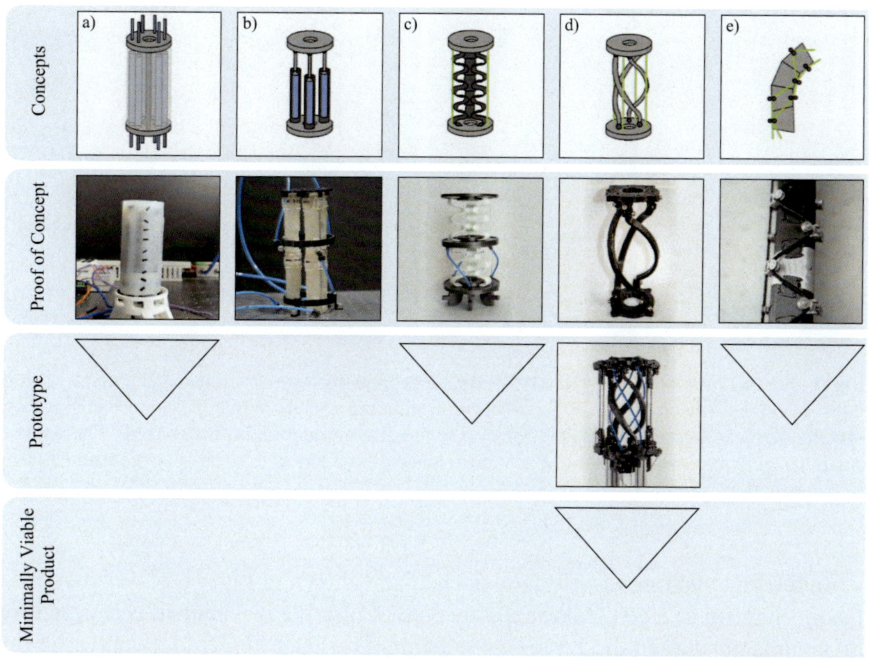

Fig. 5 For the various prototypes, two types of actuation concepts were considered: seawater hydraulic and tendon-driven designs. The structures and materials designed range from classic soft actuators to compliant structures, and even completely rigid structures. Each concept was developed into a proof of concept. If a concept proved feasible during this initial stage, it will now be developed into a prototype. Concept **a** is derived from the classic silicone soft actuator and incorporates additional strain-limiting channels filled with low melting point alloys (LMPA). In concept **b**, a hybrid design is presented that combines origami-inspired soft actuators with a bistable mechanism. Concept **c** resembles a spring-loaded continuum robot featuring fiber-reinforced polymer springs. The joint described in **d** is the only discrete joint developed and retains the damping characteristics of a virtual rolling contact. Concept **e** is entirely rigid, composed of rotating angled cylinders, and the rotation of these cylinders induces bending in the overall structure

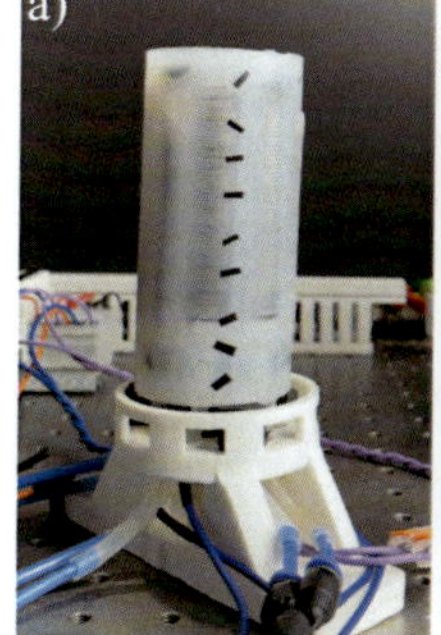

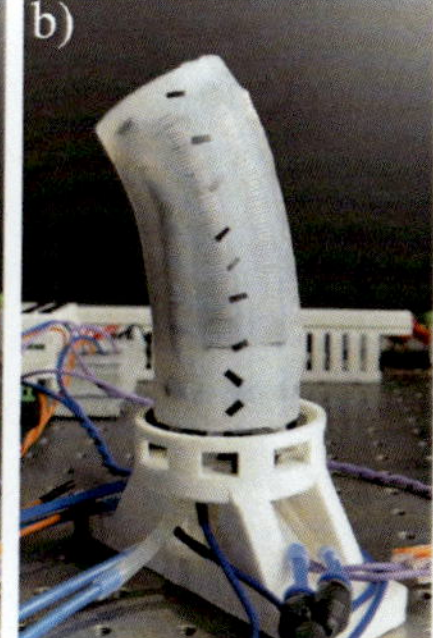

Fig. 6 The single hydraulic channel in this prototype reduces the number of necessary valves. Bending is achieved by using chambers filled with low melting point alloy (LMPA). When the LMPA is melted, the actuator section can stretch, allowing it to bend in the opposite direction of the chamber when the pressure in the hydraulic chambers is increased

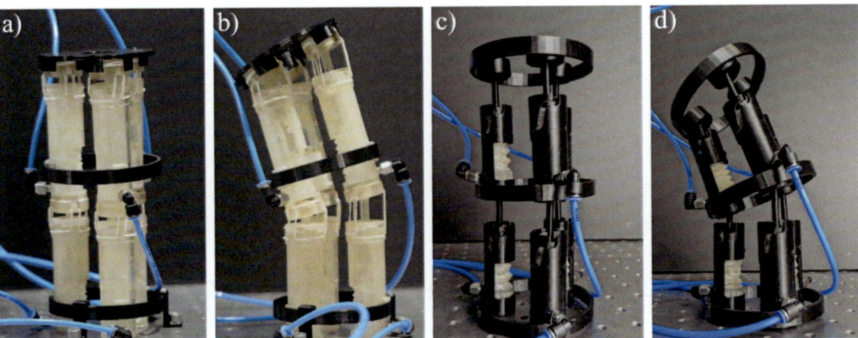

Fig. 7 For this prototype, two types of bistable mechanisms were developed and tested by combining soft, origami-inspired actuators with a rigid, stabilizing shell. While these mechanisms proved functional, manufacturing considerations led to the discontinuation of this concept. The numerous small parts involved in the bistable mechanism made it impractical to scale up to the intended length of the actuation system [7, 17]

In the first version, the bistable mechanism, shown in Fig. 7a, b, was inspired by the mechanism of a pen. Although this type of bistable mechanism performed well, its manufacturing required a high-precision 3D printer [7]. Additionally, the force necessary to activate the bistable mechanism appeared to cause long-term damage to the linear soft actuators.

For the second version, depicted in Fig. 7b, c, we simplified the bistable mechanism to reduce manufacturing costs. It consists of a groove within the outer rigid shell. Inside the groove, a pin is manipulated by the torsional origami actuator to reach either the upper or lower stable position [17].

Overall, this concept proved to be too susceptible to failure due to dirt or imprecise manufacturing. Additionally, assembly was a concern because even the simpler bistable mechanisms consisted of multiple parts that required assembly. For these reasons the concept will not be followed up on.

4.3.3 Tendon-Driven Continuum Robot with Glass Fiber Reinforced Polymer Springs

To find a middle ground between a compliant soft robot and a rigid structure, this design resembles a classic tendon-driven continuum robot with a spring backbone (Fig. 8). To avoid corrosion issues associated with metal springs, we aim to use springs made from carbon or glass fiber composite materials. Manufacturing these springs presented a significant challenge, as it typically requires specialized equipment.

To achieve low-cost production, we tested several spring designs and manufacturing processes. The resulting springs are meander springs, as they can be manufactured from fiber tubes, which eliminates the need for post-processing. In the resulting prototype, three springs are arranged equidistantly. The actuation mechanism will be

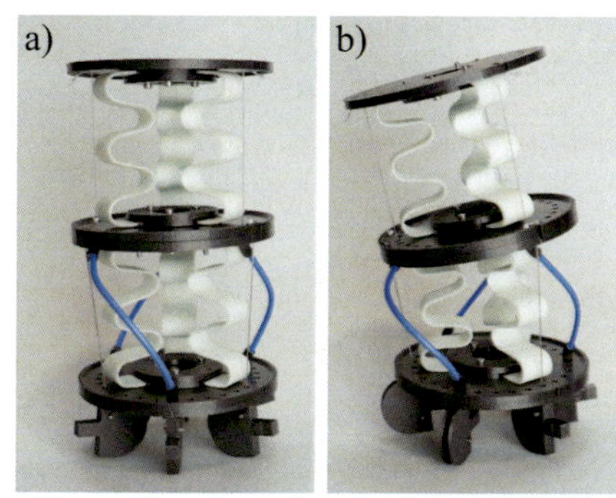

Fig. 8 This concept draws inspiration from the traditional design of tendon-driven continuum robots with a spring-loaded backbone. Since conventional springs cannot withstand the corrosive environment of the deep sea, this design utilizes glass fiber-reinforced polymer springs. This not only makes the system resistant to corrosion but also reduces its overall weight

tendon-driven [18]. With the option to now manufacture sufficiently good springs this concept will be followed up on by assembling a full scale prototype in the next step.

4.3.4 Tendon-Driven Virtual Rolling Contact Joint

The virtual rolling contact joint is the design closest to a traditional industrial robot, as it allows for the assembly of a robot with discrete joints using rigid materials. This design is particularly suitable for our current application because it enables the sampling tube to be fed through the center of the joint.

Additionally, the joint's damping characteristics help protect the ROV from sudden forces acting on the suction sampling system. The actuation is tendon-driven, which allows for a centralized actuation system within the ROV. This centralization reduces the need to pressure-proof individual motors located at the joints, simplifying the design and enhancing its reliability [19]. After completing a prototype (Fig. 9) that has been successfully tested in water, we plan to combine multiple joints to construct the complete actuation system.

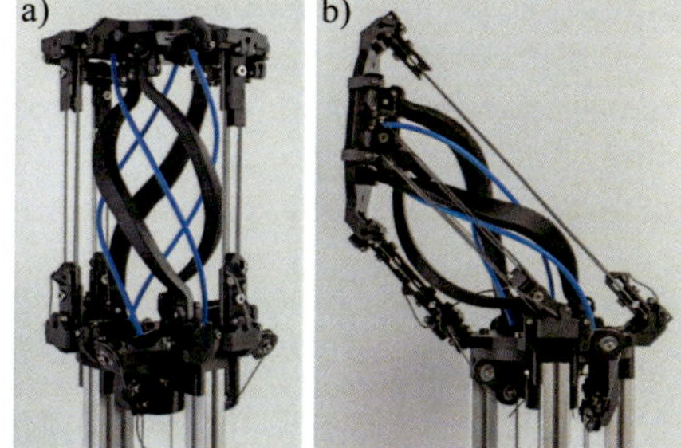

Fig. 9 The virtual rolling contact joint mimics the motion of two touching spheres, enabling a large bending radius. This design advantageously allows the sampling tube to be centrally positioned

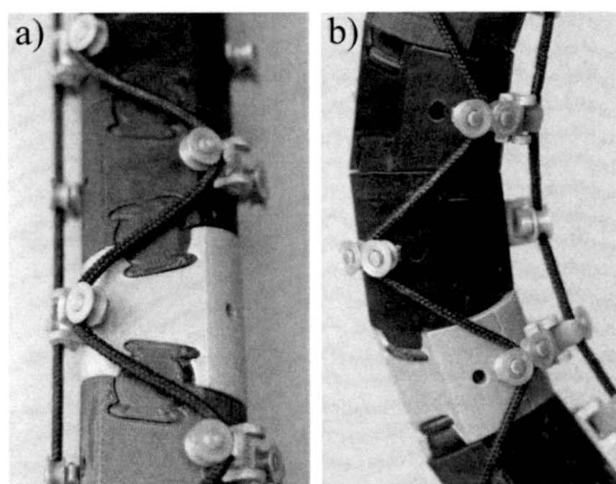

Fig. 10 This structure is composed of angled cylinders that can be rotated relative to one another to achieve bending. The design is hollow, allowing the sampling tube to pass through the center of the actuation system

4.3.5 Tendon-Driven Continuum Robot with Angular Rigid Joints

This design features a series of angled joints that enable the structure to flex when the cylinders are rotated by tendon actuation, shown in Fig. 10. The use of rigid materials enhances its robustness. However, it lacks the damping characteristics found in the previously mentioned designs.

Nevertheless, the simple design can be manufactured and assembled with minimal effort and it is lightweight and robust [20]. In the next iteration of this concept, the joints will incorporate bearing spheres instead of 3D-printed surfaces to facilitate smoother motion. Additionally, the routing of the tendons will be improved, as the current configuration, positioned on the exterior of the structure, results in high friction.

5 Conclusion and Future Work

With several options for the mechanical structure designed, the next step involves developing the actuation and power supply systems, which are essential for operating the prototypes outside of the laboratory environment. To minimize the size of the actuation system, we are considering a binary approach. Literature suggests that this approach could significantly reduce the number of sensors required to track the suction sampler's configuration, thereby allowing for a very robust [21] and lightweight structure [22]. Once the design of the suction sampling system is finalized, we aim to test it in its intended application environment: the deep sea.

Acknowledgements This research was funded by Deutsche Forschungsgemeinschaft (DFG, German Research Foundation) under grant no. 498342743. Additionally, we would like to thank Sandra Gerland for her contribution in creating several figures used in this chapter.

Competing Interests The authors have no conflicts of interest to declare that are relevant to the content of this chapter.

References

1. Huneke, H., Mulder, T.: Deep-Sea Sediments. Elsevier (2011)
2. McPhail, S.: Autosub6000: a deep diving long range auv. J. Bionic Eng. **6**, 55–62 (2009)
3. Singh, R., Sarkar, P., Goswami, V., Yadav, R.: Review of low cost micro remotely operated underwater vehicle. Ocean Eng. **266**, 112796 (2022)
4. Mc Letchie, K., Dunn, J.: Overview of Remotely Operated Vehicle Deep Discoverer's New Suction Sampler (2019)
5. Shepherd, K., Juniper, S.K.: ROPOS: creating a scientific tool from an industrial ROV. Art. Marine Technol. Soc. J. **31**, 48–54 (1997)
6. Capocci, R., Dooly, G., Omerdić, E., Coleman, J., Newe, T., Toal, D.: Inspection-class remotely operated vehicles-a review. J. Marine Sci. Eng. **5**, 3 (2017)
7. Sourkounis, C.M., Garcia Morales, D.S., Kwasnitschka, T., Raatz, A.: Hard shell, soft core: binary actuators for deep-sea applications. In: IEEE International Conference on Robotics and Automation (ICRA), pp. 9355–9361. IEEE (2024)
8. Clague, D.A., Paduan, J.B., Davis, A.S.: Widespread strombolian eruptions of mid-ocean ridge basalt. J. Volcanol. Geotherm. Res. **180**, 171–188 (2009)
9. Davies, J.B.C., Lane, D.M., Robinson, G.C., O'Brien, D.J., Pickett, M., Sfakiotakis, M., Deacon, B.: Subsea applications of continuum robots. In: Proceedings of the 1998 International Symposium on Underwater Technology, pp. 363–369. Institute of Electrical and Electronics Engineers Inc. (1998)
10. Phillips, B.T., Becker, K.P., Kurumaya, S., Galloway, K.C., Whittredge, G., Vogt, D.M., Teeple, C.B., Rosen, M.H., Pieribone, V.A., Gruber, D.F., Wood, R.J.: A dexterous, glove-based teleoperable low-power soft robotic arm for delicate deep-sea biological exploration. Sci. Rep. **8**, 12 (2018)
11. Li, G., Chen, X., Zhou, F., Liang, Y., Xiao, Y., Cao, X., Zhang, Z., Zhang, M., Wu, B., Yin, S., Xu, Y., Fan, H., Chen, Z., Song, W., Yang, W., Pan, B., Hou, J., Zou, W., He, S., Yang, X., Mao, G., Jia, Z., Zhou, H., Li, T., Qu, S., Xu, Z., Huang, Z., Luo, Y., Xie, T., Gu, J., Zhu, S., Yang, W.: Self-powered soft robot in the mariana trench. Nature **591**, 66–71 (2021)
12. Liljeback, P., Mills, R.: Eelume: a flexible and subsea resident IMR vehicle. In: OCEANS Conference 2017, pp. 1–4. IEEE (2017)
13. Xue, F., Fan, Z.: Kinematic control of a cable-driven snake-like manipulator for deep-water based on fuzzy PID controller. In: Proceedings of the Institution of Mechanical Engineers. Part I: J. Syst. Control Eng. **236**, 989–998 (2022)
14. Mazzeo, A., Aguzzi, J., Calisti, M., Canese, S., Vecchi, F., Stefanni, S., Controzzi, M.: Marine robotics for deep-sea specimen collection: a systematic review of underwater grippers. Sensors **22**, 1 (2022)
15. Aracri, S., Giorgio-Serchi, F., Suaria, G., Sayed, M.E., Nemitz, M.P., Mahon, S., Stokes, A.M.: Soft robots for ocean exploration and offshore operations: a perspective. Soft Robot. **8**, 625–639 (2021)
16. Peters, J., Sourkounis, C.M., Wiese, M., Kwasnitschka, T., Raatz, A.: Single channel soft robotic actuator leveraging switchable strain-limiting structures for deep-sea suction sampling. In: IEEE/RSJ International Conference on Intelligent Robots and Systems (IROS), pp. 6484–6490. IEEE (2023)
17. Sourkounis, C.M., Garcia Morales, D.S., Kwasnitschka, T., Raatz, A.: Exploring the deep sea: combining a bistable mechanism with origami-inspired soft actuators. In: IEEE 7th International Conference on Soft Robotics (RoboSoft), pp. 115–120. IEEE (2024)
18. Sourkounis, C.M., Kwasnitschka, T., Raatz, A.: Cost-effective manufacturing of fiber reinforced polymer springs for continuum robots. In: IEEE 8th International Conference on Soft Robotics (RoboSoft) (2025)
19. Sourkounis, C.M., Wienöbst, H., Peters, J., Kwasnitschka, T., Raatz, A.: Design of a tendon driven virtual rolling contact joint for deep-sea application. In: Procedia 35th CIRP Design Conference (2025)

20. Sourkounis, C.M., Kwasnitschka, T., Raatz, A.: Tendon-driven continuum robot for deep-sea application. In: IEEE International Conference on Robotics and Automation (ICRA), pp. 1498–1504. IEEE (2024)
21. Chirikjian, G.S., Burdick, J.W.: A hyper-redundant manipulator. IEEE Robot. Autom. Mag. (Dec 1994)
22. Sujan, V.A., Dubowsky, S.: Design of a lightweight hyper-redundant deployable binary manipulator. J. Mech. Des. **126**, 29–39 (2004)

The Future of Soft Robotics

The Future of Soft Robotics

Jan Peters, Cora Maria Sourkounis, Helge Wurdemann, Perla Maiolino, Sabine Ludwigs, Matteo Cianchetti, and Annika Raatz

Abstract At the 2025 Symposium on the Future of Soft Robotics in Hannover, a group of international experts gathered for an insightful panel discussion. The conversation addressed the significant challenges currently facing the field of soft robotics and examined potential solutions. Helge Wurdemann facilitated the discussion, while Annika Raatz, Perla Maiolino, Sabine Ludwigs, and Matteo Cianchetti contributed their expertise as panelists (Fig. 1). Drawing from diverse backgrounds including material science, engineering, and applied robotics, the panelists discussed the current grand challenges, technological trends, emerging applications, and collaborative strategies shaping the future of the field. Their discussion emphasized the importance of interdisciplinary convergence, the need for improved scalability and

Jan Peters and Cora Maria Sourkounis contributed equally to this chapter.

J. Peters (✉) · C. M. Sourkounis · A. Raatz
Institute of Assembly Technology and Robotics, Leibniz University Hannover, Garbsen, Germany
e-mail: peters@match.uni-hannover.de

C. M. Sourkounis
e-mail: sourkounis@match.uni-hannover.de

A. Raatz
e-mail: raatz@match.uni-hannover.de

H. Wurdemann
Department of Mechanical Engineering, University College London, London, UK
e-mail: h.wurdemann@ucl.ac.uk

P. Maiolino
Engineering Science Department, Oxford Robotics Institute, University of Oxford, Oxford, UK
e-mail: perla.maiolino@eng.ox.ac.uk

S. Ludwigs
IPOC—Functional Polymers, Institute of Polymer Chemistry (IPOC), and Center of Bionic Intelligence Tübingen Stuttgart (BITS), University of Stuttgart, Stuttgart, Germany
e-mail: sabine.ludwigs@ipoc.uni-stuttgart.de

M. Cianchetti
The BioRobotics Institute, Scuola Superiore Sant'Anna, Pisa, Italy
e-mail: matteo.cianchetti@santannapisa.it

A. Raatz et al. (eds.), *Soft Material Robotic Systems*,
https://doi.org/10.1007/978-3-032-22453-8_21

durability in soft systems, and the critical role of new functional materials such as stimuli-responsive polymers and self-healing composites.

1 From Materials to Market Challenges in Soft Robotics

The initial part of the panel's discussion centered on the evolving role of materials in soft robotics. While early developments relied on simple elastomers, like polydimethylsiloxane (PDMS), and commercially available silicones, like Ecoflex™ and Dragonskin™, recent advances have introduced stimuli-responsive, self-healing, and multi-functional materials. These materials now exhibit properties such as tunable stiffness, environmental responsiveness, and embedded sensing capabilities.

This evolution faces several critical challenges that impede its widespread adoption. Scaling presents a multifaceted problem - not just in physical dimensions (both miniaturization and scaling up), but also in manufacturing reliability and production volume. While soft robotic prototypes mostly function reliably in laboratory settings, most soft devices currently lack the durability required for industrial deployment, where thousands of repeat cycles are typically expected. To address these challenges, automating the manufacturing process would be a crucial step forward.

> Of course, scalability [...] involves not only dimensions but also [...] reproducibility, endurance and quality.
>
> Annika Raatz

Fig. 1 Participants in the panel discussion from left to right: Annika Raatz, Sabine Ludwigs, Perla Maiolino, Matteo Cianchetti, Helge Wurdemann

Materials that perform well in controlled environments often degrade or change properties over time in real-world conditions, affecting consistency and reliability. Meanwhile, the field suffers from "overselling," where researchers use buzzwords and make promises that raise unrealistic expectations. This creates a problematic hype cycle where, as one participant noted, people begin asking "where are the applications?" before the fundamental challenges have been adequately addressed.

> The literature is full of buzzwords like artificial muscles [...] which raises of course the expectations of their applicability. However, there are still a number of issues to be addressed to really mimic natural systems, such as an octopus or human motion.
>
> ---
>
> Sabine Ludwigs

These interconnected issues require both focused application development and continued fundamental research to bridge the gap between laboratory demonstrations and practical implementation.

2 The Unique Appeal of Soft Robotics Where Freedom Meets Function

The panelists unanimously emphasized that the soft robotics community has become a melting pot for interdisciplinary research. Initially a niche, the field now draws together mechanical engineers, chemists, material scientists, neuroscientists, and control theorists. Soft robotics has become an integral part of the broader robotics conversation. This is evident through the rising number of publications on innovative soft material robots and specialized journals like "Soft Robotics." Major conferences, such as ICRA and IROS, now include dedicated sessions on soft robotics, and the IEEE International Conference on Soft Robotics was established in 2018, underscoring its growing significance in the field. This convergence has broadened not only the scientific toolkit but also the intellectual framing of the field. Materials chemists now co-author projects with roboticists, and engineering labs increasingly train students in soft material synthesis. The shared sentiment was that disciplinary openness has become a cornerstone of progress.

What makes soft robotics truly special and amazing is its remarkable ability to break free from traditional constraints of robotic design while creating new possibilities for human-machine interaction. Unlike conventional robotics with its rigid structures and predefined movements, soft robotics embraces flexibility, compliance, and adaptability inspired by biological systems like octopuses and elephant trunks.

> Soft robotics gives us the opportunity [...] to explore things that we were not thinking could be achievable.
>
> ---
>
> Perla Maiolino

Soft robotics represents a paradigm shift that allows researchers to become creative with designs that were previously unimaginable. This interdisciplinary field brings together diverse expertise from materials science, biology, mechanical engineering, and medicine, fostering innovation through collaboration.

> What makes soft robotics so amazing is that traditional robotics comes from mechanic engineering, electrical engineering and informatics and now we have material science, neuroscience [...] and so many other disciplines coming together in this field, broadening our view [...].
>
> Annika Raatz

> What makes this community so great is that the topic is versatile and new, it is attracting many talented people.
>
> Matteo Cianchetti

While still facing challenges in sensing, durability, and scalability, soft robotics excels in applications requiring gentle interaction with delicate or irregularly shaped objects, from medical devices that safely interface with human tissue to grippers that can handle delicate agricultural products. The beauty of soft robotics lies in beyond in mimicking nature—it lies in understanding fundamental principles that allow us to develop entirely new capabilities that expand what robotics can achieve.

3 The Evolving Applications of Soft Robotics From Grippers to Artificial Organs?

Current applications of soft robotics are primarily centered around manipulation and grasping, with grippers being the most developed "low-hanging fruit" technology that's seeing commercial adoption in areas like food automation. However, the panel experts suggest that soft robotics should not be viewed as a standalone field but rather as becoming integrated into robotics as a whole, where softness is applied when beneficial.

> We will not get rid of stiff or rigid robots. [Soft Robotics] will find its niche. The soft robotics community [...] will be integrated into the general field of robotics, so [...] soft robots and traditional robots will no longer be distinguished.
>
> Annika Raatz

Future applications show promise in healthcare (though with longer development timelines due to regulatory requirements), deep-sea exploration, wearable technologies, and potentially artificial organs. The experts emphasize that while application-driven research is important for advancing technology readiness levels, the field shouldn't abandon curiosity-driven fundamental research.

> Soft robotics is not a religion. We need to see what the real need is.
>
> Matteo Cianchetti

Researchers should be pragmatic about when soft approaches are appropriate versus traditional rigid systems. The consensus seems to be that rather than seeking a single "killer application," soft robotics will likely find success in specialized niches where compliance, safety in human interaction, and adaptability to unstructured environments are essential.

> I don't see that there is a single killer application. I think that soft robotics becomes part of robotics in the sense that [...] when you need to interact with humans you need to provide safety and compliance.
>
> Perla Maiolino

The field of soft robotics faces fundamental challenges in materials development, sensing capabilities, control systems, and scaling from laboratory prototypes to real-world applications. In fact, soft robotics could transform fields ranging from medical devices to environmental monitoring to industrial automation, but the community needs sustained support to consolidate its exploratory work into practical, reliable systems that can meet the high expectations currently placed on the field.

> We are not limited by applications, it's just a matter of technology advancement. We really need time and money.
>
> Matteo Cianchetti

4 The Future of Soft Robotics

The future of soft robotics lies at the intersection of multiple disciplines working in harmony rather than in isolation. As highlighted throughout the panel discussion, promising advances can emerge when roboticists collaborate closely with material scientists, biologists, medical professionals, and engineers from diverse backgrounds. This interdisciplinary approach is not merely beneficial but essential.

> Communication and continuous interaction between experts on diverse fields is absolutely phenomenal.
>
> Matteo Cianchetti

A prominent example of interdisciplinarity is the advancement of novel materials, which is expected to be critically important. According to Sabine Ludwigs, stimuli-responsive materials, which have the ability to both sense and actuate, hold significant transformative potential.

> We can tailor materials that are self-healing [...] and biodegradable. [...] The vision would be sensing and actuation in one system by using smart and stimuli responsive materials [...].
>
> Sabine Ludwigs

The panel consensus suggests that soft robotics will not replace traditional rigid robots but will find valuable niches—in particular, healthcare, human-robot interaction, and environments requiring gentle manipulation of irregular objects. As these technologies mature beyond the current exploration phase into consolidation, we will see soft robotic components increasingly integrated into mainstream robotics, creating hybrid systems that combine the best aspects of both rigid and compliant structures.

The panel concluded on an optimistic but measured note. Soft robotics continues to evolve as a richly interdisciplinary, inventive, and impactful field. The community is aware of its limitations and actively working to overcome them. Most importantly, the spirit of collaboration—across domains, institutions, and continents—was seen as the field's greatest strength.

Acknowledgements This study was funded by Deutsche Forschungsgemeinschaft (DFG, German Research Foundation) under grant No. 405030609. The authors have no conflicts of interest to declare that are relevant to the content of this chapter.

Zeitfracht Medien GmbH
Ferdinand-Jühlke-Straße 7
99095 Erfurt, Deutschland
produktsicherheit@kolibri360.de